THE BEHAVIOR OF TEXAS BIRDS

Number Fifty-three
THE CORRIE HERRING HOOKS SERIES

THE BEHAVIOR OF TEXAS BIRDS

BY KENT RYLANDER

University of Texas Press, Austin

Requests for permission to reproduce material from this work should be sent to
Permissions, University of Texas Press, P.O. Box 7819, Austin, TX 78713-7819.

The paper used in this book meets the minimum requirements of
ANSI/NISO Z39.48-1992 (R1997) (Permanence of Paper).

Library of Congress Cataloging-in-Publication Data
Rylander, Michael K.
  The behavior of Texas birds / by Michael Kent Rylander. — 1st ed.
    p.   cm. — (The Corrie Herring Hooks series ; no. 53)
Includes bibliographical references (p.   ).

  ISBN 0-292-77120-7 (pbk. : alk. paper)

  1. Birds — Behavior — Texas.   I. Title.   II. Series.
    QL698.3 .R65   2002
    598.15′09764 — dc21   2001008472

CONTENTS

PREFACE

Twenty or so years ago a book with this title would have been entirely descriptive. Now, thanks to numerous, televised nature documentaries, as well as to illuminating scientific articles in magazines such as *Natural History, Audubon,* and *Smithsonian,* bird enthusiasts are increasingly curious about the meaning of the behaviors they observe and are looking beyond mere descriptions of the natural world.

As with any area of inquiry, we interpret the natural world in terms of laws or principles that provide a conceptual framework for our observations. Like historians who relate past events to historical models, or physicists who explain matter and energy in terms of specific theories, students of animal behavior interpret their observations in accordance with concepts or principles that apply to the behavior of animals in general. Indeed, some of the conceptual frameworks that organize and codify our perceptions of how animals behave rank among the finest and most ingenious products of human thinking.

Although the bulk of this book is a descriptive account of behaviors commonly observed in Texas birds, I have attempted to interpret many of these behaviors in terms of the basic principles of animal behavior. These principles are summarized in the Introduction. For example, some behaviors have a strong genetic component; these are the so-called innate or instinctive behaviors such as courtship displays and parental care. Other behaviors are heavily influenced by experience and fall under the rubric of "learning." Learned behaviors include habituation, classical conditioning, and operant conditioning, as well as the knotty problem of insight learning in animals. The boundary between innate and learned behavior is quite blurred; yet this interface is itself a rich source of ideas concerning why animals—including humans—behave as they do.

The Introduction also points to some issues in the field of behavioral ecology, a lively discipline that addresses questions of how an animal's behavioral responses to the environment affect its survival. Surprisingly, many such questions were not asked until twenty or thirty years ago, perhaps because too few people were pausing long enough on their nature walks to wonder about what they saw. For example, does flocking in birds make them more—or less—vulnerable to predators? Is it more efficient for a Cedar Waxwing to eat all berries regardless of size, or to select certain sizes when given a choice? Are there circumstances in which a female bird leaves more surviving offspring if she mates with more than one male?

Space limitations have been critical in the organization of this book. I chose to elaborate on the behaviors of the most commonly observed Texas birds at the expense of entries for less common species. The excluded species are listed at the end of the book. With few exceptions, breeding behavior is included only for species that nest frequently in Texas. Additionally, I have referenced articles in the text only if their relevance to a specific behavior is not obvious from the title of the original article (as listed under the species' family or order in the References section).

If this book encourages the reader to become more engaged in the experience of watching the behavior of birds, as well as to wonder about the meaning of what they see, then my goal in writing it will have been realized.

ACKNOWLEDGMENTS

I would like to express my deepest gratitude to the following persons who took time from their busy schedules to read parts of the manuscript and give helpful suggestions: Keith Arnold, Guy Baldassarre, Eric Bolen, Lee Drickamer, Amy Gangula, Robert Huber, Mark Lockwood, Michael Marsden, Craig Rudolph, Rod Rylander, Jace Shaw, R. Douglas Slack, Misty Sumner, Moira van Staaden, Roland Wauer, and Clayton White. I am especially grateful to Sam Braut, Lady Falls Brown, and Ellen Roots, who spent an extraordinary amount of time reading and editing the manuscript; to Julie Lemma and Yisenia Delgado, who helped considerably with the references; to Wyman Meinzer and Shannon Davies, who reassured me from the very beginning that writing this book was a worthwhile task; and to my wife, Laura M. Ronstadt, who continually encouraged me throughout the writing of this book.

THE BEHAVIOR OF TEXAS BIRDS

INTRODUCTION

Earlier students of animal behavior found it convenient to classify behaviors as either *innate* or *learned*. "Innate" refers to behaviors that have a strong genetic basis, behaviors that are influenced very little by experience. Fighting, copulation, and parental behavior seem to fit into this category.

Perhaps none of the numerous definitions of "learning" fits all behavior that we refer to as learned behavior. However, the definition of a learned behavior as "a behavior that has been changed by experience" will suffice for our purposes here.

The middle of the last century witnessed lively debates between classical ethologists, who stressed the importance of innate behaviors, and behaviorists, who argued that only learned behaviors are scientifically valid or even worthy of scientific attention. Currently, most students of animal behavior feel comfortable interpreting most behaviors as products that result from the interaction of genes and environment.

For example, the basic movements of food-begging in newly hatched sparrows are certainly innate, since hatchlings could hardly have had time to learn these movements by trial and error. On the other hand, food-begging becomes more effective with experience, indicating that this essentially innate behavior clearly changes with experience. It is difficult to imagine an innate behavior that cannot be changed in some way by experience, but it is equally difficult to visualize a behavior totally uninfluenced by innately programmed movements.

Some Major Categories of Bird Behavior

All attempts to define and organize behaviors are to some degree arbitrary, but the following categories are useful when dealing with the behaviors treated in this book.

Reflexes

Reflexes are innate (genetically programmed) muscular contractions that are elicited by a stimulus. The most obvious reflexes are responses to mechanical, visual, and auditory stimuli, and to sudden movements of the body. Reflexes are easily overlooked because they are usually quick and subtle.

A hawk gliding to a perch makes numerous, small reflexive wing movements to counter the effects of shifting winds. Upon landing, the hawk regains its balance with reflexive movements in the leg, wing, and body. These small reflexes are of the same type as when one taps the tendon beneath the knee and the lower leg automatically kicks forward.

The innate character of reflexes is evident in newly hatched birds, which, like older birds, blink in response to blowing dust or crouch when they hear a sudden noise.

The direction and extent of a limb movement is normally beyond our control during a reflex — a fact that is obvious when we touch a hot surface and watch our arm take its own course. On the other hand, all reflexes are probably modified to some extent by experience. With training, most animals can learn to increase or decrease the intensity of a reflex and, in some cases, to abolish it.

In one common reflex, exposure to cold causes microscopic muscles at the base of the feathers to contract so that the feathers fluff up and insulate the bird. However, a different stimulus, for example, the appearance of a bird's mate, can also cause feathers to fluff up, especially the feathers on the top of the head that form the crest. In this case feather erection has nothing to do with temperature but instead most likely functions as a signal.

Some reflexes are hidden from view, permitting us to see only their effects. In gulls, when fledglings peck at their parents' bills, the parents involuntarily regurgitate partially digested food that the young birds eat. Another hidden reflex is the contraction of the muscles that surround the salivary glands in ant-eating birds such as flickers. The stimulus for this reflex is the formic acid found in ants, and the response is the secretion of saliva that neutralizes the formic acid.

Fixed Action Patterns

Perhaps the most widely studied innate behavior in animals is the fixed action pattern (FAP). A particularly instructive example of a FAP is egg retrieval in geese: an incubating goose extends her head and neck and with her bill pulls back an egg that has rolled out in front of the nest. Even geese raised in isolation do this.

Several features characterize FAPs. Besides being innate, they are stereotyped in that they are relatively invariable (i.e., geese never retrieve eggs except in this manner). FAPs also have an obvious steering component. When the egg rolls off course because of irregularities in the substrate, the goose adjusts to the egg's changing locations by modifying her head and neck movements.

The stimulus that triggers a FAP is called a releaser. An egg is a visual releaser, but releasers can also be auditory, tactile, or olfactory. Moreover, releasers are effective only in specific contexts. For example, a goose will respond to an egg placed in front of her as long as she is sitting on her nest (the appropriate context), but she is unresponsive when she is away from the nest.

FAPs differ from learned behaviors because they are innate, and they differ from reflexes because reflexes have no steering component. Interestingly, if the stimulus (the displaced egg) is removed while the FAP is in progress, the FAP continues until it reaches completion; for example, the goose continues making egg-retrieving movements even though the egg is now in the hand of the experimenter. Although theoretically all FAPs continue to completion when the stimulus is suddenly withdrawn, it is very difficult or even impossible to demonstrate this characteristic except in special cases like egg retrieval.

FAPs are always short in duration (usually only a few seconds) because the series of muscular contractions that make up a FAP must follow a particular sequence. It is not possible for the brain to hard-wire instructions for a sequence of muscular contractions that lasts more than a few seconds.

Common FAPs in birds include pecking at a seed, reaching out for prey with the talons, tearing a piece of flesh from a carcass, drinking, bill-wiping, preening, inserting food into a nestling's mouth, and copulating. All of these FAPs are triggered by an appropriate releaser. Another behavior that is possibly a FAP (or at least has components that are FAPs) is anting, during which birds stroke their wings, bodies, and tails with ants that they hold in their bills.

Although some investigators question the usefulness as well as the validity of the FAP as a scientific concept, FAPs as classically defined here are useful for understanding more complex behaviors like those described below.

Innate Behaviors in Conflict Situations

As a rule, birds respond to threats with innate rather than learned behaviors. For example, when a threatened goose reaches out and bites an attacker, it uses basically the same hard-wired program for muscular contractions as it does for pulling at submerged vegetation (i.e., it utilizes a FAP). One reason FAPs, rather than learned movements, are the usual building blocks of defense behavior is that birds cannot afford to err during the trial-and-error process of learning a crucial defense tactic. Moreover, there is always the risk that the bird might forget a learned response at a critical moment. The same line of reasoning explains why attack behaviors are essentially innate.

Sometimes, when birds are threatened by predators but seem unable to choose between attacking or fleeing, they respond by attacking—but by attacking an object other than the predator. Thus, they redirect their attack to a substitute object, hence the term "redirection" for this type of behavior. For example, if a turkey is threatened by a mountain lion that is standing close enough to produce anxiety but not close enough to incite fleeing, the turkey may resolve the conflict by pecking a smaller turkey nearby or even an inanimate object.

Frequently, threatened animals exhibit FAPs that appear, at least to human observers, to be irrelevant or inappropriate responses to a threat. Incubating terns face a dilemma when a person slowly approaches the nest: they could risk injury to themselves by fighting, or risk injury to their eggs by fleeing. Curiously, terns often preen vigorously under these circumstances. The act of preening in this context appears irrelevant or inappropriate.

Such irrelevant or inappropriate responses to threats are called displacement behaviors. Preening, eating, bill-wiping, stretching, and drinking movements are common displacement behaviors when they are responses to threats. Bill-wiping as a displacement behavior is commonly observed in flushed birds immediately after they alight on a perch.

Intention movements are very common responses to threats. Birds begin an attack or a fleeing response, then abruptly halt the movement.

Displays and Ritualization

Displays are innate (genetically programmed) stereotyped movements that have a communicatory or signal function. In terms of their function, they may be compared to culturally acquired human gestures, which are also stereotyped movements used to communicate. In both displays and gestures, ambiguity is reduced by exaggerating the movements as well as by performing them in a more stereotyped manner. For example, the bathing movements in a gander's precopulatory display appear to be an exaggerated and more stereotyped version of movements he uses when he actually bathes. (Likewise, the gesture of saluting in humans — although culturally acquired rather than inherited — appears to be an exaggerated and more stereotyped version of a noncommunicatory behavior, perhaps raising one's hand to shade the eyes.)

Many displays are unique to the species and are determined to some degree by the bird's anatomical and behavioral characteristics. Thus, prairie chickens would be expected to evolve terrestrial displays and Common Nighthawks, aerial displays.

Courtship displays often differ markedly among species that are closely related, since a bird that confuses its own display with that of a similar species might waste time and energy courting and inseminating the wrong species. Territorial displays tend to be more general and accordingly are recognizable by other species that could potentially invade the bird's territory.

The evolutionary origin of displays intrigued early ethologists, who reasoned that it is more efficient for a display to evolve from existing FAPs, reflexes, or intention movements than if they evolved de novo. The term "ritualization" is applied to this evolutionary process. For example, erecting the feathers — originally a reflex in response to cold — might acquire, through evolution, a signal or communicatory function to become part of a territorial display in which the bird raises its crest. Likewise, the FAP for drinking (lowering the head for water and then pointing the bill toward the sky) could evolve into a courtship display that employs the same basic movements.

Not all components of a display are ritualized reflexes, FAPs, or intention movements. The inflated neck pouch in the prairie chicken's display probably evolved uniquely for that purpose.

Habituation

Obviously a bird has a greater chance of surviving if it responds appropriately to stimuli that indicate danger. Thus, a towhee feeding on the forest floor should flee when it hears a sudden rustling of leaves that might signal an approaching predator. On the other hand, the bird will waste valuable time and energy if it flies away each time the wind noisily blows leaves across the forest floor. Clearly the towhee must learn to distinguish between harmful and innocuous stimuli.

Birds learn to ignore harmless, repeated stimuli by responding less and less to the stimulus each time it is presented. Eventually they do not respond at all. A diminished response to a harmless, or innocuous, stimulus is termed habituation. In the above example we would say that the towhee habituated to the repeated rustling of leaves caused by the wind, or, in everyday language, that it tuned out the rustling sounds.

Birds are constantly habituating to the ocean of innocuous stimuli they encounter each day. For example, early in the morning, House Sparrows feeding near a highway take flight when the first automobile passes by, but as traffic continues to flow with no harmful effects, their uneasiness diminishes, and soon they are virtually unresponsive.

Dishabituation is a disruption of the state of habituation. An animal that habituates to a noise or other stimulus will again become aware of that stimulus if a different stimulus is presented to it. Sparrows that are habituated to a constant sound like an idling automobile could once again respond to this sound and flee if there is a sudden gust of wind.

Sensitization

Sensitization refers to a process whereby an animal, immediately after responding to one stimulus (for example, food), now responds to a neutral stimulus, one to which it is normally unresponsive. One of the first sensitization experiments dealt with octopuses. Octopuses normally ignore a glass rod (a neutral stimulus) that is inserted into their aquarium. However, if the glass rod is placed into the aquarium immediately after the octopuses have been fed, they will attack the rod. Thus, they become sensitized by the process of feeding and as a consequence respond to a neutral stimulus (in this case, a glass rod).

Generally the consequences of sensitization in the wild can only be inferred, but surely sensitization must be a common occurrence. Perhaps birds that have just been frightened are more readily disturbed by a neutral stimulus, such as an airplane flying overhead, than they would be otherwise.

Classical Conditioning and Operant Conditioning

Classical conditioning was first studied in detail by the eminent Russian physiologist Ivan Pavlov. Pavlov took advantage of the fact that dogs naturally salivate when presented with a piece of meat. Just before presenting the meat to the dog he presented a second stimulus (for example, the sound of a buzzer) to which dogs normally do not respond. Thus, he paired the buzzer (called the conditioned stimulus) with meat (called the unconditioned stimulus). Soon the dog was salivating every time he heard the buzzer, which was never more than a second or two before the meat appeared. The innate response (salivation) to the unconditioned stimulus (meat) is called the unconditioned response, and the learned response (also salivation) to the conditioned stimulus (the buzzer) is called the conditioned response.

We rarely witness classical conditioning while it is occurring in the wild; more often we only infer that it has occurred. For example, the first time a person drives up to a lake and throws grain to a duck, the duck responds by becoming excited. The excitement, which is a response to the grain (not to the appearance of the person), can be compared to salivation in dogs that are presented meat.

After a few days, the duck associates the approach of the person with the appearance of the grain, in the same way that the dog associates the buzzer with the appearance of the meat. Predictably, the duck becomes excited, indicating that it has become classically conditioned to the approach of the person. The stimuli and responses can be compared to those in the classically conditioned dog. The unconditioned stimulus is the grain, which corresponds to the meat. Likewise, the unconditioned response (corresponding to salivation in the dog) is the duck's excitement upon first being presented with the grain. The conditioned stimulus (the buzzer in the experiment with the dog) is the approach of the person. Finally, the conditioned response (salivation in the dog), like the unconditioned response, is the duck's excitement.

Like classical conditioning, operant conditioning (also called instrumental conditioning or trial-and-error learning) is perhaps more easily understood by example than by definition. When a pigeon is placed in a cage containing a lever that dispenses food, it pecks randomly because it is unaware of the significance of the lever. When it accidentally pecks the lever and food drops into the cage, a contingency (a connection or relationship) is set up between pecking the lever and receiving food. At this point we say that the pigeon is operantly conditioned to peck the lever whenever it wants food. In a variant of this experiment, the pigeon is not rewarded with food when it presses the lever; instead, it is punished with a shock from the electrified floor of the cage. In this case the pigeon has been conditioned to *avoid* pressing the lever.

In both experiments the act of pecking the lever is called the operant. It is important to emphasize that an operant — usually a body movement — is required for operant conditioning to take place: the animal must do something to initiate conditioning. In classical conditioning it is the conditioned stimulus rather than the animal's behavior that initiates the conditioning process; to be conditioned, the animal need do nothing but respond.

Numerous behaviors in birds are no doubt the products of operant conditioning. Turkey Vultures that wander randomly over the countryside looking for carrion can be compared to pigeons that peck randomly in their cage. By chance the vultures may fly over a highway, just as by chance the pigeon pecks at the lever. Since flying to the highway and pecking the lever are rewarded (in both cases with food), then both behaviors are operants. After a few trials, both the pigeon and the vulture are conditioned to repeat their respective behaviors when they want food.

Innate behaviors, like FAPs, are commonly shaped or modified by operant conditioning. Consider a chick a few minutes after hatching. Almost immediately it engages in an innate feeding behavior, specifically, the FAP for pecking indiscriminately at any small object that is in front of it. The releaser for this FAP could be a tiny pebble, seed, or other small object. Soon the chick picks up only the seeds and ignores the pebbles. Thus, feeding behavior has been modified by operant conditioning: the chick was obviously rewarded for picking up seeds.

Body movements in virtually all behaviors are modified by operant conditioning. For example, young birds inherit the ability to make basic flying movements with their wings, but those movements alone do not enable the bird to fly. Young birds must learn, through operant conditioning, how to modify their movements so as to achieve flight.

Animals conditioned to a particular stimulus also respond (though less intensely) to a second stimulus, as long as the second stimulus is more or less like the first one. This phenomenon is termed generalization because the animals seem to be generalizing that similar stimuli produce the same rewards and punishments. For example, pigeons conditioned to peck round levers will also peck oval levers, and vultures that have been rewarded for foraging along highways will also forage along smaller farm roads.

Extinction of a learned response occurs when the reinforcer is withdrawn. For example, pigeons trained to peck a lever for food eventually cease pecking the lever if food is not delivered. Extinction differs from forgetting. An Acorn Woodpecker might forget the location of a nut it has hidden, but this would have nothing to do with whether a reinforcer is withdrawn.

Imprinting

In imprinting, an animal, usually a very young one, establishes a bond with an animal or inanimate object that it faithfully follows for the next few weeks or months. Young animals normally imprint on their parents, but in the absence of their real parent, they will imprint on a surrogate mother, which can be another animal, including humans, or an inanimate object. Birds imprint during a brief critical period after hatching, a window that normally lasts a few hours or days.

Imprinting is characteristic of precocial birds, those that move about and feed almost immediately after birth or hatching. Waterfowl and quail are common examples of precocial birds. The close bond established by imprinting helps insure that young birds follow their parents during this vulnerable period of life. Imprinting is essentially absent in altricial species (in particular, songbirds), which hatch in a helpless condition. Because altricial birds have no opportunities to stray from parental care, they have no need to bind so closely with their parents.

Learning and the Development of Songs

The enormous literature dealing with song development in birds can hardly be summarized here, but generally, passerines acquire their songs by learning how to sequence innately produced sounds, termed the subsong, correctly. This process has been compared with language acquisition in humans, during which a baby correctly sequences innate sounds (babbling) into speech (although learned sounds are of course also incorporated into speech). Thus, song development, like human language acquisition, can be considered the result of both innate and learned processes.

Usually a young bird learns the correct sequence by listening to an adult bird. When a juvenile male hears an adult male sing the notes in the correct sequence, he learns this sequence and correctly arranges the elements of his subsong to produce the song that we hear. Learned songs also may vary geographically, giving rise to dialects. Dialects could be adaptive because when a female chooses a male with a familiar dialect, she might be selecting a male adapted for surviving in the region where her offspring will live. That dialects can influence mating is demonstrated by the observation that female White-crowned Sparrows often assume the precopulatory position when they hear the dialect of their own region, but rarely do so when they hear other dialects.

Learning is not important for song development in all species. In general, nonpasserines, such as ducks and quail, do not require exposure to the adult song in order to vocalize correctly.

Another type of learning, latent learning, is so designated because the knowledge that is acquired presumably remains latent until it is needed at a later time. Latent learning can be demonstrated in a well-known experiment with mice. A satiated mouse is allowed to explore a maze containing food pellets that are concealed in a particular part of the maze. Eventually the mouse discovers the pellets but ignores them because it is not hungry. Later, after being deprived of food, the mouse is allowed to reenter the maze. At this time it locates the food quickly, much more quickly than a mouse that has never been in the maze. Evidently the mouse learned the location of the food when it first entered the maze but did not use the knowledge until later.

Latent learning is probably impossible to demonstrate in the field. However, satiated birds may occasionally learn the location of a food source that they do not exploit at the moment, only to return to it later when they are hungry.

Insight learning is the most difficult type of learning to demonstrate in animals. Humans employ insight learning when they solve a problem—a mathematical problem or logical puzzle—in a novel way, when the solution comes suddenly as an insight. This type of learning is sometimes referred to as "Aha!" learning.

Decades ago, insight learning was proposed to explain how chimpanzees managed to rearrange tables and put together sticks to reach a banana suspended from the ceiling. It was argued that the chimps had a sudden insight as to how to reach the banana, in effect, that they figured out a novel solution to a problem they had not previously encountered. Neither classical nor operant conditioning seemed adequate for explaining this feat.

On the other hand, it has also been argued that chimps, being playful animals, accidentally solve the problem through their normal playful antics and never really require an insight to arrange the tables and sticks appropriately. Indeed, chimps experienced in play solve the problem more quickly than inexperienced chimps.

Whether birds experience such insights is debatable, but an experiment using Common Ravens certainly provokes thought along these lines. A bird standing on the top of a table was shown a string that was attached to the tabletop and hung over the side. Tied to the other end of the string and suspended about halfway down was a peanut. Most of the birds that were tested looked down at the peanut and seemed incapable of figuring out how to retrieve it. A Common Raven, however, stood on one foot, reached down with its bill, grabbed the string, pulled it up part of the way, held that part of the string with its foot, then repeated the process until the peanut was within reach.

Some Topics in Behavioral Ecology

Dominance Hierarchies

Commonly called pecking orders, dominance hierarchies are social systems in which certain members of a group dominate others. Everything else being equal, animals that maintain a dominance hierarchy are more likely to survive than those that do not. Their interactions with each other are relatively stable and predictable, making it unnecessary to waste energy fighting among themselves for food or other resources.

The simplest dominance hierarchy is a linear hierarchy, in which one male (designated here as A) dominates for food resources over all other males. The next most dominant male (B) dominates over all males except A, and so forth, so that the order of dominance can be expressed as ABCD, etc.

Dominance hierarchies can be complex and sometimes puzzling. For example, some take the form of ABCA, indicating that whereas A dominates over B and B dominates over C, C actually dominates over A. Also, ranks sometimes shift within the hierarchy depending on the location of the individuals, their state of health, their reproductive state, or any other factor that might influence the outcome of a conflict.

What seems more remarkable than shifting ranks is that sometimes dominant individuals retain their dominance even when disease, malnutrition, injury, or age makes them incapable of winning a contest should their dominance be challenged. In these cases, contests for dominance do not occur, and physically superior subordinates continue acting as subordinates. They could be doing this because they learned a subordinate role as a young bird and retained this bias throughout life.

Dominance is usually less stable and less frequently linear in females. Females engage in far fewer confrontations than males, and when they do confront each other, their mates usually step in and terminate the contest before a hierarchy has been established.

Advantages of Flocking in Birds

Several reasonable hypotheses have been proposed to explain why feeding, roosting, and moving about in flocks could increase a species' chances of surviving in nature. Unfortunately, too few experiments (like the one with pigeons described below) have been conducted to test those hypotheses. Furthermore, it is not always obvious why, or even if, natural selection should favor flocking in extreme cases, such as the hundreds of thousands of blackbirds that roost together in winter.

The following are possible ways flocking might benefit birds:

1. Improved detection of predators (Many eyes spot predators better than fewer eyes.)
2. Improved defense of food resources
3. Improved care of the young through communal feeding and protection
4. More efficient predation for larger prey

The hypothesis that flocking improves detection of predators was tested by releasing a trained Goshawk a set distance from different-sized flocks of European Wood Pigeons (*Columba palumbus*). Increased vigilance was inferred because pigeons in large flocks took flight sooner than those in small flocks. Consistent with those results is the observation that Rock Doves feeding alone feed at a slower rate and thereby have more time to look around for predators.

Altruistic Behavior

Altruistic behavior may be defined as behavior that benefits another animal at the expense of the altruist. It is generally assumed that birds are compelled to act a certain way—altruistically or not—but as long as an act benefits another individual and is detrimental in some way to the altruist, then the definition of altruism as used here is satisfied. Therefore, we need not consider questions of motivation or how to distinguish truly altruistic acts from those that are self-serving and disguised (martyrs possibly acting to satisfy a selfish need).

Acts that are detrimental to the individual can be beneficial to the species, as when a bird in a flock sees a predator and risks its life by giving an alarm call. Also, altruistic behavior directed to a family member helps insure that some of the family genes are passed along to future generations. Natural selection would be expected to favor families that have genes for altruistic parental behavior over families that do not.

The same reasoning applies to the extended family, as when sisters, brothers, or older offspring help parents raise their young. These relatives pass along some of their own genes as well as those of the parents, since they share a certain number of genes with the parents. For example, female Mexican Jays that have lost their mates are sometimes assisted by their sisters in raising their offspring. In addition to learning and practicing parental skills before they nest the following year, these young females also pass along genes they share with their sisters. Moreover, in cases where nesting birds retain their territories for life, young birds might remain with their parents a year or two because they have no choice—the habitat is saturated. With no space to establish a new territory, it is best to wait in a territory of proven quality until a death provides room for a new territory.

Finally, altruistic behavior may be a consequence of living in groups, a by-product of parental care. In this case, natural selection favors individuals that feed *any* individual.

Mobbing, such as when screaming jays swarm around a hawk or owl, has been interpreted as benefiting some birds at the expense of others. Because flying close to a predator would seem to increase the mobbing bird's risk of being captured, natural selection should favor individuals that do not mob. However, some Australian owls capture almost nine times as many nonmobbing species as mobbing species. Moreover, to avoid being mobbed, these Australian owls tend to roost in dense forests, away from the area where most mobbing species occur. Thus, in spite of its risks, the function of mobbing could simply be to drive away predators from a preferred foraging area.

Optimal Foraging Strategies

We can reasonably assume that efficient foragers reproduce more successfully (leave more viable offspring) than inefficient foragers, everything else being equal. How efficiently an animal obtains food depends to a large extent on its foraging strategy. A common foraging strategy in flycatchers is to capture insects by darting out from a perch, whereas swallows fly continuously in search of insects. A bird's foraging strategy also includes variables such as the time of day it feeds, the habitat it utilizes, and the type and size of food it selects, all of which bear on the manner in which it obtains nourishment.

A particular foraging strategy cannot be equally efficient under all circumstances. For example, a hawk's best strategy in a habitat that is densely populated with rodents may be to wait and pounce, whereas where rodents are less common, perhaps it is best to soar and stoop. Although vitamins, minerals, and perhaps taste preferences are undoubtedly important in food selection, it is easier to visualize the main points of optimal foraging theory if we consider a simple case that has only calories as its currency.

Consider a hypothetical situation in which a quail walks through a meadow where two kinds of seeds, A and B, are randomly distributed on the ground. Seed A is relatively tender and takes only 5 seconds to peel and ingest. We say that this seed has a handling time of 5 seconds. Seed B, in contrast, has a tough, acornlike covering that requires a handling time of 25 seconds. Both A and B have the same number of calories.

Three possible foraging strategies are possible. In Strategy 1, the quail selects only A. Since A has the shortest handling time, the quail will spend less time peeling and ingesting these seeds, but travel time will increase because in selecting only A the quail passes up all seeds of type B. In Strategy 2, the quail selects both seeds A and B. Travel time is shortened because the quail does not pass up any seeds, but total handling time necessarily increases because of the 25 seconds of handling time that seed B requires. In Strategy 3, A is passed up and only the tough seed B is selected.

Common sense suggests that Strategy 3 is the worst of the three strategies if calories are all we consider, since this strategy requires a longer handling time as well as a longer travel time. But how about the other two strategies? Strategy 1 turns out to be the optimal foraging strategy under some circumstances, and Strategy 2 is optimal in others.

For example, if seeds are spaced closely together, as in a bird feeder, travel time from one seed to the next is essentially zero and the bird would do best to pass up seed B, which takes 25 seconds to husk, and select only seed A (Strategy 1).

On the other hand, in a desert habitat where the two seed types are so far apart that it takes an hour just to get from one seed to the next, it would be foolish to pass up B just because it requires 25 seconds to husk. Under these circumstances, the best strategy is Strategy 2 (that is, select both seed types A and B).

In reality, of course, birds usually select from several food items instead of two; moreover, those food items are rarely distributed randomly. Nonetheless, an optimal feeding strategy can be calculated based on a specific bird's caloric requirements, the caloric content of the food items it eats, and the travel and handling times required for the food items. In many cases, these calculated values approximate those observed in the field.

Predation

Predatory behavior would not seem to be especially puzzling or complex. A hungry hawk locates a prey species and attempts to capture it while the pursued animal tries to escape. Although variations of this cat-and-mouse game are almost limitless and depend to a large extent on the morphology of the predator and prey species (osprey versus fish, flycatcher versus insect, etc.), the basic pattern in all predator-prey interactions appears, at least superficially, to be more or less identical.

On the other hand, predator-prey interactions become more complex when we consider how much predator populations depend on prey availability. When prey is abundant, predators such as owls enjoy greater reproductive success and leave larger numbers of offspring for the following year. The increased hunting pressure exerted by more predators, however, reduces prey numbers, and in subsequent years, predator populations decline because prey populations are diminishing. When the population of predators is sufficiently reduced (from lack of prey), the prey population increases and the cycle begins again.

Although that scenario seems logical enough, evidence suggests that factors other than predation pressure can influence population cycles in prey species. For example, in some cases prey populations cyclically increase and decrease in the absence of predators, in response to disease, food supply, or other factors unrelated to predation. When we add predatory pressure to this predator-independent cycle, it is obvious that models describing the dynamics of predator-prey interactions must balance a number of ecological variables, including reproductive rate and energy requirements. Such models usually require relatively complex mathematical functions to describe the interactions between these variables.

ABBREVIATIONS

c.	central
e.	east(ern)
M	migrant
n.	north(ern)
pers. comm.	personal communication
PR	permanent resident
Princ. distrib. Texas	Principal distribution in Texas
s.	south(ern)
SR	summer resident
w.	west(ern)
WR	winter resident

SPECIES ACCOUNTS

LOONS: ORDER GAVIIFORMES, FAMILY GAVIIDAE

Loons — aptly called divers in Europe — are goose-sized birds that propel themselves underwater with extraordinary efficiency, diving to depths in excess of 200 feet and remaining below the surface 5 minutes or longer. Perhaps these extreme figures need verifying; nonetheless, loons are supremely adapted for chasing and capturing fish and other aquatic animals underwater.

For example, their legs are attached toward the back of their bodies so that their feet, with which they propel themselves, are positioned for maximum thrust. Their tarsi (the lower, unfeathered part of the leg) are flattened from side to side — a shape that minimizes water resistance while swimming. Buoyancy is reduced by their mostly solid bones, as well as by well-developed muscles at the base of the feathers that enable the birds to press air from their plumage.

On land they are basically helpless; to take flight, they must run across the water for a considerable distance, sometimes 100 yards or more.

Common Loon, *Gavia immer.* Princ. distrib. Texas: Uncommon M throughout most of Texas; common WR on the coast, locally inland. Inland they usually favor larger bodies of water.

FEEDING BEHAVIOR. See the introductory notes for the family. Although occasionally they eat plants, these powerful swimmers typically concentrate on fish, which they locate visually during underwater dives that may last a minute or more. When diving, they propel themselves with their feet, which are set (as in many diving birds) far back on their bodies. This arrangement contributes to their diving efficiency but also forces them to run across the water when taking flight.

They often swim on the surface of the water, partially submerging their heads while looking for prey in the water below. Their eyes are adapted for vision both in and out of water.

VOICE. A rich, resonant, melancholy yodel.

GREBES: ORDER PODICIPEDIFORMES, FAMILY PODICIPEDIDAE

Grebes, like loons, are superbly adapted for an aquatic life: they feed, sleep, court, and care for their young on water. They share with loons many features that facilitate swimming underwater: their legs are placed back on the body, their bones are generally solid, they have flattened tarsi (the lower part of the leg), and their thick, waterproof plumage can be compressed to decrease buoyancy. They dive by plunging or by slowly sinking, in both cases hardly producing a ripple.

One price grebes pay for having limbs positioned posteriorly is that they must run across the water when taking off. They are incapable of taking flight from land, and occasionally they become stranded during rainstorms when they land on highways or wet metal roofs that they mistake for water.

They generally feed on fish, but they also take other aquatic animals, in particular insects. Curiously, they consume large quantities of their own feathers; about half of the stomach contents in the Horned Grebe and Pied-billed Grebe are feathers. They also feed feathers to their young. It was formerly thought that feathers protect the intestinal walls from being injured by fish bones, but more likely grebes ingest feathers as a way of recycling oils and other nutrients. Perhaps it is coincidental to their feather-eating behavior that grebes possess more feathers than any other group of birds.

Some grebes that breed on large lakes have courtship displays that require running long distances across the water, but the two grebes that breed in Texas (Pied-billed Grebe and Least Grebe) favor marshy habitats; predictably, their courtship behavior is more vocal than visual.

Least Grebe, *Tachybaptus dominicus.* Princ. distrib. Texas: Uncommon PR in s. Texas, in ponds, marshes, and ditches. They readily take advantage of temporary bodies of water, especially large rain pools, thanks to their light wing loading and greater flying ability.

FEEDING BEHAVIOR. These dainty birds dive for underwater insects, mollusks, small fish, and other aquatic organisms. They have been observed suddenly emerging from beneath the water's surface to snatch flying dragonflies.

COURTSHIP BEHAVIOR. An invitation posture is apparently unique to the species. In one unverified display, the pair rise to an upright position and rapidly glide across the water. The pair bond is presumably strengthened by reverse mounting, during which the female mounts the male (with no cloacal contact).

NESTING. Sometimes loosely colonial. Nest: in shallow water and built from decaying vegetation that is anchored to more secure plants. The pair always seem to be adding some bit of material to the nest, even throughout incubation. Eggs: incubated by both parents, who cover them with vegetation when they leave the nest. Both parents feed the young. The young, which can swim almost immediately after hatching, often ride on their parents' backs.

VOICE. Metallic trills, sometimes given in duet by a pair; when alarmed, a ringing *peeet.*

OTHER BEHAVIORS. Least Grebes swim without nodding when in open water, but they move their head back and forth when swimming in dense vegetation. They appear to take flight more readily than most grebes. They bathe vigorously, sometimes spraying water 1–2 feet high, and they take frequent sunbaths.

In one threat display, the bird twists the head, neck, and body into a Z configuration.

Pied-billed Grebe, *Podilymbus podiceps.* Princ. distrib. Texas: Common M and WR in most parts of the state; locally common PR except in far w. Texas, in freshwater marshes, playa lakes, and (less frequently) salt bays and other saline or brackish habitats.

FEEDING BEHAVIOR. They search for food underwater, where they take food items that include small fish, insects, and other aquatic animals, as well as a few aquatic plants.

COURTSHIP BEHAVIOR. Courtship is relatively simple: birds call to each other and sometimes duet. They do not often engage in distinctive visual courtship displays.

NESTING. Nest: a floating structure situated in shallow water, but deep enough to be approached by birds swimming underwater. Both parents incubate the eggs and cover them with vegetation when they leave the nest.

Both parents feed the young and often carry them on their backs (even during dives). Chicks exhibit noticeable dominance hierarchies that are based on body size. These hierarchies may be advantageous during times of low food availability, because the largest chick monopolizes the food resources and insures that at least one of the brood survives. Also, size differences among the young potentially allow parents to utilize a wider range of prey types and sizes (i.e., parents feed chicks of different sizes different food items).

VOICE. A loud *cuck cuck cuck cow cow cow ah cow.* Sometimes the female answers with a low grunt.

OTHER BEHAVIORS. Although secretive during the nesting season, these grebes are among the most easily spotted waterbirds during migration and winter, even if seen only briefly between dives.

In disputes at territorial borders, males first turn away from each other, hold their heads high, point their bills to the sky, and call loudly. They then quickly swing around and face each other. When anxious, but not alarmed enough to dive, a bird often will sink slowly beneath the surface and swim with only its head above water. Pied-billed Grebes also sunbathe.

Horned Grebe, *Podiceps auritus.* Princ. distrib. Texas: Locally common M and WR on the Upper Coast and on larger bodies of water inland.

FEEDING BEHAVIOR. Their dives, which usually begin with an upward and forward leap, may last up to 3 minutes. They forage for aquatic insects and other invertebrates.

VOICE. The whinnies, gobbles, and liquid *coos* associated with nesting are rarely heard in Texas.

OTHER BEHAVIORS. On their nesting grounds they seem to be the least wary of our grebes. They are less gregarious than the Eared Grebe and are capable of taking off and landing on relatively short spaces of water.

Eared Grebe, *Podiceps nigricollis.* Princ. distrib. Texas: Common M throughout Texas; locally common WR on the coast and on inland reservoirs. They quickly avail themselves of temporary bodies of water — perhaps an adaptation to living in arid regions.

FEEDING BEHAVIOR. These smart-looking birds propel themselves with their feet in pursuit of insects, crustaceans, and occasionally fish.

VOICE. Call: a soft *poo eep,* as well as several abrupt, wheezy notes; none of these sounds are heard very often in Texas. Experiments show that courting males are able to distinguish the advertising calls of unpaired females from those of unpaired males.

OTHER BEHAVIORS. This is a gregarious species that often forms large flocks. Although they usually dive to avoid danger, they are capable of rising easily from the water and flying away swiftly. They appear to be bolder than the other small grebes, usually reappearing close to the place where they submerged to escape danger.

Occasionally migrants are trapped and starve to death on small bodies of water that suddenly freeze.

Western Grebe, *Aechmophorus occidentalis.* Princ. distrib. Texas: Rare M throughout Texas; locally common WR in w. part of the state.

FEEDING BEHAVIOR. They forage underwater for fish and, to a lesser extent, for other aquatic animals. This species and Clark's Grebe are unique among grebes in that the structure of their neck allows them to thrust it like a spear.

VOICE. Both the Western and Clark's Grebe utter a loud advertising call, a rasping *crrik* that has two syllables in the Western, one in Clark's. Possibly these different call notes, in addition to superficial differences in facial pattern and bill color, are important isolating mechanisms that prevent hybridization, thereby justifying splitting the former single species into the current two.

OTHER BEHAVIORS. There are four types of dives, each differing markedly in their movements: alarm, feeding, springing, and surface.

In some parts of the United States, Clark's forages farther from shore than the Western, but this difference may not be evident in migrating Texas birds. Both species tuck one or both feet under their wing and flank feathers, possibly as a mechanism to conserve heat.

In the Western Grebe, males and females differ with respect to bill length (the male's is longer) and shape (it is upturned in the female). These morphological differences may reflect subtle differences in prey selection that reduce competition for resources between the members of a pair.

Clark's Grebe, *Aechmophorus clarkii.* Princ. distrib. Texas: Locally uncommon WR. *See* Western Grebe for a discussion of its behavior, and Lockwood (1992) for an interesting discussion of the nesting behavior of the two *Aechmophorus* grebes in w. Texas.

TUBE-NOSED SWIMMERS: ORDER PROCELLARIIFORMES

SHEARWATERS AND PETRELS: FAMILY PROCELLARIIDAE

Audubon's Shearwater, *Puffinus lherminieri.* Princ. distrib. Texas: Fairly common offshore in deep water. The narrow-winged, fast-flying shearwaters maneuver with remarkable skill even in high winds. They capture fish, squid, and other marine animals by dropping down, with wings held high, then running over a patch of seaweed while flapping their wings. They also dip the head underwater while swimming and sometimes dive. Although at sea they may form large flocks, our birds are usually encountered either alone or in small groups.

STORM PETRELS: FAMILY HYDROBATIDAE

Band-rumped Storm-Petrel, *Oceanodroma castro.* Princ. distrib. Texas: Fairly common offshore in deep water. They fly low over the water, apparently locating prey by scent as well as by sight. They consume fish and crustaceans, picking their prey from the surface while hovering. They less frequently swim and dive.

TOTIPALMATE BIRDS:
ORDER PELECANIFORMES

BOOBIES AND GANNETS: FAMILY SULIDAE

Boobies and gannets superficially resemble large gulls. They are distinguished by having straight, sharp bills, long, pointed wings, and spindle-shaped bodies. All dive for fish, often from spectacular heights. The impact with the water is softened by air sacs located beneath the skin.

Masked Booby, *Sula dactylatra.* Princ. distrib. Texas: Locally uncommon M and SR on the coast, usually seen offshore.

FEEDING BEHAVIOR. Masked Boobies sometimes follow ships hundreds of miles from land. They fly with steady wingbeats that alternate with glides, and they make vertical dives from 80 or more feet that may plunge them as deep as 10 feet below the surface. They consume many flying fish and lesser numbers of small squid and other invertebrates.

Frigatebirds sometimes grab boobies by the tail and force them to disgorge.

VOICE. Because these birds are generally silent at sea, the male's whistles and the female's quacking sounds are infrequently heard in Texas.

Northern Gannet, *Morus bassanus.* Princ. distrib. Texas: Uncommon to locally common M on the coast, often seen offshore.

FEEDING BEHAVIOR. Gannets feed on small fish, especially those in schools, by seizing them with the bill and swallowing them underwater. They plunge into the water headfirst, from a height of 100 feet or more. They also forage by swimming and picking food from the surface or by submerging the head to locate prey just beneath the surface. They sometimes take offal from around fishing boats and steal food from other birds.

VOICE. Guttural barks and other harsh sounds, probably not heard away from the nesting area.

OTHER BEHAVIORS. Flight: swift (45 mph), with rapid wingbeats and short glides. Most gannets observed in Texas are immatures wandering at sea, a typical behavior during the first 3 years of life.

PELICANS: FAMILY PELECANIDAE

To most of us, pelicans are recognized by their enormous throat (gular) pouch—an expandable sac that holds about 3 gallons of water. Pelicans use this pouch as a scoop to capture fish and other aquatic animals that they swallow after they tilt their head to drain out the water. They do not carry fish in the pouch.

The pouch is made up of naked skin that has numerous blood vessels close to the surface. It pulsates to cool birds in hot weather.

These highly social birds nest in colonies, some of which are quite large. They lay their eggs in scrapes on the ground or on stick platforms that they build in trees.

Displays emphasize the pouch and the large bill, as when a bird opens its mouth wide and faces the opponent (Fig. 1). Other behaviors that have been interpreted as displays (for example, pointing the bill up or spreading the gular pouch) may be simply stretching movements that maintain flexibility in the pouch.

Pelicans are large, heavy birds. They expend considerable energy running across the water to take flight, but once aloft, they fly with powerful wingbeats and, as someone remarked, "with great solemnity and dignity."

American White Pelican, *Pelecanus erythrorhynchos.* Princ. distrib. Texas: Common M in e. half of Texas, becoming less common to the w. Common WR in s. Texas, especially on the coast. Uncommon, local SR on the coast.

FEEDING BEHAVIOR. They feed principally on fish that have little commercial or recreational value. They dip their bill into the water and scoop up fish with their distensible gular pouch. Sometimes before a pelican can empty the water and swallow the fish it is spotted by a gull that soon lands on the pelican's head and deftly snatches the fish from the pouch.

White Pelicans feed cooperatively, not a common behavior among birds. Up to several hundred swimming birds form a large circle in the water. They approach each other to make the circle smaller and smaller while violently flapping their wings to frighten the fish to the center. When the fish are sufficiently concentrated the pelicans dip down in concert and scoop up their meal.

In another version of this feeding strategy, pelicans form a line parallel to the shore and drive fish to shallow water.

Most likely it is not possible for White Pelicans to make the high dives characteristic of Brown Pelicans, as they generally weigh at least twice

FIGURE 1. Defensive posture of Brown Pelican, emphasizing large mouth. (From a photograph by Rod Rylander.)

as much as the latter species. Also, their buoyant bodies make them ill-equipped to forage in this manner.

During the breeding season they regularly feed at night. (In Canada, two to three times as many pelicans were observed foraging at night as during the day.) Usually their feeding areas are distant from their nesting sites.

COURTSHIP BEHAVIOR. The female bows low to the male, who approaches her, extends his neck over hers, and sways. Groups containing both sexes circle high in the air during courtship flights, and on land they strut about with their heads held erect and their bills pointed downward, a posture that emphasizes the nail on the bill.

NESTING. Colonial, in coastal locations, including islands in shallow bays, and generally on flat terrain. Nest: on open, bare soil, in grassy areas, or under trees. The two eggs are incubated by both parents, who hold them on or under the webs of their feet. Unless food is abundant, the second nestling to hatch usually dies within a couple of weeks. If it does not die of starvation, the older sibling may kill it. Both parents bring food to the nestlings.

Embryos communicate with the parents through the shell. One sound they produce prompts the parents to move the egg so as to regulate the temperature of the embryo.

VOICE. Adults are generally silent, although they sometimes make a low clucking sound. Young birds in the nest bark and squeal.

OTHER BEHAVIORS. These birds soar at great heights, especially in stormy weather. They may do this to avoid the storm, or perhaps simply as a form of play. In spite of their weight (normally 10–15 pounds), they soar easily because of their extraordinary wingspan — the longest (9 feet) of any Texas bird. They fly in vee formation or in lines, alternately flapping and gliding with slow, powerful wingbeats. They fly lower over water than over land.

On water they swim buoyantly with the help of air sacs distributed in their breast tissues. On land they waddle clumsily.

Unlike many birds in this order, they do not perch in trees but instead roost at night on the ground. Yearlings form pods that wander northward before they begin their migration south. They fly over virtually any habitat, including deserts.

They shed the nail (a small, vertical plate on the upper surface of the bill) after the eggs are laid, supporting the suggestion that nails play a role in courtship. Piles of nails accumulate on the ground around roosting areas.

There is a report of a group of White Pelicans that kept a blind member alive by regularly feeding it.

Brown Pelican, *Pelecanus occidentalis.* Princ. distrib. Texas: Uncommon to locally common PR on the coast, where it nests in colonies.

FEEDING BEHAVIOR. Brown Pelicans plunge into the water from as high as 60 feet, submerging themselves violently, and resurfacing with a fish in their pouch. The force of the bill hitting the water expands the gular pouch and fills it with 2–3 gallons of water, along with the prey items. Evidently the bird realizes immediately if prey has been caught. If nothing is in the pouch, the bird rapidly raises its head out of the water, with bill open, so as to drain the pouch; if prey is caught, it slowly raises its head, with the bill closed, so that the water drains out and the prey remains inside to be eventually swallowed.

Brown Pelicans sometimes scavenge for food. They habituate readily to humans who feed them fish.

COURTSHIP BEHAVIOR. While the female is squatting on bare ground, the male silently circles around her, lifting his wings slightly and throwing his head back. When she rises from the ground and flies away, he follows her.

NESTING. Nest: in Texas, a simple scrape in the ground or a depression in a pile of debris. The female builds it with materials brought to her by her mate. Both parents incubate the eggs by holding them on or under the webs of their feet, and both feed the young regurgitant.

Brown Pelicans may practice brood reduction when food is scarce, the smaller chicks succumbing to the aggression of the larger ones.

After a few weeks, juvenile birds leave the nest and spend the day together in groups called creches. Evidently, parents returning to the creche quickly recognize their own offspring.

VOICE. Although young birds make grunts, barks, and squeals, their parents are generally silent.

OTHER BEHAVIORS. One of the more captivating sights in nature is a hundred or so Brown Pelicans flying low over the water en route to their evening roosting sites. They stretch out into a long line of evenly spaced birds, each individual coordinating its flaps and glides with the bird immediately in front.

CORMORANTS: FAMILY PHALACROCORACIDAE

Cormorants are black, crowlike waterbirds easily recognized by their large heads and spindle-shaped bodies. In flight they hold their heads above the horizon.

By squeezing air from their plumage they are able to swim partly submerged. They pursue fish underwater with considerable skill, propelling themselves with their feet. Yet they must monitor the length of their dives carefully, as their plumage is extremely absorbent—a plus because wet plumage increases specific gravity and facilitates underwater swimming. On the other hand, cormorants become waterlogged if they remain in or under the water too long; thus they cannot rest in the water like ducks, but rather must come to shore to dry out, which they do by assuming an eagle-spread posture.

Apparently, cormorants utilize this spread-wing behavior only to dry their plumage, as only wet individuals assume this posture. In contrast, the closely related Anhinga apparently spreads its wings not only to dry the wet plumage but also to warm its body, as Anhingas are less able than cormorants to endure cold ambient temperatures.

Cormorants capture aquatic animals in addition to fish. They bring prey to the surface before swallowing it, later regurgitating indigestible bones and scales as pellets.

Cormorants nest socially in colonies that include other waterbirds.

Neotropic Cormorant, *Phalacrocorax brasilianus.* Princ. distrib. Texas: Uncommon to locally common PR on the coast, less common inland.

FEEDING BEHAVIOR. They dive for fish, frogs, aquatic insects, and other animals, usually in protected waters. They very rarely plunge from the air. These cormorants sometimes forage in groups, beating the water with their wings to drive fish into shallow water. (Compare the feeding behavior of the White Pelican.) Adults capture prey more efficiently than do immatures, indicating a strong learned component in their foraging behavior.

COURTSHIP BEHAVIOR. The male sits with his tail elevated and his bill pointed toward the sky as he raises and lowers the tips of his folded wings. Both males and females stretch their necks upward, open their bills, and wave their heads back and forth. This courtship behavior contrasts strongly with the Double-crested Cormorant's and no doubt is an important mechanism for preventing the two species from interbreeding.

NESTING. Colonial. Nest: in bushes, trees, or on the ground, and sometimes on man-made structures; it is probably built by both parents. Both parents incubate the eggs by placing them on the webs of their feet, and both parents feed the nestlings.

VOICE. Generally silent. Call: guttural, piglike grunts (Pough, 1951).

OTHER BEHAVIORS. They perch on slender twigs and wires, assuming the typical eaglelike posture, and repeatedly make squatting or bouncing movements. They are generally unresponsive to moderate human activity.

Double-crested Cormorant, *Phalacrocorax auritus.* Princ. distrib. Texas: Uncommon to abundant M in most parts of Texas; common to locally abundant WR on the coast and on larger bodies of water inland. They nest in colonies at scattered locations.

FEEDING BEHAVIOR. These birds fish in both clear and muddy waters, but they typically avoid the bottom of the lake or bay. They feed alone or in loosely formed groups that sometimes contain a thousand or more birds. When foraging in a group, all birds face the same direction and dive more or less in synchrony. They do not dive from the air.

COURTSHIP BEHAVIOR. The male splashes the water with his wings, swims around in a zigzag course, then dives to the bottom for bits of vegetation that he brings to the surface. At the nest he calls loudly while vibrating his wings, and during courtship he frequently assumes "grotesque" postures (Pough, 1951).

NESTING. In colonies. Nest: usually on the ground or in a tree. The female builds a platform from sticks and debris brought by her mate. Both parents incubate the eggs and bring food to the nestlings. Young birds wander through the colony a few weeks after hatching but return to the nest periodically to be fed by their parents.

VOICE. Virtually silent, but at the nest they make a variety of vocalizations, some sounding like *tik*s and *ook*s; when alarmed, they utter a grunt that suggests a bullfrog.

OTHER BEHAVIORS. When taking flight from a limb, they drop slightly to gain flight speed, but thereafter their flight is strong and direct.

They seem to favor dead trees as perches; here they assume their characteristic drying posture with the bill pointed up and the wings spread in eagle fashion. They habitually perch on wires—dozens can be observed any winter day perched on power lines near Aransas Pass (Michael Marsden, pers. comm.).

DARTERS: FAMILY ANHINGIDAE

See the family account for cormorants.

Anhinga, *Anhinga anhinga.* Princ. distrib. Texas: Uncommon to locally common M and SR in e. and c. Texas, as well as on the Coastal Bend, in quiet, sheltered waters. They favor freshwater ponds and marshes over brackish habitats.

FEEDING BEHAVIOR. In contrast to cormorants, which swiftly and tenaciously pursue their quarry underwater, Anhingas swim slowly either just beneath or on the surface of the water, waiting for fish to pass near. A special mechanism in their neck equips them to stab prey with remarkable speed. They sometimes dive into the water from flight.

COURTSHIP BEHAVIOR. After selecting a nesting site in the colony, the male attracts a mate to the site by waving his wings, lifting his tail, pointing his bill upward, and bowing deeply.

NESTING. In isolated pairs or, more often, in mixed colonies with cormorants and herons. Nest: a platform built by the female from sticks and green cypress foliage brought by her mate. Sometimes the pair appropriate an old heron nest. Both parents incubate the eggs and bring food to the nestlings.

Birds climb at an early age, working their way up tree limbs using their feet and bill. When disturbed they drop to the water but soon climb back up to the nest.

VOICE. Generally silent; at the nesting site they are quarrelsome and challenge each other with numerous clicking and croaking sounds.

OTHER BEHAVIORS. They dry their plumage by spreading their wings like cormorants, but they carry the process a step further by alternately turning their wings to expose both upper and lower surfaces.

They assume the spread-winged posture more frequently when temperatures are low; at these times they also turn their back to the sun to expose more surface area. They respond to high ambient temperatures by facing into the sun and fluttering their gular pouch (as a form of panting).

Anhingas sometimes swim with only their slender head and neck emerging from the water (hence the colloquial name "snakebird," a rather thin allusion, as snakes hardly swim in this manner). They rarely take flight from the water, typically ascending to a perch first. Their flight is strong and graceful. During migration and especially during the nesting season they spread their long tails and soar at high altitudes on outstretched wings, often in company with Turkey Vultures; under these circumstances they appear as tiny, dark crosses.

They spread the tail—a unique structure because of tiny ridges that cross from side to side—in a turkeylike fashion, giving rise to yet another colloquial name, "water turkey."

When they emerge from the water, Anhingas flap their wings and spread their tails. They move around clumsily when perched on limbs.

FRIGATEBIRDS: FAMILY FREGATIDAE

Magnificent Frigatebird, *Fregata magnificens.* Princ. distrib. Texas: Uncommon visitor on the coast in summer, sometimes occurring far offshore.

FEEDING BEHAVIOR. Their ability to fly gracefully and with agility accounts for their unparalleled skill and efficiency in foraging. Flying low over the water, they daintily dip their long bill (and sometimes their head) into the water to capture fish and numerous species of invertebrates; over land they pick up both dead and live fish that have washed up on shore, all the while remaining in flight. While circling around docks and fishing boats they consume offal thrown into the water, and under these circumstances they seem quite fearless of humans. Why frigatebirds soar inland over coastal cities, as they often do, is not clear.

Frigatebirds regularly rob gulls and other birds, forcing them to drop their catch or harassing them until they helplessly disgorge their partially digested stomach contents.

Apparently frigatebirds forage more successfully when high winds and turbulent water cause prey either to jump out of the water or to be thrown into the air. Accordingly, it is said that stealing food from other birds intensifies during calm weather.

VOICE. Apparently always silent in Texas.

OTHER BEHAVIORS. They soar for hours on seemingly motionless wings, or fly with deep, strong wingbeats, sometimes opening and closing their forked tail. They often take flight from an elevated perch and into the wind.

When flying low over water they do not let their highly absorbent breast plumage touch the surface. Unlike all other members of this order, they never swim, mainly because their long wings (and probably their absorbent plumage) prevent them from taking off while in water.

Their feet, which are tiny and out of proportion to their bodies, make them walk awkwardly on land.

During his courtship display, the male inflates his red gular pouch to the size of a large balloon, a well-known behavior (seen in so many Galápagos Islands films) that does not occur in Texas.

HERONS, IBISES, STORKS, AMERICAN VULTURES, AND ALLIES: ORDER CICONIIFORMES

HERONS, BITTERNS, AND ALLIES: FAMILY ARDEIDAE

Most herons and egrets prefer open water to vegetated areas, with lakes being a favorite habitat. In general, longer-legged species favor deeper water, although both they and short-legged species occur in shallow water. High overlap in habitat preference occurs between Reddish Egrets and Great Egrets, between Reddish Egrets and Tricolored Herons, and between Great Blue Herons and Snowy Egrets. Great Egrets and Tricolored Herons show almost total overlap.

Herons and egrets stalk or ambush fish and other aquatic animals; in both cases they capture prey by rapidly thrusting their spearlike bills and grasping (not spearing) the animal. A specialized vertebral arrangement allows them to straighten their S-shaped necks with astonishing swiftness.

Capture success depends on the striking angle, which becomes more acute as a fish moves away from the egret. Capture success for Little Egrets (*Egretta garzetta,* an Old World close relative of the Snowy Egret) increases when the striking angle becomes more acute, possibly because, from the fish's vantage point, egrets appear closer to the horizon and therefore are more difficult to detect.

Most members of this family can alight on water and take off with little difficulty.

Indigestible parts are regurgitated as pellets. Unlike most waterbirds, these birds leave the water to defecate.

American Bittern, *Botaurus lentiginosus.* Princ. distrib. Texas: Locally uncommon WR on the coast, and locally common M throughout the state, in marshes, bogs, and other dense wetland habitats. They rarely leave the protection these plants afford.

FEEDING BEHAVIOR. Standing alone and motionless on the shore or in shallow water, these highly camouflaged birds wait for fish, frogs, tadpoles, salamanders, snakes, and other animals to pass by, which they capture with a sudden thrust of the bill. Sometimes they pursue animals by walking slowly and stealthily, like a heron or egret. They forage throughout the day but are most active at dawn and dusk.

VOICE. Call: an unforgettable pumping sound uttered at dusk or at night. This deep, guttural croak, rendered as *pump er lunk,* is repeated two to seven times and carries up to half a mile.

OTHER BEHAVIORS. Alarmed birds freeze with the neck outstretched and the bill pointed to the sky; in this posture they blend well into the surroundings because of the vertical stripes on their plumage. When threatened at the nest, a bird aggressively points its bill at the intruder and makes itself appear larger by spreading its wings and puffing out its feathers.

When flushed from a feeding area — reluctantly, as they are more likely to hold their ground — they take flight with wings flopping loosely and feet dangling, dropping down after a short distance to find concealment in dense vegetation. When they fly farther distances they do so with strong, regular wingbeats.

Unlike some smaller bitterns, they never land on cattails or other emergent plants. American Bitterns are excellent mimics of wetland vegetation; they have been observed waving gently in concert with cattails and sedges that sway back and forth in the breeze.

Least Bittern, *Ixobrychus exilis.* Princ. distrib. Texas: Uncommon to locally common M and SR in most parts of Texas, in freshwater marshes and shallow ponds. Apparently they favor deeper water than the American Bittern. On the Coastal Bend they also breed in coastal salt and brackish marshes (Michael Marsden, pers. comm.).

FEEDING BEHAVIOR. For the most part they eat small fish, but they also take crayfish, frogs, leeches, and other aquatic animals. Their foraging behavior contrasts sharply with the American Bittern's: they climb about in emergent vegetation, often over deep water, and from their perch they jab down with their bill to capture animals on the surface. They flick their wings, presumably to startle prey, and sometimes bend vegetation to form a feeding platform next to the water.

One method of capturing prey is distinctive. After spotting an animal, a bird will inch its head slowly toward the water while slowly undulating its neck, possibly to mimic moving vegetation. At the critical distance it suddenly thrusts its head forward and seizes the victim. Afterward the bird usually leaves the area to resume foraging in a nearby site.

COURTSHIP BEHAVIOR. Although the predominant courtship behavior is vocal, the pair also engages in contact and noncontact bill clappering, and possibly courtship feeding. Territories are advertised by assuming characteristic postures.

NESTING. Isolated pairs or loose colonies. Nest: hidden in tall marsh vegetation. It is a platform that the male builds (with help from his mate) by bending down plants and adding sticks and grass. Both parents incubate the eggs and feed regurgitant to the nestlings.

When predators approach the nest, parents simulate a larger bird by fluffing out their feathers and partially spreading their wings. Young birds leave the nest in less than a week, thanks to rapidly developing legs and feet that allow them to climb with ease; however, they generally remain around the nest for another week or so.

VOICE. Call: generally heard at dawn and dusk. The male produces a cooing *uh uh uh uh uh oo oo oo oo oo oouh*, to which the female replies with *uk uk uk*.

OTHER BEHAVIORS. These birds are elusive, but in well-visited places, such as the wildlife preserve at Port Aransas, they readily habituate to human intruders and can be approached in full view at less than 15 feet.

When flushed, they fly away weakly, with legs dangling, then quickly drop back down into the vegetation. When alarmed, they freeze with their neck and head pointed toward the sky.

Great Blue Heron, *Ardea herodias.* Princ. distrib. Texas: Uncommon to common PR throughout most of Texas; in winter, less common inland than on the coast. They frequent fresh- and saltwater wetland areas, in particular, marshes, lakes, and estuaries, as long as the water is calm.

FEEDING BEHAVIOR. On the shores of almost any lake or impoundment, where the noisy palavers of Killdeer and the restless flights of blackbirds always seem to dominate the landscape, one's attention is eventually drawn to the calm, stately Great Blue Heron, poised like a statue with neck extended and erect, patiently surveying the still water for passing prey. Whether standing motionless or walking slowly in the shallow water, it strikes quickly and accurately at fish, frogs, or snakes that pass within its reach.

Thus, it surprises us when these birds employ feeding strategies that seem out of character, such as diving from the air or from a branch in pursuit of a swimming fish, or capturing rodents and birds concealed in wet grasslands. In one instance a Great Blue Heron repeatedly stabbed a cottontail until it died.

These herons feed at night as well as during the day, but they are most active at dawn and at dusk.

COURTSHIP BEHAVIOR. The male stretches his neck upward and points his bill to the sky; he also circles above the colony with his neck extended. On the ground he erects his feathers and snaps his bill. He has a larger repertoire of displays than most egrets, yet he displays less often. He continues to display after the pair bond is established.

NESTING. In colonies, often with other wading birds, and in the general vicinity of foraging areas. Nest: in a tree 20–60 feet above ground or water; other sites, including the ground, are sometimes selected. The male brings sticks to his mate, who uses them to construct a large platform. Both parents incubate the eggs and feed regurgitant to the nestlings.

VOICE. Normally silent, except for gooselike honking noises in flight and a hoarse, guttural croak when alarmed.

OTHER BEHAVIORS. Flight is strong and deliberate, yet leisurely in pace. The frequently applied term "majestic" aptly describes this bird's demeanor both on the ground and in the air.

Great Egret, *Ardea alba.* Princ. distrib. Texas: Locally common SR in e. and c. Texas, and common PR on the coast. Locally uncommon WR in scattered locations, especially in e. and c. Texas. They favor lake shores, shallow coastal lagoons, large marshes, and other open areas, but they also occur on rivers in forested areas.

FEEDING BEHAVIOR. Their foraging posture is characteristic: standing in shallow water, the bird holds its neck erect and leans the neck and body forward. When prey (in particular, fish) approach, it extends the neck with lightning speed and captures the animal with the spearlike bill.

Great Egrets also feed on crustaceans, salamanders, snakes, and other aquatic animals. In pastures, where they often associate with cattle, they consume many insects. Over water they fly low and drag their feet on the surface, possibly to lure fish.

They often forage in mixed groups that include White Ibises or other waders. Field studies in Texas show that Great Egrets feeding in groups made up of other Great Egrets have higher capture success than solitary foragers. However, overall prey intake rates (grams per minute) are similar between group and solitary foragers, probably because solitary birds capture larger prey. On the other hand, foraging in flocks may be more efficient because the birds expend less energy chasing conspecifics (Wiggins, 1991).

Great Egrets steal prey from smaller species (kleptoparasitism), a feeding strategy shown to be five times as efficient as this species' typical for-

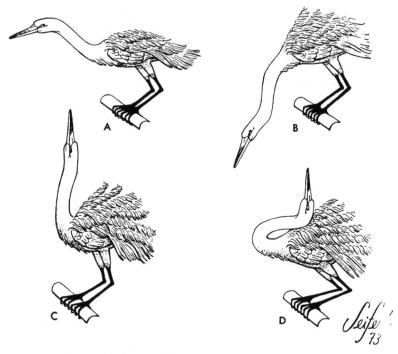

FIGURE 2. The stretch display of the Great Egret as employed during courtship. (D. A. McCrimmon, 1974, *Wilson Bulletin* 86: 165–167. With permission.)

aging behavior. Unlike many members of this family, these birds do not feed at night.

COURTSHIP BEHAVIOR. After selecting a nest site, the male engages in displays to drive away competitors and to court the female. These displays can be grouped as either reactive or spontaneous. During reactive displays the bird maintains a relatively fixed orientation toward another bird. Reactive displays tend to be aggressive responses to threats and include a fluffed neck and bill clappering. Spontaneous displays have a much less rigid orientation and apparently function mainly to attract the female. They include the wing preen, twig shake, and stretch display (Fig. 2).

NESTING. Usually in colonies of up to a thousand birds, with herons, egrets, Anhingas, or cormorants. Favored areas in Texas are *Scirpus* marshes or coastal islands dominated by mesquite, huisache, prickly pear, and salt cedar (Oberholser and Kincaid, 1974). Nest: in a shrub or tree. Both parents construct a flimsy platform of sticks.

Both parents incubate the eggs and feed regurgitated food to the nestlings. A few weeks after hatching, the young leave the nest, climb around in nearby branches, then return to the nest.

VOICE. Call: a deep, harsh, rattlelike *cuk cuk* or *fraaawnk*.

OTHER BEHAVIORS. In the evening they fly to large communal roosts that they share with other waders. Their flight is buoyant and graceful, in part because their wings are proportionately longer and broader than those of most white herons.

In their greeting ceremony, the birds erect their plumes and raise their wings.

Snowy Egret, *Egretta thula.* Princ. distrib. Texas: Uncommon M in most parts of the state; uncommon SR in e. half of Texas and common PR on the coast. They inhabit swamps, marshes, lakes, bays, and other wetland habitats.

FEEDING BEHAVIOR. Although superficially resembling the larger Great Egret (which weighs about three times as much), Snowy Egrets are much livelier and have more diverse foraging techniques. They sometimes stand quietly in still water and wait for passing prey, yet their activity level more closely approaches that of the Reddish Egret. They run in shallow water with partly raised wings, stir the bottom sediments with their feet, hover over the water like a petrel and drop for prey, feed next to other waders (to take advantage of aquatic animals that are stirred up), or follow cattle in pastures to catch insects that are flushed.

Fish comprise a relatively small proportion of their diet, the more important items being crustaceans, insects, and other invertebrates. When feeding on crayfish, they select recently molted individuals that have a softer exoskeleton.

COURTSHIP BEHAVIOR. The nuptial plumage of most herons and egrets is outstanding, but that of these well-proportioned birds is unparalleled: the delicate plumes on the head and neck are transformed into an elegant, fan-shaped arrangement that is quickly raised and lowered throughout courtship displays.

The male selects a nest site and engages in displays that both advertise his territory and attract a mate. He points the bill toward the sky, pumps his head up and down, flies in circles around the nest site, or flies high and tumbles in a spectacular descent.

NESTING. Usually in colonies with other waders. Nest: in a tree or shrub, but sometimes in marsh vegetation or on the ground. Both parents build a platform from sticks fetched by the male, and both incubate the eggs and bring food to the nestlings. Eggs hatch asynchronously, and often the last bird that hatches receives less food and starves to death. A few weeks after hatching, the young leave the nest and climb around in nearby branches.

VOICE. These birds are very noisy at the beginning of the nesting season but at other times are remarkably quiet. Call: a harsh, grating, and generally unpleasant *raah* or hiss.

OTHER BEHAVIORS. They roost communally outside the breeding season, going to and from the roost in small flocks. Their flight is light and graceful.

Little Blue Heron, *Egretta caerulea.* Princ. distrib. Texas: Common SR and uncommon WR in e. and c. Texas, fairly common PR on the Coastal Bend. They occur in marshes, ponds, swamps, lakeshores, and occasionally dry fields, and favor (in Texas, at least) freshwater over saltwater habitats.

FEEDING BEHAVIOR. Little Blues consume basically any animal they can capture and handle. Food items include crabs, crayfish, grasshoppers, lizards, and spiders, as well as fish and crustaceans (probably their main food item). In shallow water or wet meadows they wait patiently for prey or walk along methodically, changing feeding sites frequently. They capture prey with a rapid thrust of the bill.

Their deliberate manner of feeding can be a useful identifying field mark in mixed groups of herons and egrets. Their gait has been described both as measured and dainty, and they generally do not wade as deeply as many herons. Occasionally one will wade up to its belly in water. They select rain pools, mud puddles, and ditches as feeding sites. When pools dry up, they readily take to dry pastures where they seem quite capable of surviving on grasshoppers.

COURTSHIP BEHAVIOR. The formation and maintenance of the pair bond is complex (see *Birds of North America,* No. 145) and includes behaviors in which the male displays by stretching his neck, snapping his bill, preening his wing, and vigorously driving away competing males. Courting pairs cross and intertwine necks, mutually clap their bills, and nibble at each other's plumage.

NESTING. In heronries that contain other nesting herons and egrets. In mixed colonies, Little Blue Herons generally occupy the outer edges of the colony, where they sometimes compete with Cattle Egrets for nesting sites. Nest: in a tree or shrub. Both males and females build the nest, which they complete in 3–5 days. It is a platform of sticks with a depression in the middle. The eggs are incubated by both parents, and both parents feed regurgitated food to the nestlings. A few weeks after hatching, the young leave the nest and climb around in nearby branches.

VOICE. Usually silent; sometimes they make a clucking or croaking sound. Sounds made by quarreling birds suggest a flock of noisy parrots.

OTHER BEHAVIORS. These wary birds are usually difficult to approach. They roost in trees at night and fly in small groups to feeding areas in the morning. Their flight is strong, graceful, and unhurried. When alighting, they are able to descend almost vertically.

Many birds, especially the white-plumaged juveniles, move northward in Texas in late summer, then later migrate south.

Tricolored Heron, *Egretta tricolor.* Princ. distrib. Texas: Uncommon to common PR on the coast, frequenting swamps, streams, marshes, and bays. They are second only to the Reddish Egret in their preference for saltwater habitats, but it is not unusual for small numbers to nest inland regularly.

FEEDING BEHAVIOR. As a Tricolored Heron moves from one feeding area to another, almost always alone, it establishes and defends a small feeding territory at each location. It stands quietly as it waits for approaching prey, but it also forages actively, dashing quickly, with wings partly spread, perhaps to confuse the prey. During fishing maneuvers these herons are more graceful and less frenetic than Reddish Egrets.

Some observers have noted Tricolored Herons flying relatively long distances in order to forage in natural rather than disturbed habitats — in contrast to Snowy Egrets and Great Egrets, which readily accommodate themselves to highway ditches and other small bodies of water that are man-made.

They mainly consume fish, sometimes dashing after those that swim in schools. They also take salamanders, crustaceans, frogs, insects, and other animals and often stir the bottom sediment with their feet to disturb prey.

COURTSHIP BEHAVIOR. After selecting a site within the heronry, the male displays by stretching his neck, bowing deeply, hopping around the female, raising his crest feather, calling, clappering his bill, and flying overhead in circles. Prior to the stretch display, his throat swells (as in Fig. 3) and he utters an *unh* call (Rodgers, 1977).

NESTING. Gregarious, nesting in colonies with other wading birds. Nest: in trees, thickets, dry scrub, or mangroves, and sometimes on the ground. The male brings sticks to his mate, who builds a platform and lines it with softer materials. Both parents incubate the eggs and feed

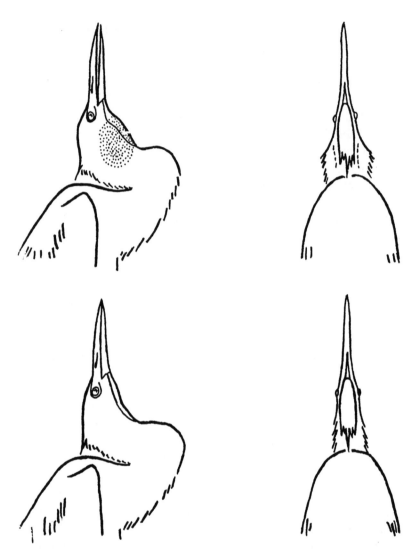

FIGURE 3. Prior to the stretch display, the throat of the Tricolored Heron swells (stippled) and the bird utters an *unh* call. (J. A. Rodgers, Jr., 1977, *Wilson Bulletin* 89: 266–285. With permission.)

the nestlings. Young birds only a few weeks old climb about in nearby branches.

VOICE. Croaks and squawks heard mainly on the nesting areas; however, quarreling birds make a variety of other sounds, two of the harsher ones rendered as *aaah* or *raah*. Compared to other herons and egrets, these birds are exceptionally noisy.

OTHER BEHAVIORS. The members of a pair greet each other by raising their crests and engaging in bouts of bill nibbling (bill clappering). Bill nibbling, which consists of opening and then gently closing the mandibles to create a rattling sound, apparently reduces aggressive behavior.

Reddish Egret, *Egretta rufescens.* Princ. distrib. Texas: Locally common PR on the coast, infrequent in freshwater habitats.

FEEDING BEHAVIOR. The Reddish Egret's foraging antics are unexampled. After assuming an ungainly posture, the bird tilts its head so that one eye sees the water and the other is aimed at the sky. Then, without warning, it dashes through the shallow water with exaggerated strides while raising and lowering one or both wings. As if its behavior at this point does not already appear dysfunctional, the bird now abruptly changes directions, leaps excitedly into the air, and finally stabs at a fish. Following capture, it may jump sideways, flap its wings back and forth, pause, and repeat the whole performance, usually executing the above movements in another sequence.

Precisely how such behavior facilitates prey capture is not clear. Sudden and unpredictable movements might surprise and confuse fish, or shadows cast by spread wings might entice fish to seek shelter beneath the wings.

Interestingly, its foraging behavior at other times can be quite placid; also, like several other herons and egrets, it uses its foot to stir up prey. In whatever manner it chooses to feed, it almost always defends its feeding territory with vigor.

COURTSHIP BEHAVIOR. Courtship is as spirited as foraging. In one display (for which there are several variants), the male raises his shaggy back feathers, stretches out his head and neck, and repeatedly thrusts his head forward. In another, he circles around the female while tossing his head and raising his wings.

NESTING. In arid coastal islands covered with thorny vegetation, on a site selected by the male. Nest: a platform of sticks built on the ground by both parents. Both parents incubate the eggs and feed the nestlings. About a month after hatching, but before they can fly, young birds leave the nest and wander around the area.

VOICE. Call: a guttural *squawk,* as well as chickenlike notes during the nesting season. At other seasons this is generally a quiet bird.

OTHER BEHAVIORS. In the greeting ceremony, head tossing is accompanied by bill clappering, during which birds touch bills.

Compared to many herons and egrets, this is a relatively unwary bird. It flies with strong, graceful wingbeats.

Cattle Egret, *Bubulcus ibis.* Princ. distrib. Texas: Common to locally abundant SR in many parts of the state, especially coastal areas; less common to the w. They are a locally abundant WR in extreme s. Texas. They frequent pastures, plowed fields, flood fields, and other open habitats. Cattle Egrets are native to Africa. They appeared in the United States in 1952. Apparently they occupy a largely unexploited niche and compete very little, if at all, with native birds.

FEEDING BEHAVIOR. Cattle Egrets seldom wade in water to feed; indeed, their success rate in capturing submerged prey is lower than that of other herons and egrets. They forage near the head or feet (sometimes on the back) of cattle and other ungulates, or follow tractors for insects that are flushed. They jealously guard a feeding territory that moves with the horse, cow, tractor, etc.

These birds also forage independently of large animals, and they readily capture and eat frogs, small snakes, young birds, and other animals of comparable size. They make undulating movements with their necks when stalking prey. These neck movements may help the egret estimate its distance from its prey.

They have been known to fly to distant grass fires to take advantage of fleeing animals.

COURTSHIP BEHAVIOR. After the male selects a territory in or near a colony, he displays to the female by stretching his neck, raising his plumes, and swaying from side to side. He also flies around the area with deep, exaggerated wingbeats. Courting birds mutually preen each other's back feathers and engage in contact bill clappering. Extrapair copulations are fairly common.

NESTING. Colonial, with other herons and egrets. Nest: a platform in a tree or shrub usually built by the female from sticks brought by her mate. Sometimes sticks are stolen from another nest. Both parents incubate the eggs and feed regurgitant to their offspring. The nest is always attended by one of the parents.

As long as his mate is fertile, the male constantly remains at or near the nest, an anticuckoldry strategy that insures that his own genes are passed on. A few weeks after hatching, the young leave the nest and climb around in nearby branches.

VOICE. Normally silent except during the breeding season, when they utter barks and croaks. When disturbed, they sometimes produce a harsh gargle.

OTHER BEHAVIORS. Cattle Egrets greet each other by lowering their crests and raising their back plumes. Each morning they depart together from their communal roosts in trees. Their flight is quicker, shallower, and less graceful than most egrets. They readily habituate to human activity.

Green Heron, *Butorides virescens.* Princ. distrib. Texas: Common SR in e. two-thirds of the state, in most saltwater and freshwater habitats; they favor small bodies of water that are lined with vegetation.

FEEDING BEHAVIOR. These intriguing herons engage in several different foraging strategies. Birds standing by the water's edge waiting for fish to pass by "often freeze motionless in odd positions and hold them with remarkable patience" (Pough, 1951); at other times they walk along the shore "stealthily, catlike, putting each foot down carefully" (Terres, 1980). They have been observed dropping a feather or twig on the water's surface to bait fish. In one extraordinary case of baiting, a Green Heron placed bits of bread in the water, captured fish that were lured by the bread, and even defended the bread against birds that tried to eat it.

Green Herons also rake the bottom mud with a backward stroke and examine the water for moving prey, as well as dive for fish from logs or low limbs.

Besides fish, they take crustaceans, aquatic and terrestrial insects, tadpoles, and frogs. They feed primarily at dawn and dusk.

COURTSHIP BEHAVIOR. After selecting the nesting territory, the male calls repeatedly from a tree limb or other prominent perch, even if a female is not present. In the presence of a potential mate, his displays intensify into the full forward display: he stretches his neck forward and down, audibly snaps his bill shut, and points it upward while swaying his body back and forth. He also erects his neck plumes, swells his throat, and hops from foot to foot.

NESTING. Usually in small groups or as isolated pairs. Proximity to water is not critical. The male begins building the nest, then brings nesting materials to his mate, who completes it. Both parents incubate the eggs and feed regurgitant to the nestlings. Before allowing his mate to relieve him at the nest, the male performs the stretch display.

About 2 weeks after hatching, the young leave the nest and climb about in nearby branches. They can also swim at this age.

VOICE. Call: a penetrating and far-carrying *skeow* that has an almost mechanical quality; also several clucks and grunts.

OTHER BEHAVIORS. The full forward display described under "Courtship behavior" is the most effective of their threat displays, as it intimidates most opponents.

Because of their size and their strong, deep wing strokes, Green Herons in flight are sometimes mistaken for American Crows. They sunbathe by partially opening their wings and facing the sun. When walking, they nervously flick their tail and raise and lower their crest.

Black-crowned Night-Heron, *Nycticorax nycticorax.* Princ. distrib. Texas: Common PR on the coast, and locally common SR throughout most of Texas. Locally uncommon WR inland. They inhabit both fresh and salt water, including rivers, tidal pools, ponds, and marshes. They also do well in urban habitats, especially lakes in parks where at night prey are attracted to the shoreline illumination.

FEEDING BEHAVIOR. They usually forage at dusk, dawn, and night. Competition with diurnal herons and egrets may be responsible for some of their nocturnal foraging behavior. Although their larger eyes probably enable them to discern objects over a wide range of light levels, their binocular vision is no better than that of diurnal herons and egrets such as Cattle Egrets.

They feed mainly on fish, but they are consummate opportunists that readily eat what is at hand, including insects, rodents, frogs, and carrion. They readily consume eggs and young birds of colonial nesters, especially gulls, terns, and blackbirds.

A characteristic foraging strategy is to walk slowly and deliberately while searching for prey. They also stand motionless in shallow water, then quickly capture prey that come within reach. Unlike most herons, they swim well, and to procure food they sometimes alight on water or dive headfirst into the water.

COURTSHIP BEHAVIOR. After selecting a nesting site, the male erects his breast feathers and back plumes, stretches his neck upward and forward, and bows. As he does this he treads from one foot to the other. At the lowest point in his bow he utters a hissing sound.

NESTING. In colonies with other wading birds, and often initiated slightly earlier than other herons. Nest: in trees and shrubs, frequently on islands or above water (perhaps to avoid predators). The female builds a platform 10–40 feet above ground, using sticks brought by her mate. Both parents incubate the eggs and feed regurgitant to the nestlings. A few weeks after hatching, young birds leave the nest and clamber about in nearby branches. They are aggressive and defend themselves by regurgitating and defecating on intruders.

VOICE. Call: a throaty *wok,* heard from birds flying overhead at dusk; when alarmed, they make a more rapid and higher pitched sound. Fledglings are surprisingly noisy.

OTHER BEHAVIORS. Their flight is slow, steady, and more labored than that of other herons. These birds are gregarious at all seasons. They adapt well to urbanization and frequently fly low over busy cities at dusk.

Yellow-crowned Night-Heron, *Nyctanassa violacea.* Princ. distrib. Texas: Uncommon to locally common SR in e. and c. parts of the state; locally common WR in s. Texas on the coast. They are found in bayous, cypress swamps, streams, rivers, and shallow tidal pools. These herons are less likely to inhabit marshes than are Black-crowned Night-Herons.

FEEDING BEHAVIOR. Compared to other herons, these are feeding specialists. On the coast, they feed predominantly on crustaceans; inland, they capture crayfish emerging from burrows. Thanks to a stout bill, they can manage crabs that other herons ignore. They only rarely capture fish.

Interestingly, they transport most of their larger prey items, such as crabs and eels, to areas away from the water before subduing them; here, captured animals have less opportunity to escape and return to the water. They feed both at night and during the day, apparently being slightly more diurnal than the Black-crowned Night-Heron.

COURTSHIP BEHAVIOR. The male gives a loud call as he erects his plumes, stretches his neck upward, and points his bill to the sky. He advertises his territory with circling flights. Juveniles that have not yet molted into breeding plumage sometimes pair with adults, but not with other juveniles.

NESTING. Isolated pairs or small groups, rarely in the large colonies characteristic of many waders. Nest: a platform of sticks in a tree about 30 feet above ground. Both parents incubate the eggs and feed the nestlings.

VOICE. Call: a *quack,* higher pitched than the Black-crown's call, and not as harsh or as guttural.

OTHER BEHAVIORS. Compared to Black-crowned Night-Herons, Yellow-crowns are generally more solitary and more secretive. They greet each other by raising their crests, nibbling at each other's feathers, and by contact and noncontact bill clappering.

IBISES AND SPOONBILLS: FAMILY THRESKIORNITHIDAE

These gregarious wading birds probe the mud or water and detect food items by means of touch-sensitive sense organs in their decurved or spatulate bills. They are strong fliers and always fly with their necks extended.

White Ibis, *Eudocimus albus.* Princ. distrib. Texas: Locally common PR on the coast, in salt and freshwater marshes, mudflats, flooded pastures, and grassy fields.

FEEDING BEHAVIOR. They walk leisurely through the shallows, sweeping their long, decurved bills from side to side as they probe the bottom mud for crustaceans (especially crayfish), worms, and other animals.

In general, this ibis is a nonvisual, tactile forager: it places its partially opened bill in the water or bottom sediment, then snaps it shut when it detects prey. Prey taken from the water's surface, mud, or short grass habitats are generally located by sight. White Ibises also steal food items from one another.

COURTSHIP BEHAVIOR. The male leans over, grasps a twig in his bill, points it to the sky, and lowers his head onto his back. He also engages in ritualized preening, with movements that are exaggerated and more stereotyped than in normal preening.

NESTING. In colonies with other wading birds. Nest: in trees or shrubs, 2–15 feet above ground or water; it is a platform built by the female from sticks brought by her mate. Sometimes they usurp the nest of a neighbor. Both parents incubate the eggs and feed the nestlings. To obtain their meal, young birds reach into the parent's mouth and pull out partially digested food.

When ibises on high salt diets are experimentally given fresh water, they quickly gain weight. This may explain why birds that breed in coastal colonies fly considerable distances inland to capture freshwater prey for their young.

Parents are sensitive to intruders and readily abandon their nests if disturbed. Young birds leave the nest when they are a few weeks old and climb around on nearby branches before finally abandoning the area.

VOICE. Call: a loud *hrunk hrunk hrunk,* when taking flight, as well as several grunting notes.

OTHER BEHAVIORS. White Ibises are one of the most sociable of the waders. They not only nest in colonies, some very large, but they also feed, loaf, and roost in flocks. Egrets and other waders follow White Ibises to capture animals that the latter species stirs up from the bottom sediment.

They fly swiftly and strongly, alternating rapid wingbeats with glides. Sometimes flocks ascend high in the sky and circle about. At dusk, birds as far as 15 miles away congregate at special roosting areas; to get there they fly in lines a mile or more in length.

Glossy Ibis, *Plegadis falcinellus.* Princ. distrib. Texas: Rare visitor on the coast, increasing in numbers and rarely nesting locally. Apparently their behavior differs little from the White-faced Ibis's. Their call is a grunt followed by four bleating notes.

White-faced Ibis, *Plegadis chihi.* Princ. distrib. Texas: Locally uncommon to common M in most of Texas, and locally common PR on the coast, in marshes, irrigated fields, and other freshwater (rarely saltwater) wetlands. They take advantage of changing water levels in temporary ponds and grasslands that result from flooding.

FEEDING BEHAVIOR. Birds fly considerable distances to locate good feeding sites. Groups of these highly gregarious birds wade in shallow water, probing the soft mud for crustaceans, insects, crayfish, and other animals. As they advance over a feeding area, birds at the rear leapfrog over those in front. They also pick insects from aquatic plants or from the surface of the water.

COURTSHIP BEHAVIOR. Apparently not recorded.

NESTING. In colonies. They select habitats that may change with varying water levels, for example, temporary habitats created by flooding. Nest: a low platform constructed by both parents from marsh plants or materials stolen from other nests. White-faced Ibises are less aggressive toward members of their own species than are most colonial waterbirds, and females sometimes lay eggs in the nests of other ibises.

Both parents incubate the eggs, and one parent is always on the nest. During their nest relief display, birds preen and coo. Both parents feed partially digested food to the nestlings.

VOICE. Call: a low croak or grunt (*ka oonk*), to some listeners suggesting a pig.

OTHER BEHAVIORS. At all seasons, flocks fly high and circle around in what might be play, although this behavior appears to be more common during the breeding season.

They fly in diagonal lines, with strong, rapid, and deliberate flaps that alternate with glides. They roost communally at night and depart together each morning for their feeding grounds.

Roseate Spoonbill, *Ajaia ajaja.* Princ. distrib. Texas: Locally common PR on the coast, in fresh, brackish, and saltwater lagoons, marshes, and mudflats.

FEEDING BEHAVIOR. Small flocks wade in shallow, muddy water, sweeping their heads from side to side as they sift through mud with their partly open spatulate bills. They detect small animals with minute sense organs inside the bill. They also visually locate shellfish and other invertebrates. Less frequently they consume roots and stems of aquatic plants.

COURTSHIP BEHAVIOR. Initially, the male and female are aggressive toward each other; later they present sticks to each other and cross and clasp each other's bills.

NESTING. In colonies on coastal islands. Nest: in a low tree or dense shrub, occasionally on the ground. The female builds a bulky, deeply cupped platform from sticks brought by her mate and lines the center of the depression with softer materials. Both parents incubate the eggs and feed the nestlings. They relieve each other at the nest with special calls. A few weeks after hatching, young birds leave the nest and clamber about in nearby shrubs.

VOICE. They are usually silent, but they make grunting sounds while feeding, as well as when alarmed.

OTHER BEHAVIORS. They fly with deep, slow wingbeats, gliding only occasionally. Flocks move in a line or in vee formation. Sometimes they fly up suddenly and circle around, especially at the beginning of the nesting season.

STORKS: FAMILY CICONIIDAE

Wood Stork, *Mycteria americana.* Princ. distrib. Texas: Postbreeding visitors (from nesting sites in Mexico) are locally common on the coast and on larger bodies of water in e. and c. Texas, sometimes occurring locally in large numbers. They favor freshwater habitats.

FEEDING BEHAVIOR. The Wood Stork wades in shallow water, walking carefully and deliberately with its partially open bill submerged in the water. When it detects an animal, by sight or touch, it snaps the bill shut.

VOICE. Generally silent, except on its nesting grounds. In flight, the wings make a *hwoof* sound.

OTHER BEHAVIORS. Their flight is slow, but their wingbeats are strong; they often soar on warm days. They frequently perch high in dead branches.

NEW WORLD VULTURES: FAMILY CATHARTIDAE

Until recently the New World vultures were considered raptors (Order Falconiformes), but currently they are regarded as more closely related to the storks. They eat mainly carrion; indeed, capturing and killing larger animals is almost precluded by their weak feet (which are, however, nicely adapted for walking around carcasses). Because their feet cannot carry food to the nest, they transport it in the stomach and feed regurgitant to their young.

Maintaining the correct body temperature requires no little effort. They defecate on their legs and feet, a behavior generally interpreted as a cooling mechanism (although no one seems interested in assessing the efficacy of this behavior). Vultures also pant when hot. At night their body temperature lowers, and in the early morning they spread their wings to capture the sun's warmth with their black, heat-absorbing plumage. On the Coastal Bend, Black Vultures become active at least half an hour later in the morning than do Turkey Vultures (Michael Marsden, pers. comm.).

Virulent bacteria ingested with the carrion are destroyed in the vulture's digestive tract; therefore, birds do not spread potential pathogens from one carcass to another or to young birds—a belief used to justify killing these birds in the past.

Black Vulture, *Coragyps atratus.* Princ. distrib. Texas: Common PR in lowlands throughout Texas except the High Plains and the Trans-Pecos; they rarely venture into the thick forests of e. Texas. Birds in n. part of the range move south in winter if carcasses remain frozen for long periods of time.

FEEDING BEHAVIOR. Black Vultures consume carrion, especially larger carcasses. Occasionally they eat live birds, mammals, plant material, and even young livestock if they can subdue them. They sometimes locate carrion by watching Turkey Vultures, which, being lighter on the wing, are able to fly lower and search the ground more closely. Turkey Vultures also locate food by scent. However, Black Vultures are the more aggressive of the two vultures and displace Turkey Vultures at carcasses. See Buckley's (1997) study of these two vultures in Texas.

Black Vultures that nest near heronries readily rob unguarded nests, as well as eat young birds that fall from the nest. Having good night vision, they are capable of foraging nocturnally.

There are reports of cooperative hunting in Black Vultures. Several birds — as many as 25 in one reported instance — surround a skunk, opossum, or other animal of comparable size and kill it by tearing off the skin with their stout bills. (The Turkey Vulture's predatory behavior is considerably less effective because of its weaker bill.)

COURTSHIP BEHAVIOR. One or several males chase a female, sometimes engaging in complex aerial displays that include spiraling downward. On the ground, the male circles the female with lifted wings and outstretched neck while exhaling noisily.

Most pairs probably remain together for several years, and evidently for life in some cases. The pair bond is reinforced by mutual preening, during which birds nibble at each other's lower neck feathers.

NESTING. Eggs are laid on cliffs, in caves, in tree cavities, on the ground in clearings, or in barns or other man-made structures. Both parents incubate the eggs and feed regurgitated food to the nestlings. There is no evidence that they defend a territory.

VOICE. They are typically silent, as their vocal apparatus is rudimentary, but they occasionally make muffled barks and hisses.

OTHER BEHAVIORS. Black Vultures regurgitate their stomach contents when threatened. The lighter body helps the birds take flight, and the unpleasant regurgitant may repel attackers. Young birds also disgorge food when annoyed, reingesting it without hesitation when danger is past. If threatened, they stretch out their wings, hiss, and lunge at intruders. This aggressive response contrasts markedly with the passive behavior of threatened young Turkey Vultures.

When trapped, adult birds struggle, bite, regurgitate, and throw the head violently from side to side. In spite of such vigorous defensive behavior, adult Black Vultures are less wary than Turkey Vultures and many become quite tame around urban areas if not molested.

Because their body weight and proportions reduce lift, Black Vultures require strong thermals for soaring and must regularly flap between glides. They do not wobble in flight like Turkey Vultures because their body weight stabilizes them, nor do they deviate from their course like the latter species. Black Vultures tend to hunt from great heights and to circle while hunting, whereas Turkey Vultures (which can also soar high) regularly course back and forth close to the ground.

The family unit usually remains intact throughout the year. Families often associate with other families at roosts, where information about food resources is possibly shared (in a manner not clearly understood). Black Vultures are more gregarious than Turkey Vultures and associate in larger groups when soaring, feeding, and roosting.

Turkey Vulture, *Cathartes aura.* Princ. distrib. Texas: Common SR in open habitats throughout the state, in both mountains and lowlands; much less common during the winter except in s. Texas (Michael Marsden, pers. comm.). They rarely overwinter in the Panhandle and Trans-Pecos, probably because carcasses there remain frozen for long periods of time. It has been said that in Texas during the nineteenth century young Turkey Vultures were eaten by humans.

FEEDING BEHAVIOR. *See* Black Vulture. Turkey Vultures typically locate carrion by sight; however, unlike most birds (including Black Vultures), their olfactory apparatus is functional and permits them to locate food items by scent. They rarely if ever kill live animals because of their relatively weak bill and feet. They have been known to eat vegetal matter (pumpkins) and feces (including human) when food is scarce.

Flying low over the ground, and often alone, Turkey Vultures search for carrion that cannot be detected at greater heights, including small toads and snakes smashed on the highway. (Black Vultures are rarely able to locate such small items.) Turkey Vultures are unique among North American vultures in being able to forage efficiently beneath the forest canopy. They frequently soar with Black Vultures, but more often they fly beneath them.

Black Vultures displace Turkey Vultures at the carcass, but both vultures give way to Crested Caracaras, which chase them and force them to regurgitate what they have just eaten. When looking for food, Turkey Vultures pick up cues from other scavengers, including American Crows, ravens, and even cats and dogs.

COURTSHIP BEHAVIOR. The male and female follow each other in flight, occasionally diving at one another. Later, on the ground, several males, with partly spread wings, rapidly bob their heads and strut and hop around a female.

Before they copulate, and sometimes immediately afterward, both members of the pair spread their wings, a display possibly derived from their sunning posture. They are monogamous during the mating season, and many birds form life-long bonds. Once the pair bond is established, birds strengthen it by touching their mate's bill.

NESTING. They usually do not make a nest, but when they do, they prepare a rudimentary structure of leaves and stones in a cavity or under a boulder. Some observers suggest that what appears to be nest-making by the incubating bird (i.e., drawing materials around the eggs in a haphazard manner) is nothing more than a response to boredom.

If the clutch is destroyed, the pair does not renest that year.

VOICE. They are essentially silent because of an undeveloped syrinx (vocal apparatus), but occasionally they produce a hiss or grunt.

OTHER BEHAVIORS. *See* Black Vulture. After feeding, Turkey Vultures perch nearby to digest their meal. They usually roost near water, sometimes in enormous groups of several hundred birds, and often alongside Black Vultures.

They conserve energy by lowering their body temperature at night. Undoubtedly, the hour or so they spend preening and sunning at the roost each morning enables them to regain their normal body temperature more quickly. Turkey Vultures generally do not defend a territory.

At roosting sites, as densities increase, Black Vultures displace Turkey Vultures at the upper levels.

SCREAMERS, SWANS, GEESE, AND DUCKS: ORDER ANSERIFORMES

DUCKS, GEESE, AND SWANS: FAMILY ANATIDAE

Some Comments on Waterfowl Displays

Waterfowl displays pique our curiosity because they are complex, varied, and often bizarre. Texas provides numerous opportunities to view waterfowl displays, especially those related to pair formation, because in winter and spring many ducks and geese en route to their nesting grounds are already selecting a mate.

Given the anatomical constraints that physically limit the number of movements a bird can make, it is not surprising that ducks sometimes incorporate the same movement into more than one display. For example, Wood Ducks pump their head backward and forward during one pair-forming display, yet the same movement (though more rapid) is also part of a flight intention display.

In ducks it is generally the female rather than the male that selects the mate, and she usually selects a new male each year. Therefore, to prevent wasteful hybridization, it is important that she distinguish the courtship displays of her own species from those of other species. That is probably why many male ducks have invested in unique and complex courtship displays. Conversely, monogamous species — geese, swans, and whistling ducks in particular — usually do not form a new pair bond each year and their courtship displays are relatively simple.

Selection pressures that produce monogamy and long-term pair bonds in whistling ducks, swans, and geese probably also account for their reduced sexual dimorphism (i.e., males and females have similar plumages and vocalizations). It would be costly for a male to develop an elaborate plumage or distinctive vocalization if he mates with the same female each year.

There are other consequences of monogamy: the young develop more slowly, stay with their parents longer (up to 3 years in swans), and learn more from them. This luxury is unavailable to ducklings whose parents change mates each year.

The following waterfowl behaviors can be observed in Texas in winter and spring.

In many ducks, courtship begins with a *social phase,* when drakes engage in group displays—energetic bouts of posturing and chasing that probably function as much to bring them into a state of sexual readiness as to repel males of other species. The social phase is especially important for species that form a new pair bond each year, such as dabblers and pochards.

When females begin approaching, males shift to new displays that not only announce their availability for mating but also anticipate the *pair-forming displays* that follow. Prominent among pair-forming displays is the hen's *inciting display,* an example of which is vigorous head pumping. This display intensifies the drake's responses and causes him to engage in still other displays (many quite complex) that eventually lead to pair bonding.

Precopulatory behaviors encourage hens to mate with drakes of their own species, besides helping to bring both drakes and hens into sexual readiness. *Copulation* (called treading in waterfowl) tends to be a brief and invariable event. Because it functions to inseminate rather than communicate, there is no reason why this behavior should differ much among species.

The function of *postcopulatory behaviors*—some quite complex and many unique to the species—is somewhat puzzling. By definition, displays communicate, but it is not obvious exactly what is communicated or why there is a need to communicate immediately after copulation. Perhaps postcopulatory displays serve to reinforce the pair bond. It is possible that they are not displays at all but just elaborate displacement behaviors triggered by the stress of the copulatory event.

In swans, geese, and whistling ducks, the pair bond is periodically reinforced by a *triumph ceremony,* a mutual display that frequently follows an encounter in which a male successfully chases away an intruder. He rushes back to his mate, and both place their heads near to each other and vocalize loudly and excitedly.

Whistling-Ducks

Both of our whistling-ducks generally mate for life and invest very little time and energy on courtship displays, which, accordingly, are relatively simple. Also, males freely associate with other whistling-ducks during the breeding season, precisely the time of year when most male birds are expending considerable energy defending their mates and territories.

Though quick to defend their young, whistling-ducks are relatively casual about their nesting behavior and frequently lay eggs in the nests of other whistling-ducks (dump nesting). They do not surround their eggs with down, a behavior that may reflect a tropical origin (where warmth from down is unnecessary). Both parents incubate the eggs and care for the nestlings.

Black-bellied Whistling-Duck, *Dendrocygna autumnalis.* Princ. distrib. Texas: Locally common SR in s. third of Texas, wintering in varying numbers on the coast and in extreme s. Texas. They frequent freshwater marshes and shallow lakes.

FEEDING BEHAVIOR. These handsome birds forage on land at night. When they do feed in water, they wade in the shallower parts rather than swim in deeper areas. (*See* Fulvous Whistling-Duck.) They are better adapted than Fulvous Whistling-Ducks for terrestrial life, as well as for feeding at night: they have relatively longer legs and smaller feet, they walk in a more upright posture, and they have a higher concentration of rods (receptors for night vision) in their eyes. Their shrill whistling call and bold white wing patches may be adaptations for locating other whistling ducks in dim light.

COURTSHIP BEHAVIOR. The few pair-bonding and courtship displays include mutual head dipping, neck stretching, and diving. They seem to have no precopulatory displays. After copulating, males strut about with breasts puffed up and head and neck held in a tight S-shaped position. Although typically monogamous, birds choose another mate when widowed, and sometimes even when their mate is still alive.

NESTING. Nest: normally in tree cavities, commonly hackberry, Texas ebony (*Pithecellobium*), live oak, and mesquite; less frequently on the ground, and sometimes a mile or more from water. They normally place no down or other nesting material in the cavity, perhaps because their ancestors were from tropical regions or because the pair continually incubate the eggs. Eggs: usually one per day for about 2 weeks; dump nesting is common. (*See* Fulvous Whistling-Duck.)

The male helps to incubate the clutch. He has a brood patch, a patch of skin on the abdomen that touches and warms the eggs (normally found only in female birds). On the second day after hatching, young birds scramble up to the rim of the cavity, pause, and in response to calls from their parents, release their hold on the rim and fall to the ground. They land without injury, immediately pick themselves up, then follow their parents to the nearest body of water.

VOICE. Both sexes: a descending *peeeeee djee djee djee,* less penetrating than the squeals of the Fulvous Whistling-Duck. Also, a *djeet djeet djeet,* given in flight, as well as other calls.

OTHER BEHAVIORS. Being exceptionally gregarious, they form large postbreeding flocks that sometimes number into the thousands. Of more interest is that they congregate during the breeding season, a time when rivalry among the males of most bird species is at its peak. In mixed groups, Black-bellies are dominant over Fulvous Whistling-Ducks.

These birds rarely alight on, or swim in, deep water, but when young birds swim, they sandwich themselves between one parent in the front and the other behind. Occasionally ducklings ride on a parent's back. When threatened, one parent may harass the intruder while the other leads the brood to safety. Both juveniles and adults respond to threats by holding the head low and forward.

Fulvous Whistling-Duck, *Dendrocygna bicolor.* Princ. distrib. Texas: Common SR and uncommon WR on the coast, frequenting marshes, irrigated lands, and rice fields. Although difficult to see in dense vegetation because they blend so well with their surroundings, they are actually easily approached when located.

FEEDING BEHAVIOR. They consume enormous quantities of rice (Hohman et al., 1996). They are also fond of corn, alfalfa, and other crops. In water they generally swim or dabble rather than wade, and they locate food items by tipping up, by diving (up to 15 seconds), or by lowering the head into the water. They spend much time feeding at night. Young birds dive with considerable skill and dexterity.

COURTSHIP BEHAVIOR. Courtship displays are rarely observed because of the species' strong pair bonding and life-long monogamy. During the prenesting period small groups fly together with considerable turning, twisting, and banking. Prior to copulation both males and females head-dip; after copulating they rise out of the water, place their bills in their puffed-out breasts, raise the wings opposite their partner, and tread water (the so-called postcopulatory step-dance).

NESTING. Nest: neatly woven in or near water, often on rice-field levees. Females frequently deposit eggs in the nests of other whistling ducks. The record appears to be 62 eggs in one of these dump nests. Dump nesting may be related to the parents' tendency to leave nests unattended for long periods of time when they are associating with other whistling ducks.

Although the young rather quickly swim and find their own food, they accompany their parents for several months after hatching.

VOICE. In flight, a thin, high-pitched whistle (*ke whee oo*), suggestive of a plover, and given by both sexes; also, a double-syllabled *pit CHEE*.

OTHER BEHAVIORS. In one interesting display, the bird holds its head and neck stiffly erect, rapidly walks toward another bird, puffs out its breast while holding the neck and head in an S-shape, and utters a *weep* call. It then simulates feeding by holding the bill to the ground. This display probably functions as a greeting or to reinforce the pair bond. For defense or threat, Fulvous Whistling-Ducks employ the head back display, in which the head is held stiffly over the shoulders.

They fly at low altitudes and with slow wingbeats, suggesting a heron or ibis, and they often circle prior to landing. Unless suddenly threatened, they exhibit a flight intention display (lateral head-shaking), which signals to others that they are not taking flight to escape danger.

Birds feed and loaf together even during the breeding season, a time when most birds isolate themselves as pairs.

Geese and Swans

Behaviorally, geese more closely resemble swans than they do ducks. In both geese and swans, courtship displays are few and simple, and defense and threat behaviors are frequent, distinctive, and emphatic. Both groups typically mate for life, and both groups are primarily vegetarian upland grazers.

Swans and geese lack the boldly patterned wings that enhance many duck displays; thus, they commonly rely on head and neck movements to communicate. Geese also vibrate the erect feathers, and in some species, the vibrating feathers are arranged in vertical furrows that intensify the signal by making the neck more conspicuous.

The voice and plumage of geese are essentially alike in males and females, although males vocalize at a slightly higher pitch than females.

Swans use their long neck to feed on submerged plants. Being heavy, with short legs that prevent a full wing downstroke when taking off, they must run over water 15–20 feet before they can take flight. When migrating, they fly at high altitudes in a characteristic vee formation. The vee formation in waterfowl was once thought to give an aerodynamic advantage to all but the leading bird, but more likely this arrangement is efficient because it gives each bird a broader range of view.

The male (cob) and the female (pen) are identical in plumage. Because the plumage in North American species is almost entirely white, visual signals are appropriately concentrated on the unfeathered parts of the head, especially the color patterns on the bill, where they are readily seen when the bird moves its head. Both parents incubate and care for the young (cygnets).

Many species, including the Tundra Swan, produce very loud sounds that carry long distances, thanks to an elongated trachea that loops through the sternum. The call of the Trumpeter Swan of the western United States is, in fact, the loudest of all waterfowl.

Our geese are divided into the following two groups:

1. *Pale-breasted Geese.* In Texas, the pale-breasted or gray geese (genera *Anser* and *Chen*), include the Greater White-fronted Goose, Snow Goose (Blue Goose), and Ross's Goose. The European greylag (*Anser anser*), the wild ancestor of the domestic goose commonly seen in parks, also belongs to this group. Pale-breasted geese have high, narrow, and laterally serrated bills that support terrestrial feeding habits. They graze by cutting off the tops of grasses, grub by digging out roots and stems, glean seeds in cultivated fields, or shear leaves and seeds from plant stems. They also forage in aquatic habitats. Their brightly colored feet and bills probably serve as visual signals.

Gray geese vocalize and shake their heads laterally prior to flight. Threat displays include the diagonal neck and head-forward displays, both commonly accompanied by vibrating the neck feathers.

The pair bond in pale-breasted geese is durable, and families stay together much longer than in most other waterfowl. Gray geese are highly gregarious during migration and winter, when they form large, discrete flocks. Because these flocks tend to be made up of a single species, gray geese are sometimes viewed as clannish.

2. *Dark-breasted Geese.* The dark-breasted or black geese (*Branta*), including the Canada Goose and Brant (rare in Texas), have black feet, black bills (not as serrated as in gray geese), and relatively darker plumages. Contrasting head and neck patterns—for example, the white chin patch in the Canada Goose—serve as visual signals in the absence of striking markings on the body. These geese usually graze and glean, and inland birds generally do not feed in water. Black geese move their bodies more rapidly than gray geese and also have harsher and higher-pitched voices.

The bent neck posture (or display) in black geese may be homologous to the diagonal neck display in gray geese. Both gray and black geese exhibit similar neck-forward and erect postures, the same triumph ceremony, and basically the same copulatory behavior.

Greater White-fronted Goose, *Anser albifrons.* Princ. distrib. Texas: Common M throughout much of c. and e. Texas, and locally abundant in winter on the coast.

FEEDING BEHAVIOR. Wintering birds feed in habitats that include stubble fields, pastures, and aquatic and marshy areas. They consume cultivated grains, especially rice, as well as native grasses. They typically graze, but sometimes glean or grub for roots and stems. Occasionally they feed by submerging the head in water.

COURTSHIP BEHAVIOR. The displays that lead to pair formation are rarely observed because they probably occur only once, during the second year when pairing occurs. Thereafter, pair bonds are reinforced with the triumph ceremony, an arresting encounter during which both birds stretch their necks low over the ground and call loudly into each other's ears.

VOICE. Both sexes: rapid, high-pitched paired notes, which would suggest laughter except for their melancholy quality.

OTHER BEHAVIORS. They walk and run adeptly, and swim with strong leg movements. Other than during migration, they rarely form the large flocks typical of other North American geese. They sometimes associate with Canada Geese, less frequently with Snow Geese. They tend to be more wary than the other two species.

Quite likely individuals recognize each other on the basis of the shape of the white forehead patch and the extent of barring on the belly. When threatened, a bird will stretch its neck forward or diagonally and vibrate the strongly furrowed neck feathers.

Before taking flight (provided danger is not present), they shake the head laterally to expose the white forehead patch; or they make one to three quick wing flaps, then freeze in a high-alert position. Greater White-fronted Geese are considered the most agile of North American geese; they can rise almost vertically into the air when taking flight, and when landing, they accelerate their descent by performing a slip-sliding or falling-leaf maneuver.

Migrating birds form large, noisy flocks that fly in vee formation.

Snow Goose, *Chen caerulescens.* Princ. distrib. Texas: Common M in e. part of the state, and abundant WR on the coast, sometimes forming extraordinary flocks. More than 300,000 birds were observed one winter in Chambers and Jefferson counties.

Snow Geese occur in two genetically determined color phases, the white morph and the blue morph. They were formerly regarded as two distinct species (Snow Goose and Blue Goose, respectively), but because they are now known to interbreed, they have been lumped into a single species.

Because breeding birds normally select a mate of the same color, off-spring are usually the same morph as the parents, even though some broods are mixed. Such preferential mate selection may be due to slightly different geographical distributions and breeding times for the two morphs. One interesting hypothesis proposes that sexual imprinting biases a bird to select a mate of its own color. Thus, juvenile geese, because they imprinted on their parents at an earlier age, are biased toward selecting a mate of their parents' color.

FEEDING BEHAVIOR. Snow Geese feed on rice, waste grain, and seeds; in pastures and wet areas they eat native grasses and shear off stems. They also use their stout, serrated bill to grub roots from the ground. When large numbers concentrate in marshy areas, mainly in the northern United States, the odor from their droppings can be detected several miles away. On their Texas wintering grounds they seem to forage mainly in low prairies and brackish marshes.

COURTSHIP BEHAVIOR. The displays that lead to pair formation occur only once in the lifetime of a bird, usually during the second year of life. The pair bond is maintained by the triumph ceremony: the two birds approach each other with their necks stretched forward and near the ground, each calling loudly into the other's face.

VOICE. Both sexes: high-pitched, falsetto-sounding, musical honks or barking that can be rendered as *hounk* or *lauk!*

OTHER BEHAVIORS. Considered one of the most social of our geese, Snow Geese form large flocks that exclude all other species except Ross's Goose. Pair bonding is strong and permanent, with remating occurring only when one of the pair dies or is killed.

When threatened, they stretch the neck forward or diagonally and vibrate their strongly furrowed neck feathers. In the absence of danger, Snow Geese exhibit flight intention by shaking the head laterally. They migrate in diagonal lines or broad vee formations.

Ross's Goose, *Chen rossii.* Princ. distrib. Texas: Uncommon WR or winter visitor on the coast, sometimes in numbers greater than 100, and on the Coastal Bend more common than the Canada Goose (Michael Marsden, pers. comm.). Ross's Geese are becoming more common throughout most of the state, except the forested parts of e. Texas.

Once abundant, but later rare, they have been reported regularly east of the Rockies since the 1950s. They are only slightly larger than a Mallard. They so closely resemble the Snow Goose, with which they associate, that they are sometimes killed by careless or inexperienced hunters. When in large numbers, they usually segregate themselves from Snow Geese.

FEEDING BEHAVIOR. Apparently Ross's Geese feed exclusively on plants, including roots and stems from grasses and sedges; they forage both on land and in shallow water.

COURTSHIP BEHAVIOR. Most courtship behavior concludes during winter and spring migration. Pair bonding is facilitated and maintained by the triumph ceremonies. (*See* Snow Goose.) Pair bonds appear to be permanent.

VOICE. Both sexes: a high-pitched, double *keeek* and a weak, repeated grunt.

OTHER BEHAVIOR. The flight of these highly maneuverable birds is rapid and powerful.

Canada Goose, *Branta canadensis.* Princ. distrib. Texas: M and WR throughout the state, common on the coast and abundant in the High Plains. Some populations on the High Plains may have permanently shifted their winter ranges from croplands to urban areas after the harsh climatic conditions in 1990–1991 (Ray and Miller, 1997).

The several subspecies (geographic variants) that regularly winter in Texas differ in size and plumage, but often so slightly that they cannot be reliably separated in the field. Moreover, views differ as to which North American populations warrant subspecific designation, as well as which of these populations winter in Texas.

Nevertheless, we can appreciate that the variation we see in a flock of wintering Canada Geese is attributable in part to two behavioral traits: monogamy and persistent family bonds. Because wintering families generally return to their previous nesting grounds, related individuals are more likely to interbreed in that region, thereby producing populations that are genetically adapted to the environment. (In contrast, most ducks that winter in Texas show no significant geographic variation because each year wintering drakes follow a different hen back to her nesting area.)

FEEDING BEHAVIOR. Canada Geese are basically terrestrial grazers, feeding on grasses, sedges, waste grain, and marsh plants. They typically do not feed in water except on the coast.

COURTSHIP BEHAVIOR. Pair bonds are formed during the second year and are typically strong and permanent. They are regularly strengthened by the triumph ceremony described below. In one common display the male approaches the female with his head held low and his neck weaving back and forth.

Prior to copulation the male begins head-dipping movements that are probably ritualized bathing movements. He is soon joined in this display by the female. After copulating, the birds partially extend their wings, raise their breasts, extend their necks, and point their bills upward (all possibly derived from intention movements).

In the spectacular triumph ceremony—triggered by the gander's victory over an intruder—both gander and goose extend their necks, hold their heads low, open their bills, and honk loudly.

VOICE. Both sexes: a deep, musical double honk, *kuh lonk;* also, a high-pitched cackle. Larger subspecies generally have low-pitched calls (honking), whereas the calls of the smaller subspecies tend to be more of a cackle. At least ten different sounds have been identified in this species, each being a response to a different situation.

OTHER BEHAVIORS. These highly social geese form large flocks on their Texas wintering grounds. Within these flocks family units remain intact throughout the winter.

Canada Geese show several threat behaviors. A bird may (a) hold the neck erect while flipping the bill upward, (b) stretch the neck forward and low while honking, (c) bend the neck in an S-shape, (d) perform a head-pump or head-roll display, or (e) vibrate the neck feathers (which are not furrowed as in the gray geese). In the presence of a dominant bird, this goose curls its neck, ruffles its feathers, and flees with the neck held erect.

Prior to taking flight in nonthreatening situations, Canada Geese signal to each other by tossing the head or by wagging the head vertically and sideways—movements that expose the white throat, which no doubt increases the effectiveness of the display. They take flight rapidly, and in flight they are highly maneuverable. Like many waterfowl, they accelerate their descent by performing a slip-sliding or falling-leaf maneuver.

These highly gregarious birds readily habituate to humans, quickly making themselves at home in most urban situations such as parks and golf courses, especially if they are regularly fed. When possible, they spend the night in groups in open water.

Tundra Swan, *Cygnus columbianus.* Princ. distrib. Texas: Wintering birds are reported infrequently and in widely scattered localities. They eat mainly plants but also consume invertebrates such as aquatic insects. Although the courtship display is unlikely to be observed in wintering birds, it deserves mention: the male performs a high-stepping walk while lifting his neck into an arch and stretching out his wings. Both sexes bow and call repeatedly.

Three to five birds observed together probably represent a family, as most yearling birds apparently remain with their parents throughout migration.

VOICE. Both sexes: a barking sound suggesting a Canada Goose, but higher and more musical, *whoop WHOOOOOP whoop,* with a lengthened middle syllable. There are unverified reports that Tundra Swans shot in flight emit a special sound heard only when the bird is falling to its death, the so-called swan song. (*See* "Swan Song" in Terres, 1980; the idea of a swan song goes back to ancient Greece.)

Surface-feeding Ducks

Traditionally, all surface-feeding ducks except the Muscovy Duck and the Wood Duck have been referred to as dabblers. Dabblers swim well, but their relatively short tails, short legs, and small feet do not allow them to dive as effectively as pochards and other diving ducks. They prefer shallow water that is fresh or brackish, and in this habitat many of them feed by upending. They feed primarily on plant material, including seeds. Many have died from lead poisoning by mistaking lead shot for seeds.

Their ability to take off directly from the water (some almost vertically) appears to be an adaptation to living in small, shallow ponds.

Males and females differ in voice and plumage. Pair bonds are seasonal, and the male deserts the female soon after the eggs hatch.

In his pioneering study of waterfowl behavior, Konrad Lorenz identified the following components that he considered basic to the displays of surface-feeding ducks: bill shake, head flick, tail shake, grunt whistle, head-up-tail-up, turning toward the female, nod swim, turning the back of the head, bridling (drawing the bill backward along the back while giving a whistle), and down up. Not all are present in every dabbler, but most can be seen in Texas by patiently watching wintering birds, especially on sunny days.

There are reports of intraspecific nest parasitism (laying eggs in the nests of other birds) from seven North American dabblers: Northern Shoveler, Green-winged Teal, Cinnamon Teal, Mallard, American Black Duck, Mottled Duck, and Gadwall.

Muscovy Duck, *Cairina moschata.* Princ. distrib. Texas: Rare PR in the lower Rio Grande Valley. Most individuals in Texas are domesticated birds that mingle with (and sometimes hybridize with) domesticated Mallards, domestic geese, and other waterfowl in urban parks.

The male's agonistic and sexual behavior is distinctive: he hisses repeatedly while vigorously shaking his tail and pumping his head backward and forward. Males do not form pair bonds or engage in precopulatory displays; reproductive behavior is limited to chasing one or more females and forcing copulation.

Wood Duck, *Aix sponsa.* Princ. distrib. Texas: Locally common SR in c., s., and e. part of the state, wintering locally in e. Texas.

FEEDING BEHAVIOR. These extraordinarily colorful ducks feed on aquatic seeds and plant parts, as well as crustaceans, insects, amphibians, and small fish (which they sometimes capture by diving). They generally forage from the surface or dip the head and neck into the water. They rarely upend like dabblers.

On land they walk erect and with a fast gait; they also run. They are very much at home on the forest floor, where they consume more acorns than any other species of waterfowl.

COURTSHIP BEHAVIOR. They have numerous and complex courtship displays, many of which incorporate movements that emphasize the drake's highly ornamental plumage. These displays are initiated as early as fall and winter. A few are described below.

The hen incites the male with a series of rapid, forward-pointing movements or by pointing her bill upward. The drake responds by turning the back of his head toward her. He may also engage in ritualized drinking movements, perform the burp display (raising the crest while stretching the neck vertically and making a squeaky sound), or shake his head while producing a whistle. One common display is a back-and-forth pumping of the head while aiming the bill forward. Prior to copulation the drake bill-dips or makes drinking movements.

Wood Ducks are not territorial in the usual sense of the word, but males do defend their mates in a moving territory.

NESTING. Nest: in a tree cavity, approximately 20–50 feet from the ground, and placed over water if possible. These ducks compete for cavities with European Starlings, owls, and squirrels, and they readily take advantage of nesting boxes.

Females sometimes deposit eggs in the nest of another female (dump nesting). In south Texas, where both Wood Ducks and Black-bellied Whistling-Ducks compete for nesting cavities, mixed clutches of the two species have been reported. In northeast Texas, Wood Ducks have shared nest boxes with Hooded Mergansers.

When they arrive at the nest, parents cling to the cavity entrance by bracing themselves with their tail, much in the manner of woodpeckers.

On the day after hatching, the ducklings respond to parental calls from below by scrambling up to the rim of the nesting cavity and plunging to the ground or water. Young Wood Ducks are notably independent, even at an early age, and some voluntarily abandon the care of their parents when less than 2 weeks old.

Wood Ducks are seasonally monogamous. The drake deserts the hen at the beginning of incubation and pairs with a different female the following year.

VOICE. The male's *khoo eek* whistle contrasts with the female's owllike *hou eek* or *krrrek*. They have a remarkably large and varied repertoire.

OTHER BEHAVIORS. Most of the year Wood Ducks stay together in groups of fewer than a dozen birds, but sometimes during migration and in winter several hundred individuals roost together at night. When swimming or when walking, they point by moving the head forward and backward, like pigeons. Their flight intention movement resembles these pointing movements but is more stereotyped and more rapidly executed. Their aerial skills are singular among our waterfowl, permitting them to fly swiftly and directly through the crowns of trees, as well as to take off from the water almost vertically. They readily perch and walk on tree branches.

These unwary ducks habituate easily to humans and become quite tame when not molested.

Gadwall, *Anas strepera.* Princ. distrib. Texas: Common to abundant M and WR throughout most of Texas.

FEEDING BEHAVIOR. Gadwalls feed by tipping up, usually in shallow marshes and ponds where submerged plants lie just beneath the surface. They dive only rarely.

Their diet includes more stems and leaves than seeds. They take considerable plant material from the water's surface and occasionally forage for insects, mollusks, and small fish.

COURTSHIP BEHAVIOR. Gadwalls select a new mate each year. The social phases prior to pairing, as well as pair-forming behaviors, are similar to the Mallard's. In Texas, most males display in winter and spring. They respond to the female's inciting behavior with displays that include chin-lifting and turning the back of the head. Pursuit flights are common, and at this time the female attempts to thwart an undesired male with the gesture of repulsion — flying away with her head drawn into her neck, her head and back feathers ruffled, and the upper mandible bent upward.

Before copulating, the pair engage in mutual head-pumping and courtship flights. Afterward, the male gives the burp display. The pair bond is reinforced by mutual chin-lifting.

VOICE. During courtship the male croaks, whistles, and trills, producing sounds sometimes rendered as *raeeb tzee tzee raeeb raeeb*. The female utters a series of quacks that fall in pitch and volume and suggest the Mallard's call.

OTHER BEHAVIORS. This is a gregarious species that forms small, compact flocks.

American Wigeon, *Anas americana.* Princ. distrib. Texas: Common to abundant M and WR throughout Texas.

FEEDING BEHAVIOR. The American Wigeon, more than any other North American dabbler, consumes the vegetative parts of aquatic plants, its small, compact bill being nicely structured for grazing. Many of the food items these birds consume are plant parts dislodged from the bottom by Canvasbacks and other pochards; wigeons have learned to follow pochards and other diving ducks to take advantage of this food source. In winter they rely heavily on cultivated crops.

COURTSHIP BEHAVIOR. Courtship behavior is commonly observed in winter, as most pair bonds have already formed before the birds begin their northward migration in spring. Prior to pairing, 5–15 drakes display together by preening behind the wing and making short jump flights. Later, the female incites by swimming toward the selected male and turning the back of her head toward the other males. The pair engage in mutual head-pumping prior to copulating, and afterward the male engages in bridling.

VOICE. The male makes a pleasant, mellow *whee whee wheu* in flight or on the water. The female is silent most of the time, occasionally producing a weak, guttural *qua ack, qua ack, qua ack.*

OTHER BEHAVIORS. Threatened females exhibit the gesture of repulsion: they fly away with their head drawn into their neck, head and back feathers ruffled, and the upper mandible bent upward. When alarmed, these alert and nervous ducks rise almost vertically from the water, producing a rattling noise with their wings. They fly swiftly and erratically, with much twisting and turning.

American Wigeons form flocks during migration, but other than when feeding, they normally do not congregate in large numbers.

Mallard, *Anas platyrhynchos.* Princ. distrib. Texas: Uncommon to common PR in Texas, especially in n. parts, where they nest locally. Mexican Ducks, which nest in the Trans-Pecos, are considered a subspecies of the Mallard. As far as is known, their behavior is indistinguishable from the Mallard's.

FEEDING BEHAVIOR. Mallards consume agricultural grain crops as well as grasses, aquatic vegetation, and aquatic invertebrates, feeding frequently at night and sometimes altering their feeding schedules to avoid hunters. In the western United States (possibly including west Texas) Mallards that feed on exposed mudflats sometimes suffer from alkaline poisoning and from botulism. They feed by tipping up, but they are capable of diving for food if necessary.

COURTSHIP BEHAVIOR. Courtship is initiated in September, and by January about 90 percent of the females are already paired. Courtship behavior is elaborate; however, it is not entirely clear why such an extraordinarily complex repertoire of displays is necessary for successful reproduction in this species. Important courtship behaviors include the following.

During the social phase, which may function to exclude other species, groups of males display in nonthreatening postures and with head feathers raised. At this time the white ring around the neck is partially concealed, suggesting that when exposed the ring signals aggression. Clearly it would be counterproductive to signal aggression during the social phase when males behave in concert (cf. the role of the red shoulder patches in male Red-winged Blackbirds). Apparently the ring also is important in male-female interactions, as females select males with wide rings over those with narrow ones.

When females approach this group of displaying males, both sexes engage in the bill-shake display. The female incites by swimming toward one male and turning her head away from others. Even then, one or more males may pursue a female and attempt forced copulation. Occasionally females drown as a result of these chases.

The speculum-flashing display was probably derived, through ritualization, from preening. The drake orients himself laterally toward the female, raises his wing in a preening motion to expose the speculum, and runs his bill behind the wing.

If, after she is paired, a female is pursued by drakes other than her mate, she may exhibit the gesture of repulsion, a particularly graphic response to male aggression. She flies away with her head drawn into her neck, the feathers on her head and back ruffled, and her upper mandible bent upward.

The pair bond is maintained by mutual head-pumping. The hen may incite the drake to chase an intruder, after which the pair, upon being united, engage in a noisy palaver that perhaps functions like the triumph ceremony in geese. Copulation is preceded by mutual head-pumping. The drake's postcopulatory displays include bridling, nod swim, and turning the back of the head.

NESTING. Nest: constructed of grasses, cattails, and other plant materials, and situated in fairly dry areas near water. The hen continually lines the nest with down and turns the eggs about twice an hour by paddling them with her feet. Eggs: incubated by the hen, who also leads the young to feeding sites.

Males mate with a different female each year. Midway through incubation they abandon their mates to join groups of other males.

VOICE. Male: a soft, high-pitched *yeep* that contrasts sharply with the calls of most dabblers. Female: a loud *quack,* the sound commonly caricatured for a quacking duck. When undisturbed, females utter a series of notes that decrease in volume (the decrescendo call).

OTHER BEHAVIORS. Territoriality—in the sense of defending a plot of ground—is weak in this species, if it exists at all. Males defend their mates both inside and outside the nesting area, and the term "home range" (an area occupied but not defended) probably better represents the area Mallards inhabit during the nesting season.

Mallards signal flight intention by raising and lowering the head while turning it laterally.

Mottled Duck, *Anas fulvigula.* Princ. distrib. Texas: Locally common all year long on the coast, nesting locally inland. Preferred habitats on the coast include salt marshes, bluestem meadows, prairies, and fallow rice fields.

FEEDING BEHAVIOR. Like the Mallard, Mottled Ducks feed on grain, seeds, aquatic vegetation, and grass, much of which is taken in ungrazed fields. Animal food items include fish, insects, snails, and crustaceans.

COURTSHIP BEHAVIOR. Their courtship behavior closely resembles the Mallard's (although their pair bond lasts longer). Territorial defense by the male is weak, and agonistic encounters are infrequent.

NESTING. Time of nesting in spring depends on October–February rainfall totals. Nest: in the drier parts of marshes, on sandy ridges, and in other secure places, sometimes far from water; it is concealed under vegetation and constructed of aquatic plants, matted grass, and other plant materials. Intraspecific nest parasitism has been reported.

They line their nests with down, a practice more typical of northern ducks, where the advantages of added warmth to the eggs outweigh the disadvantages of overheating. Possibly this behavior in Mottled Ducks indicates an evolutionary origin from a northern ancestor. Eggs: incubated by the hen, who also leads the young to feeding sites where they forage independently. When the mother gives the alarm call, ducklings quickly scatter and hide in vegetation. Apparently this is the only duck that has been observed carrying an egg.

Because Mottled Ducks are very sensitive to disturbance by humans and other intruders, they will probably desert their nest if disturbed.

VOICE. Similar to the Mallard's.

OTHER BEHAVIORS. Most behaviors probably resemble those of the closely related Mallard. They walk adeptly on land and fly with strong wingbeats.

Blue-winged Ducks

The following three species are known as blue-winged ducks. Ornithologists formerly regarded the Blue-winged, Cinnamon, and Green-winged Teals as closely related (at least close enough to call them all teal), and they considered the Northern Shoveler sufficiently distinct in appearance to place it in a genus to itself.

The four species are viewed somewhat differently now. In spite of its small size and teal appearance, the Green-winged Teal has been removed from its former group and replaced with the Northern Shoveler. Thus, bill length, formerly considered a very important taxonomic character, is now regarded as much less important for classifying these ducks.

Blue-winged Teal, Cinnamon Teal, and Northern Shovelers do, in fact, share a number of characters that justify grouping them together, including an extensive blue wing patch that gives rise to the name "blue-winged duck." Moreover, Blue-winged and Cinnamon Teals have spatulate (spoon-shaped) bills, although hardly as exaggerated as the shoveler's.

Behaviors of the Green-winged Teal and the blue-winged ducks are compared in Table 1. They engage in the same behaviors at the beginning and end of the social phase (male grouping and pursuit flight); in pair formation, the males all engage in burp display and head-up-tail-up; and both groups practice precopulatory head-pumping.

Blue-winged Teal, *Anas discors.* Princ. distrib. Texas: Common M and WR throughout most of Texas, WR in s. Texas; they nest locally in scattered locations throughout much of the state.

Table 1: Behaviors of Green-winged Teal and Blue-winged Ducks

	Shared Behavior	Green-winged Teal Behavior	Blue-winged Ducks Behavior
Social phase prior to pairing	males group, display around female	raise crown feathers grunt whistle tail shake bill up head-up-tail-up nod shake	head pumping turn back of head preen behind wing jump flight (Shoveler)
Pair formation	pursuit flight	F: moves head sideways head-up-tail-up	F: head pumps chin lifts (Shoveler raises head, jerks bill downward)
	M: burp display	raise crown feathers grunt whistle tail shake bill up	lateral dabbling preen behind wing preen dorsally chin lifting
	head-up-tail-up	nod shake	body shake swimming shake
Precopulatory Postcopulatory	MF: head pumping	M: bridling	M: bill pointed down head feathers erect waggles tail
Defense or threat		?	M: hostile pumping head held high

FEEDING BEHAVIOR. In the shallows of small bodies of water, they skim the surface with their bill or reach below the surface with their head and neck. They also tip up to feed. Often they forage next to American Coots, as well as other ducks, to take advantage of debris stirred up. Their diet includes seeds and vegetative parts of aquatic plants.

COURTSHIP BEHAVIOR. (*See* Table 1.) New pair bonds are formed in winter and early spring. Because female Blue-winged Teal and Cinnamon Teal are virtually identical, males court females of both their own and the other species, placing the burden on the hen to ensure that mating occurs with the proper species. Even then, from time to time the two species hybridize.

VOICE. Male: a weak, peeplike whistle heard mainly in spring. Female: a weak, high-pitched *quack.*

OTHER BEHAVIORS. By the time fall migrants reach Texas, these highly social ducks have often formed enormous flocks. They seem overly cautious when landing, and repeatedly circle over the place from which they were flushed, only to alight again at that place. In spite of their apparent wariness under these circumstances, they tend to become tame and unsuspicious when not persecuted.

Cinnamon Teal, *Anas cyanoptera.* Princ. distrib. Texas: Common M in w. Texas; nests locally in scattered localities.

FEEDING BEHAVIOR. Apparently Cinnamon Teal consume the same basic foods as Blue-winged Teal, including seeds of grasses and sedges, as well as some animal matter, such as insects and mollusks. They tip up in typical dabbling duck fashion and strain food items with their lamellate bill (somewhat in the manner of the Northern Shoveler). They frequently follow ducks that are agitating the water and take advantage of food brought to the surface by all the activity.

COURTSHIP BEHAVIOR. Like most dabbling ducks, they form a new pair bond each year, usually on their wintering grounds. By the time migratory birds reach Texas, their courtship displays have intensified considerably and can be anticipated on warm, sunny days in late winter. These displays are very similar to those of the Blue-winged Teal and Northern Shoveler. (*See* Table 1.)

VOICE. Not a particularly vocal species. Female: a weak *quack,* virtually indistinguishable from the Blue-winged Teal's. Male: a low, guttural, squeaky chatter, as well as a rattling sound similar to that of the closely related Northern Shoveler.

OTHER BEHAVIORS. Although gregarious, they seldom congregate in large flocks. They walk easily on land and take flight by leaping up at a sharp angle. In hostile encounters, males head-pump with the bill tilted upward. As a species they are relatively tame.

Northern Shoveler, *Anas clypeata.* Princ. distrib. Texas: Very common M and WR in most of the state, with many nonbreeding individuals lingering throughout the summer in scattered localities. Food availability and gut morphology data suggest that freshwater wetlands provide higher quality habitat than saltwater wetlands for Northern Shovelers that winter on the Texas coast (Tietje and Teer, 1996).

FEEDING BEHAVIOR. Shovelers feed in small, compact groups, swimming together in a circular pattern. They rarely tip up like other surface-feeding ducks. They use their highly specialized bill like a sieve to strain food items from the surface of the water, so they compete very little with other dabblers for food resources.

They probably consume more crustaceans and other small aquatic animals than any other North American dabbling duck. The remainder of their diet consists of seeds and vegetative parts of aquatic plants.

COURTSHIP BEHAVIOR. See Table 1.

VOICE. Usually silent. Male: an occasional deep croaking or rattling (*dookh dookh*) during courtship. Female: a weak *quack* and a decrescendo call about five notes long.

OTHER BEHAVIORS. Migrating and wintering birds associate with Gadwalls, Lesser Scaups, Blue-winged Teal, and American Wigeons.

These highly maneuverable ducks rise quickly and easily from the water, then dart off swiftly and erratically. Like teal, they drop suddenly from the air to alight on the water, often at the place from which they just took flight. They walk adeptly on land.

Northern Pintail, *Anas acuta.* Princ. distrib. Texas: Common M and WR throughout Texas. This is the most widely distributed and perhaps the most abundant duck in North America.

FEEDING BEHAVIOR. They feed on grasses, seeds, and grains as well as insects and aquatic invertebrates, sometimes diving deep for food items.

COURTSHIP BEHAVIOR. Pintails form a new pair bond each year and begin displaying in Texas as early as December. During the social phase prior to pairing, males form groups and engage in displays that include chin-lifting, burping, and turning the back of the head. Before copulating, the pair mutually head-pump; afterward, the male performs several displays, including bridling, burping, and turning the back of the head. The intense aquatic and aerial courtship activity observed in Texas during spring migration has been attributed to the larger proportion of unpaired males in the population.

VOICE. Male: generally silent, but they may produce a low, wheezy, teallike whistle, suggesting *pfeeh* or *prreep prreep.* Female: a series of low quacking sounds that diminish in volume.

OTHER BEHAVIORS. Territorial defense is apparently very weak, with little hostility among males in the same area. These fast, graceful fliers make acrobatic turns with ease and often hover when alighting. They swim buoyantly and walk and run adeptly on land.

Green-winged Teal, *Anas crecca.* Princ. distrib. Texas: Locally common to abundant M and WR in most of Texas.

FEEDING BEHAVIOR. The vegetal diet of these smart-looking birds — the smallest of the North American dabblers — is limited by bill size to small seeds, grass, grains, and plant shoots. They also consume many insects. On the coast they forage both night and day in tidal creeks, as well as in marshes near rice fields. They walk considerable distances for food and sometimes wander into forested areas in search of acorns. When feeding in water they tip up more than the other teal.

COURTSHIP BEHAVIOR. (*See* Table 1.) Pair-forming behaviors should be looked for in late winter and spring. Because food resources on their breeding grounds are scattered, it is relatively unimportant for male Green-winged Teal to defend a territory; thus, this species did not evolve the territorial hostile-pumping display of the blue-winged ducks (Blue-winged Teal, Cinnamon Teal, and Shoveler).

VOICE. The male's loud, abrupt two-syllable whistle is heard in winter and spring. The female utters a simple *quack* as well as a four-note call.

OTHER BEHAVIORS. They occur in medium-sized, compact flocks that wheel and twist like sandpipers.

Pochards and Allies

Although the term "diving duck" has been applied to these birds to distinguish them from the dabbling ducks above, "pochard" is the preferred term, as many other ducks, for example, mergansers and stifftails, dive as well as or better than pochards. The designation "pochard" is derived from the Common Pochard (*Aythya ferina*), an Old World species occasionally sighted in Alaska. Pochards are also called bay ducks because many species winter in coastal bays and estuaries, often in enormous congregations called rafts.

Their feet are large and set back on the body, an arrangement that facilitates diving but also produces a waddling gait on land. The lobed hind toe increases diving efficiency. Most species run across the water when taking off.

Plumage differences between the sexes are slight, in contrast to the dabbling ducks, and displays are less elaborate, illustrating that displays tend to be simpler in waterfowl that show less sexual dimorphism (cf. geese and swans).

Our five pochards show very similar pair-forming, precopulatory, and postcopulatory behaviors; exceptions are noted in the individual species accounts. During the social phase prior to pairing, 3–10 drakes swim around a hen. As soon as a hen chooses a drake, the two fly off together.

During her inciting display, the female engages in pronounced neck-stretching and threatening movements and repeatedly tugs at the male's plumage. The male responds with displays that include head throw, neck stretching, neck kinking, coughing, and turning the back of the head. The pair bond is maintained through mutual neck-stretching or the kink-neck display. Preening of the dorsal region is also an important display component.

Prior to copulation, the male makes alternate bill-dipping and dorsal-preening movements. After copulation, as the female bathes and preens, he utters a courtship call and swims away with his bill pointing downward.

Pochards show a strong tendency to associate with individuals of their own kind.

Canvasback, *Aythya valisineria.* Princ. distrib. Texas: Uncommon M and WR throughout Texas, sometimes occurring locally in fairly large numbers, especially on the coast.

FEEDING BEHAVIOR. Because they forage almost entirely by diving, most of their diet consists of food items unavailable to dabblers. These items include underwater roots, tubers, the basal parts of bottom plants, as well as mollusks and crustaceans. They prefer to feed in wide, shallow bodies of water.

COURTSHIP BEHAVIOR. See "Pochards and Allies" above.

VOICE. The male during courtship utters a series of low grunts and a coolike *koo.* The female's rarely heard call is a *quack.* Both sexes are normally quiet in flight.

OTHER BEHAVIORS. During migration and winter they form enormous rafts. Spring rafts are smaller because many birds dissociate themselves from the group after pairing in winter. Canvasbacks associate with other pochards, especially Redheads and Lesser Scaups. Although generally wary of humans, they become tame in nonthreatening situations.

Canvasbacks have the largest wingspan of any pochard (up to 36 inches), and they probably fly faster than any other duck except the Ruddy Duck. Their flight has been described as "direct, powerful, [and] purposeful . . . , giving the impression of a squadron of bombers" (Palmer, 1976).

Redhead, *Aythya americana.* Princ. distrib. Texas: Uncommon to locally abundant M throughout the state. Common WR on the coast, in particular on the Laguna Madre, where they find protection from high wave action. Both Redheads and Lesser Scaups spend more time feeding and moving in estuaries and more time drinking and preening in ponds, with a preference for ponds with lower salinities and deeper water (Adair et al., 1996). In the 1950s about three-quarters of all North American Redheads wintered on the Texas coast.

FEEDING BEHAVIOR. On their south Texas wintering grounds, Redheads feed almost exclusively on shoal grass, *Halodule wrightii* (Mitchell et al., 1994). They also take small quantities of other aquatic plants as well as a few insects.

COURTSHIP BEHAVIOR. *See* "Pochards and Allies" above. What seems to be unique to the Redhead's courtship behavior is the hen's inciting behavior: she alternately makes lateral and chin-lifting movements with her head. These movements contrast sharply with the inciting behavior of the female Canvasback (tugging at the drake's plumage).

NESTING. The nesting behavior in Redheads is not included here except to mention that females are notorious for laying eggs in other birds' nests. At the populational level, three out of every four eggs are laid in the nests of other ducks (usually Redheads), but Redheads also parasitize the nests of other species, including American Coots and bitterns. Consequently, young Redheads do not imprint strongly on their parents.

VOICE. Male (during courtship): an unusual, catlike *meow* (to some listeners it sounds like a violin) that carries a considerable distance. Female: normally silent; occasionally she utters a *quack*.

OTHER BEHAVIORS. Redheads associate with other ducks, especially scaups. On the coast, rafts sometimes contain several thousand birds. Redheads fly swiftly, with strong and rapid wingbeats; they are noticeably more agile in flight than Canvasbacks.

Ring-necked Duck, *Aythya collaris.* Princ. distrib. Texas: Uncommon M and WR throughout Texas. They generally avoid strictly saline waters.

FEEDING BEHAVIOR. Ring-necked Ducks consume seeds, vegetative parts of pond weeds, wild rice, and other aquatic plants. Their feeding tends to resemble dabbler behavior: they prefer shallow water with denser vegetation, they frequently tip up to feed, and they dive for shorter periods of time than Canvasbacks, Redheads, and scaups.

COURTSHIP BEHAVIOR. Commonly performed in winter and spring, courtship reaches its peak in March and April. (*See* "Pochards and Allies" above.) The pair bond is stronger and persists longer than in most related duck species.

VOICE. Usually silent in flight. A weak *pur-r-r-r* is infrequently heard in Texas.

OTHER BEHAVIORS. In Texas, Ring-necked Ducks associate freely with both scaups. Flocks in fall tend to be larger than in spring — up to several thousand birds sometimes congregate at that time.

Behaviorally, these are the least typical of the pochards. They are more inclined to feed in fresh water, they forage more like dabblers, and they rise from the water much more easily than the other pochards. During their swift and vigorous flight they make a distinctive whistling sound with their wings.

Greater Scaup, *Aythya marila.* Princ. distrib. Texas: Rare to uncommon M throughout Texas; locally uncommon WR in e. and coastal parts of the state.

FEEDING BEHAVIOR. Mussels and other aquatic invertebrates make up the major part of their diet, but Greater Scaups also take a limited amount of plant material. They typically forage in less than 5 feet of water. Because they generally do not feed out of water, their food intake on the coast is affected by low tides. Competition with Lesser Scaups wintering in the same region is minimal because the two species have different water salinity preferences. (*See* Lesser Scaup.)

COURTSHIP BEHAVIOR. *See* "Pochards and Allies" above.

VOICE. Normally silent except during courtship. Male: a soft *ba woo.* Female: a low growling sound (*tuk tuk tuk churratuk*), given in response to the male's call.

OTHER BEHAVIORS. They form large rafts in open coastal waters, some containing thousands of birds. Curiously, Greater Scaups are wary of human intrusion when they are in large rafts, yet they become remarkably trusting when in smaller flocks.

Lesser Scaup, *Aythya affinis.* Princ. distrib. Texas: Common to abundant M and WR throughout Texas. Because they are among the last of the prairie-nesting ducks to nest in summer, their courtship behavior occurs relatively late in spring. This species is the most numerous pochard in the United States, even though their late nesting makes them more vulnerable to predation (more young foxes and other predators to be fed at that time).

Wintering habitat preferences in Texas need to be studied in more detail, but apparently Lesser Scaups select sheltered freshwater habitats more often than the open saline waters preferred by Greater Scaups. (*See* Red-head.)

FEEDING BEHAVIOR. They generally forage in water 3–8 feet deep, frequently near piers, pilings, and other physical structures that support an abundant supply of invertebrates. They take large numbers of aquatic animals, principally mollusks and fish. Although they sometimes upend to feed, they more often dive. They dive with extraordinary energy, leaping forward and upward before plunging into the water with closed wings and partly spread tails.

COURTSHIP BEHAVIOR. *See* "Pochards and Allies" above. Their extremely rapid head-throw (the fastest among the pochards) is truly extraordinary and appears to the observer as little more than a blur. It is accompanied by a soft *whee oouu*.

VOICE. Male (when alarmed): a growling *quawk* or *scaup*. Female: a soft purring sound, uttered mainly during courtship.

OTHER BEHAVIORS. Lesser Scaups form enormous rafts that may contain several thousand individuals and cover hundreds of acres. Although by nature they are active and restless, they readily habituate to humans, even to the point of taking food from a person's hand. They fly erratically and with considerable twisting and turning, appearing as "compact but irregular formations that dart about like squadrons of pursuit planes" (Palmer, 1976).

Seaducks and Mergansers

All the seaducks and mergansers dive for food, but otherwise they share very few field characters that are useful for recognizing them as members of the same group.

Surf Scoter, *Melanitta perspicillata.* Princ. distrib. Texas: Rare M and WR in scattered localities throughout the state. Surf Scoters eat primarily mollusks, crustaceans, and insects, only about 10 percent of their diet being plant material. The male's courting call has been described as a liquid gurgling; the female's, as crowlike. Surf Scoters fly low over the water in irregular lines and make a humming sound with their wings.

White-winged Scoter, *Melanitta fusca.* Princ. distrib. Texas: Rare M and WR throughout the state except the lower Rio Grande Valley and the Trans-Pecos. The major component of their diet is mollusks. Males make a bell-like whistle (a tinkling-ice sound) and females, a thinner whistle. White-winged Scoters commonly fly in loose flocks low over the water.

Black Scoter, *Melanitta nigra.* Princ. distrib. Texas: Rare winter visitor, most observations being on the coast. They feed principally on animal matter, especially mollusks, which they gather by diving, sometimes to great depths. The male's courting call, a melodious, bell-like whistle that suggests a curlew, is rarely heard in Texas. The female makes a croaking sound that has been compared to a rusty hinge, and both males and females produce a whistling sound with their wings.

Oldsquaw, *Clangula hyemalis.* Princ. distrib. Texas: Rare M and winter visitor in most parts of the state. These handsome ducks seem phenomenally strong and energetic, flying low over the water, twisting and turning, then dropping to the surface with a splash. Even after alighting they seem unable to keep still. They feed mainly on crustaceans and other invertebrates, sometimes diving 100 feet or more — considerably deeper than most waterfowl, and in the range of the loons.

They are exceptionally garrulous and produce two loud, unique courtship calls sometimes heard in Texas in spring: *ugh ugh ah oo GAAH* and *a oo a oo a oo GAH.* The latter call, sometimes given in chorus and audible up to a mile away, suggests a pack of hounds.

Bufflehead, *Bucephala albeola.* Princ. distrib. Texas: Common M and WR throughout Texas. They prefer more sheltered waters than do goldeneyes, and they can be observed on salt, brackish, and fresh water.

FEEDING BEHAVIOR. Small groups of these lively, restless, teal-sized ducks commonly swim together; then suddenly, with remarkable synchrony, they plunge into the water and search for aquatic insects, pond weeds, and bulrushes. In saltwater habitats, they seek crustaceans and mollusks. They compress their plumage before diving and propel themselves underwater with their feet.

COURTSHIP BEHAVIOR. The jerky, mechanical displays of the drake are unforgettable. Many movements appear to be ritualized attack and escape behaviors, including the head bob (most common), wing lift, wing flap, and head shake forward. Their displays emphasize the strikingly patterned head and neck.

VOICE. Generally silent. Male: a hoarse whistle. Female: a hoarse *quack.*

OTHER BEHAVIORS. Buffleheads rarely associate to any extent with other species. They are regarded as moderately gregarious, however, and form small groups among themselves during migration and when on their wintering grounds. They respond to threats with the head-forward posture.

Their agility in flight distinguishes them from most ducks. They take wing very easily (more like dabblers than other diving ducks) and fly low with rapid, insectlike wingbeats. They may suddenly emerge from underwater and appear to fly out of the water.

Common Goldeneye, *Bucephala clangula.* Princ. distrib. Texas: Uncommon to common WR throughout most of Texas, generally on the coast or on the larger reservoirs. Because many ducks have already paired in winter, migratory flocks are smaller in spring than in fall.

FEEDING BEHAVIOR. Their diet includes crustaceans, insects, and mollusks, but they take advantage of most plant or animal food items when available. These restless birds feed primarily during the day and commonly form rafts (congregations of swimming birds) at night. Strong swimmers and divers, they sometimes turn over stones with their bill during underwater forays for food.

COURTSHIP BEHAVIOR. Male goldeneyes probably show more diversity and complexity in their social displays than any other North American waterfowl. Often these displays are ritualized forms of aggressive behaviors. The female's most common inciting behavior is the head-forward display; others include the head flick, head up, and dip.

The male responds with fancifully named displays that include ticking (his most common display, a lateral shaking of the head), head throw, bowsprit, nodding, masthead, slow and fast head throw, head kick, head up pumping, head back, head back bowsprit, and head flick.

VOICE. The male during courtship utters a harsh, rasping, nasal *pee eenk,* suggesting a Common Nighthawk. Females are generally silent but sometimes utter a *quaak.*

OTHER BEHAVIORS. These are generally shy, wary, and restless birds; however, juveniles during their first autumn can be quite tame.

In calm weather, Common Goldeneyes must patter across the water to take flight, but a light breeze permits them to take wing directly from the water. Flight intention is signaled by ticking (lateral shaking of the head), a display also used to reinforce the pair bond. Their swift, strong flight has been clocked at more than 40 mph. When they fly they produce a vibrant, melodious whistling sound with their wings (hence their colloquial name "whistler").

Hooded Merganser, *Lophodytes cucullatus.* Princ. distrib. Texas: Uncommon to common M and WR throughout most of Texas e. of the Trans-

Pecos, frequenting lakes, brackish estuaries, and other open bodies of water.

FEEDING BEHAVIOR. These sharp-looking birds dive gracefully for aquatic animals that live near the bottom or just beneath the water's surface. They swim rapidly and use both wings and feet to propel themselves. Diet: approximately 95 percent animal, including crayfish, aquatic insects, and other invertebrates. Fish, making up only 40–50 percent of their diet, are less important for this species than for the larger Common and Red-breasted Mergansers.

COURTSHIP BEHAVIOR. Not surprisingly, the striking crest plays a prominent role and is raised or lowered depending on the display. Commonly observed displays in winter are the head throw, elliptical neck stretch, and tail cock.

VOICE. During courtship the male's froglike *krrrooooh* is audible for up to half a mile. He also chatters and makes low grunts. Females are generally silent.

OTHER BEHAVIORS. Hooded Mergansers appear nervous, often jerking their heads as they swim; they also seem to be constantly preening. They maintain a strong (but quiet) flight that can change abruptly in direction. Like Buffleheads, they take wing easily from the water, even in the absence of wind. Their wings produce a whistling sound in flight.

Although occasionally they form large flocks, they normally congregate in groups smaller than 15. They typically keep to themselves even when they join a larger, mixed flock of ducks.

Common Merganser, *Mergus merganser.* Princ. distrib. Texas: Uncommon WR in n. half of Texas. Most individuals that remain in the Panhandle in summer are migrants unable to complete their northward trip. Unlike Red-breasted Mergansers, which in winter favor salt water, Common Mergansers tend to spend the winter on bodies of fresh water.

FEEDING BEHAVIOR. Common Mergansers forage opportunistically, taking both game and rough fishes. They feed in water less than 6 feet deep, plunging forward into dives that last 10–20 seconds. Like grebes, they sometimes sink slowly into the water when disturbed.

The main item in the diets of both Common and Red-breasted Mergansers is fish. When foraging on the same lake the former species tends to follow the shoreline, where the water is shallow and trees are numerous; the latter selects more open and often deeper water. In winter, the Red-breasted Merganser occurs in more saline habitats.

Although they minimize competition for food resources by utilizing different foraging strategies, these two mergansers share some basic feeding behaviors. Both search for prey by dipping the head into the water or by swimming along with the head submerged, and both collectively drive fish into shallow water (recalling the feeding behavior of White Pelicans).

COURTSHIP BEHAVIOR. In one common display, occurring in Texas in the spring, the male stretches his neck, fluffs his head feathers, and utters a purring sound that has been likened to strumming a guitar. In another display, he suddenly stretches his head and neck (salute posture) and produces a note described as bell-like. In the sprint display, he swims swiftly across the water while flapping his wings against the surface.

VOICE. Both sexes produce a harsh croak. The male's courtship calls are described above.

OTHER BEHAVIORS. If threatened on land they run to water to dive or to take flight. When taking flight they first run across the water.

As they fly in long lines low over the water, Common Mergansers produce a whistling sound with their wings. They rarely form flocks larger than about a dozen individuals. Being a cavity-nesting bird of the northern wooded areas, they are adept at flying among tree branches and entering tree cavities. (*See* "Other behaviors" under Red-breasted Merganser.)

Common Mergansers rest and spend the night on shore. They are usually quite wary around humans.

Red-breasted Merganser, *Mergus serrator.* Princ. distrib. Texas: Common WR on the coast but rare to uncommon WR elsewhere in the state. Although their Canadian nesting grounds include both inland waters and coastal areas, their preferred wintering habitats are decidedly marine (in contrast to Common Mergansers).

FEEDING BEHAVIOR. (*See* Common Merganser.) Red-breasted Mergansers feed in close groups, one quickly following another in dives, and all regrouping when they surface. They sometimes swim abreast to drive fish into shallow water; at other times they leapfrog, with birds at the rear of the feeding group flying ahead, only to be overtaken by the birds they just flew over.

COURTSHIP BEHAVIOR. The spectacular courtship displays can be observed in Texas in winter and spring. The female incites by dashing forward with her head high and her bill pointed downward. The male responds with any of the following displays. In the knicks display he circles around the female, suddenly extends his neck and head upward, then pulls his head downward while uttering a catlike *weow weow.* He may also utter a catlike call while pulling his head downward, raising his wings, and tilting his tail downward.

VOICE. The male's courtship calls are noted above. The female's calls are similar to the female Common Merganser's. Both sexes also make a purring note and a variety of hoarse croaks.

OTHER BEHAVIORS. Where they form large flocks, they may be responding more to local concentrations of fish than to a need to associate with members of the same species. These are wary, energetic, and constantly active birds that pump their head back and forth as they swim.

When alarmed, both Red-breasted and Common Mergansers can take off directly from the water, but usually they first run across the water. Red-breasted Mergansers (which take flight more easily than Common Mergansers) "appear to be more buoyant and maneuverable in flight" (Palmer, 1976). They have one of the swiftest flights of our ducks, and they usually fly low over the water. When they fly they produce a soft whistling with their wings.

Stiff-tailed Ducks

Stifftails are represented in the United States by the Ruddy Duck and the Masked Duck. Both exploit a niche unavailable to most ducks: they forage in the bottom silt for plants and animals. They are well adapted for diving, having large feet, short legs placed far back on the body, short wings, and stiff tail feathers that serve as rudders.

The swollen neck observed during the courtship displays of both species is attributed to an inflatable saclike extension of the trachea. Clutch size is small for a duck, but the eggs are the largest of all waterfowl in relation to body size. For a detailed account of the behaviors of stifftails, see Johnsgard and Carbonell (1996).

Masked Duck, *Nomonyx dominicus.* Princ. distrib. Texas: Rare and irregular PR on the coast and in the lower Rio Grande Valley, where they have nested in freshwater ponds and pools. Water lilies, used both for food and concealment, are an important, although probably not indispensable, habitat component. Because almost all U.S. nesting records are from Texas, their nesting behavior is described below.

FEEDING BEHAVIOR. Masked Ducks dive with impressive facility. They may remain underwater 30 seconds or longer while searching for submerged seeds and plant parts. They are reported to locate food items by sweeping aside underwater plants with their tail. Important food plants besides water lilies include dodder and sawgrass.

Their bill is structured more like a dabbler's or pochard's than that of their nearest relative, the Ruddy Duck, suggesting that Masked Ducks are less well adapted for locating and extracting invertebrates from bottom silt.

COURTSHIP BEHAVIOR. In what appears to be a unique behavior among waterfowl, the male expands his upper neck to the size of a golf ball. Earlier accounts of males striking their inflated breasts (in the manner of Ruddy Ducks) need to be confirmed.

NESTING. They nest as early as September and October. They place the nest near water, often in vegetation, and cushion it with a small amount of down. Clutches larger than the usual four to six eggs apparently represent dump nesting.

Possibly some of the femalelike birds observed attending the young are males in winter plumage, as males of the closely related Ruddy Duck regularly assist in caring for the young.

VOICE. The male's courtship call (a clucking *kiri kirou*, repeated with variations) has been compared to a pheasant's call. Females make hissing noises.

OTHER BEHAVIORS. Masked Ducks can emerge gradually from their dives, sometimes remaining for a while with only the head above water. They usually patter along the water before taking off, but some observers report that when taking flight they leap from the water. They usually fly low over the marsh and quickly drop down to find concealment among emergent vegetation. Unlike Ruddy Ducks, they rarely hold their tail above water.

Masked Ducks are probably monogamous. They occur singly or in pairs, rarely in large flocks. When birds are disturbed and occasionally when they alight they pump their head up and down (displacement behavior).

Ruddy Duck, *Oxyura jamaicensis.* Princ. distrib. Texas: Common M and WR in most parts of the state, nesting in scattered localities. Most summer birds do not nest.

These ducks are consummate divers. Even young birds dive almost as soon as they get into the water (in contrast to most ducklings, which must mature for a few weeks). Adults are capable of submerging so slowly that they hardly leave a ripple in the water.

FEEDING BEHAVIOR. Ruddy Ducks use their stout bill to tear off pieces of submerged leaves and other plant parts. They also sift food items from muddy bottoms, especially midge larvae and pupae, by rapidly opening and closing the bill as they move their head back and forth. Apparently they are capable of detecting organisms entirely by touch; thus, they probably feed as efficiently at night as during the day. In Texas, insects and other invertebrates are a large component of the winter diet.

COURTSHIP BEHAVIOR. During the breeding season the male's bill turns blue and probably functions as a releaser for courtship behaviors. Apparently female Ruddy Ducks do not incite like most ducks. In the bubbling display, the male raises his head and tail to a vertical position and beats his bill on his inflated neck to produce a clucking or thumping sound rendered as *raa annhh* — a behavior unique among waterfowl. This sound, like the sounds of other marsh-dwelling birds such as rails and bitterns, carries well in highly vegetated wetland habitats. The display is called bubbling because beating his breast feathers forces out trapped air that forms a ring of bubbles on the water. Bubbling also occurs in contexts besides courtship.

Other displays performed in winter and spring include head dipping, head shaking, and cheek rolling (in which cheeks are rolled on the back, a display probably derived from preening). Prior to copulation the male engages in head flicking and head dipping. One postcopulatory display consists of bubbling followed by preening.

NESTING. Nesting behavior is not described here, except to mention that on their breeding grounds, Ruddy Ducks sometimes lay eggs in nests of other birds, including bitterns, American Coots, grebes, and several duck species. The Ruddy Duck is the only North American duck in which the male helps the female care for the young.

VOICE. Normally silent, except during the bubbling display described above. Sometimes the female produces a nasal sound and a high-pitched note.

OTHER BEHAVIORS. During migration and in winter, Ruddy Ducks sometimes congregate in flocks that contain several thousand birds. They rarely associate closely with other waterbirds, except, curiously, American Coots. They are generally unwary outside the nesting season.

They have two flight intention displays. In cheeking, the head is lowered over the back and the cheek is rubbed near the base of the tail (ritualized preening behavior). In head shaking, the bird quickly shakes its head before taking flight.

DIURNAL BIRDS OF PREY:
ORDER FALCONIFORMES

Flight in Raptors

Flight varies among raptors according to habitat and type of prey. For example, Red-tailed Hawks and other buteos survey enormous expanses of prairie by *soaring* on thermals (updrafts of hot air). Their broad wings enable them do this, but at the price of increased drag created when air from high-pressure areas beneath the wings flows over the broad wingtips to the low-pressure areas above. To increase the flow of air and reduce drag, buteos spread the tips of their primaries, an adaptation lacking in soaring birds like Ospreys, whose wingtips are small in comparison to their long, narrow wings.

Accipiters such as Cooper's Hawk *flap and glide,* which permits them to keep aloft in the absence of thermals. Thus they can fly beneath the forest canopy and prey on fast-flying birds that are essentially unavailable to the soaring buteos above. During glides accipiters maintain speed by closing their primaries and partially bending their wings.

Kites and a few other raptors maintain a fixed wing position that allows them to hang in the wind. This method of flying is aptly called *kiting,* because by suspending themselves like this they suggest the flight of a man-made kite.

When raptors *course,* they fly steadily and low over the landscape in search of prey. Coursing is generally done more or less randomly, but when done methodically, as in flying back and forth across a field, it is called *quartering.*

Quartering is a common foraging strategy in Northern Harriers. Its efficacy is evidently recognized by other raptors, some of which use Northern Harriers as *beaters.* For example, Peregrine Falcons and other falcons sometimes follow quartering Northern Harriers to locate prey in much the same way that hunters use dogs (also called beaters) to flush quarry.

Hovering—a frequent behavior in Ospreys, American Kestrels, and other hawks—allows a raptor to suspend itself over the prey just before swooping down for the capture. To hover efficiently, most raptors must face the wind (in contrast to hummingbirds, which are capable of hovering in still air).

When dropping to the ground for their quarry, hawks often *stoop*—that is, they draw their wings close to their bodies to minimize wind resistance and accelerate their descent.

HAWKS, KITES, EAGLES, AND ALLIES: FAMILY ACCIPITRIDAE

This family has been divided into the following groups on the basis of morphological and behavioral characteristics: the Osprey, or fish eagle (unique among our raptors because of its specialized manner of capturing fish); kites, whose graceful soaring suggests the buoyant movements of a well-built paper kite; harriers, which course low over the ground, searching for prey in the manner of owls; accipiters—the sprinters of the forest—whose slim bodies reduce drag and whose short, broad wings help them gain speed when chasing swift, hard-to-catch birds; and buteos (including the Golden Eagle), which soar over open areas in search of rabbits, rodents, and other prey. The Bald Eagle does not fit well into any of these groups, although on the basis of DNA it seems to be related to the kites.

Osprey, *Pandion haliaetus.* Princ. distrib. Texas: Uncommon M throughout Texas, WR in s. part of the state, and rare SR at scattered localities throughout all but w. Texas.

FEEDING BEHAVIOR. The fish eagle's unexampled plunges for fish are spectacular. After briefly hovering, it partially closes its wings to gain speed, then rapidly plunges into the water and grabs a fish with sharp, strongly curved talons. It momentarily disappears beneath the water's surface and resurfaces with a fish that it carries headfirst, in its talons, to an elevated feeding site, commonly a tree or utility pole, or to its nest. The ability to reverse the outer toe (so that two toes point forward and two backward) gives Ospreys an extremely effective tool for grasping slippery quarry.

Hovering is an important component of their foraging behavior: birds that hover before diving are about 50 percent more successful in capturing prey than those that do not. Ospreys forage most successfully in clear waters, where fish swim near the surface.

They occasionally eat amphibians, reptiles, small rodents, and other animals. In spite of their powerful, swift, and dashing flight, and the intensity with which they defend their fishing areas against other Ospreys, these birds are sometimes robbed of food by Bald Eagles (although probably rarely in Texas).

COURTSHIP BEHAVIOR. Monogamous. The pair pursue each other rapidly, circling, dodging, and swooping with dexterity.

NESTING. In a tall tree near or over a river or lake. Nest: refurbished each year. They are not choosy about nesting materials and incorporate sticks, grass, leaves, rocks, dirt, and even cow dung into a nest that often becomes very large as trash accumulates.

The male brings food to his mate while she incubates, and also brings food for her to feed the young—a behavior that probably serves to reinforce the pair bond as well as to reduce exposure of the highly visible nest to predators.

VOICE. A series of high-pitched whistled notes, *whiew whiew;* also, a *sheeap.*

OTHER BEHAVIORS. Ospreys live alone or in pairs. Because they show relatively little fear of humans, they can be difficult to control around fish hatcheries.

Hook-billed Kite, *Chondrohierax uncinatus.* Princ. distrib. Texas: Rare PR in lower Rio Grande Valley, frequenting thorn scrub and thorn forests (Brush, 1999). Their reclusive habits make them difficult to locate. They feed on tree snails (*Rabdotus alternatus*) and hunt from a perch. Call (rarely heard in Texas): a short rattle.

"In flight, [they are] quite distinctive: the longish wings are narrower at the base than toward the tip and have been described as spoonlike or paddle-shaped. The wings are flapped rather loosely and then held fairly flat, often pressed forward and sometimes raised slightly" (Brush, 1999).

Swallow-tailed Kite, *Elanoides forficatus.* Princ. distrib. Texas: Rare M throughout most of Texas except the Trans-Pecos and Panhandle. They are a locally common M on the coast and a rare and local SR in e. Texas. They prefer marshes, riverine and swamp forests, and adjacent wet grasslands.

FEEDING BEHAVIOR. These gracefully contoured raptors float just above the treetops of open forests, constantly adjusting their balance by lightly tipping back and forth. They are versatile foragers, sometimes soaring to great heights, at other times quartering low over open fields like Northern Harriers. They capture dragonflies and other large insects with their talons, and consume their prey in flight by bending their head under their body; they also pluck lizards, snakes, and other animals from trees or from the ground.

COURTSHIP BEHAVIOR. While perched on a horizontal limb, the two birds approach each other, then the female backs off under a limb. The male feeds her a snake or other prey item, a display that is followed by repeated copulations. The male also performs aerial displays, including swift chases over water.

NESTING. Nest: hidden at the top of a tall tree, and built by both sexes from twigs, leaves, mosses, and other materials. Pairs have been known to travel in excess of 800 miles in order to collect the 200 or more pieces of material they incorporate into the nest. Sometimes an old nest is refurbished. The female broods the nestlings at first, feeding them food brought to her by her mate; later both parents hunt and bring food to their young.

VOICE. Soft, plaintive whistles and loud, shrill cries (*pii pii pii pii*).

OTHER BEHAVIORS. These are gregarious birds; they migrate and feed in flocks of 4–20, occasionally (outside Texas) in considerably larger groups. They sometimes associate with Mississippi Kites.

Their flight is graceful, elegant, and effortless. They make remarkable swoops to within a few feet of the ground, executing quick turns while opening and closing the forked tail, and they bathe and drink by skimming the surface of the water like a swallow. Since they spend most of their day flying, one rarely sees perched individuals.

They play by soaring with streamers of Spanish moss in their bill, or by rapidly chasing each other through trees.

White-tailed Kite, *Elanus leucurus.* Princ. distrib. Texas: Uncommon to common PR on the Coastal Plain, frequenting damp meadows, marshes, and cultivated fields. This is one of the few Texas raptors that benefit significantly from expanding agricultural practices.

FEEDING BEHAVIOR. They generally hunt alone, coursing gracefully and low over the ground (but not so low as Northern Harriers) as they search for rodents and other small vertebrates, as well as large insects. When they locate their quarry, they hover (up to half a minute), and with feet dangling, they slowly flutter down for the capture. This method of attack contrasts sharply with the rapid swoop of falcons. These kites also hunt from perches.

COURTSHIP BEHAVIOR. Courtship displays include spectacular dives and loops. As the pair circle slowly, one bird passes below, rolls over onto its back, and interlocks feet with its mate. In the wings-up display, the male flies with his wings in an upward V position while rapidly vibrating them, making a chattering call as he does this. Often he presents his potential mate with prey.

NESTING. In treetops near water, and normally difficult to see from below. White-tailed Kites are generally tolerant of other kite nests nearby, but they become quite aggressive if raptors other than their own species nest in the vicinity.

The male brings food to the female while she is incubating, sometimes transferring it to her in flight; later, both parents drop food into the nest for the nestlings to eat.

VOICE. Shrill whistles, *keeep kee ep,* suggesting an Osprey, as well as a *gee grack* and a guttural call.

OTHER BEHAVIORS. In winter they roost communally, sometimes with more than 100 birds at one site.

Their light, graceful, and slow flight, with wingtips pointing downward, has been compared to that of Ospreys and gulls. They also remain stationary in flight (kiting) with their legs dangling. As raptors go, these birds are considered unwary around humans.

Mississippi Kite, *Ictinia mississippiensis.* Princ. distrib. Texas: Common SR in the High Plains and adjacent parts of the state, frequenting open woodlands, pastures, and urban areas. Common SR on the coast.

FEEDING BEHAVIOR. Noted for their buoyant flight and the dexterity with which they capture prey, primarily large, flying insects. They also seize animals from the ground and sometimes capture bats at entrances to caves.

Sometimes as many as 20 birds follow livestock to take advantage of insects that are flushed. Mississippi Kites may consume their prey in flight, bringing the foot forward and bending the head beneath the body. This maneuver often presses partially eaten prey against the underside of the tail to produce plumage discoloration that is sometimes mistaken for a field mark.

COURTSHIP BEHAVIOR. They are already paired when they arrive on the breeding grounds. Their courtship displays, not often observed in Texas, include a series of graceful aerial flights. Other than offering food to his mate, the male initiates copulation with little or no precopulatory behavior.

NESTING. Several pairs sometimes nest together in loose colonies at the tops of tall trees. Where tall trees are unavailable, nests are placed as low as 6 feet above the ground. Old nests are frequently reused. Although they accept other nesting kites nearby, Mississippi Kites vigorously and routinely drive owls and other hawks from their nesting areas. Both parents incubate the eggs and feed the nestlings. Helpers (usually yearlings) sometimes aid their parents at the nest, an altruistic behavior that presumably provides parenting experience and prepares the young birds to raise their own young. Also, by helping at the nest, young birds increase the likelihood that some of their own genes — those they share with their parents, brothers, and sisters — will be passed down to future generations.

VOICE. A high, thin, descending *shi chiew.*

OTHER BEHAVIORS. Mississippi Kites are clearly the most gregarious of our three kites, frequently foraging in small groups and at times traveling with Turkey Vultures. They also congregate at communal perches.

The elegance of their buoyant flight is perhaps second only to that of the Swallow-tailed Kite. In a notable feature, a bird will abruptly check its flight, spread its wings horizontally, and hang motionless in the air (kiting).

Bald Eagle, *Haliaeetus leucocephalus.* Princ. distrib. Texas: Uncommon M and WR in most parts of the state; rare local PR on the coast and in scattered locations inland.

FEEDING BEHAVIOR. In spite of their imposing appearance, Bald Eagles feed primarily on fish, carrion, small mammals, and small birds. They are capable of taking waterfowl on the wing, but they are more likely to chase down sick or wounded ducks. They commonly steal prey from Ospreys and other birds, a behavior evidently refined by practice, as older birds steal more efficiently than do young birds.

COURTSHIP BEHAVIOR. From a position high in the air, two birds lock talons and fall in a series of spectacular somersaults. During their migration northward in spring, birds circle at a considerable height, then suddenly dive at other birds.

Apparently most Bald Eagles mate for life. Their monogamous behavior has been attributed to their tendency to remain at the nest year round or to return each year to the same nest.

NESTING. Nests are generally placed in the fork of a tree or on a cliff. Most nests observed on the coast were in water oak (*Quercus nigra*) or pecan trees (Benson and Davis, 1985).

Each year the pair adds new sticks to the previous year's massive nest. In some cases the nest represents a number of nesting seasons and measures up to 7 feet across and 12 feet deep. It is thickly lined with soft materials and greenery.

Both parents incubate, fetch food, and feed the nestlings. Young birds commonly kill their siblings in the nest, but they do this less frequently than do Golden Eagle nestlings.

VOICE. A high-pitched chatter or squeaking cackle that suggests a gull.

OTHER BEHAVIORS. In Alaska, thousands of eagles congregate to feed on salmon, but in Texas rarely more than 5–10 birds are observed together. In Utah, small groups hunted cooperatively to flush and kill rabbits, a behavior that could be looked for in Texas.

Communal roosting is common on their wintering grounds.

Bald Eagles soar majestically on horizontal, flat wings, then make low swoops to capture prey or to feed on carrion. Some observers report that their maneuvering skills are so extraordinary—they turn quickly, then double back—that, unlike Golden Eagles, they are almost impossible to kill from a helicopter.

The head-toss display is commonly performed in the presence of an approaching human: the bird tosses its head backward, then forward and down, in a deliberate manner. These movements may represent displacement behavior.

Some activities seem to be play. For example, several birds spiral high into the air while following a bird that is carrying a stick. When the bird drops the stick, another bird catches the stick before it reaches the ground, and then this bird flies up and drops the stick.

Northern Harrier, *Circus cyaneus.* Princ. distrib. Texas: Common M and less common WR throughout most of Texas, where they inhabit wet meadows, marshes, pastures, and comparable habitats, often near lakes; rare SR in widely scattered localities.

FEEDING BEHAVIOR. Northern Harriers are the most owllike of our hawks, both in structure and behavior. They feed somewhat like Short-eared Owls, flying low over the ground (the search-pause-pounce strategy) in search of rodents, snakes, large grasshoppers, and sometimes carrion. They regularly quarter when foraging, and they may follow a route that remains unchanged for several weeks.

Like owls, Northern Harriers locate prey by sound as well as by sight. Sound reception is enhanced by a facial ruff that apparently functions like the facial ruff in owls. Their directional hearing is similar to that in owls and surpasses the capabilities of most diurnal raptors.

They typically flush rather than chase prey, and if they do not capture the animal on their first attempt, they pass it up. The female's larger size and different plumage color probably allow her to capture different prey species from those her mate selects.

Although in California (and probably Texas as well) adult males rarely defend a winter territory, females do—and quite vigorously. They chase away any raptor large enough to steal their prey, yet they allow smaller hawks in their territories, possibly because now they are in a position to steal prey.

COURTSHIP BEHAVIOR. Among the male's spectacular courtship dives is the sky dance: barrel rolls in multiple U-shaped loops, as though the bird were following "the path of an invisible coil spring in the sky" (Palmer, 1988). In the tumbling display, the male ascends to a great height and positions his wings over his back to produce a fall. During his descent he turns over and over, pulling out of the fall just before reaching the ground.

Often males present food to the female prior to copulation, sometimes dropping it to her in midair (aerial transfer of prey). Copulation generally occurs on the ground.

NESTING. Both parents bring food to the nestlings. On their nesting grounds parents are quick to chase away almost all intruders, including eagles and human beings.

Some males have several mates (polygyny), especially when vole populations, their preferred prey in the western United States, are high. Polygyny has not been reported in Texas birds.

VOICE. Short, high-pitched whistles, as well as shrill screams.

OTHER BEHAVIORS. Their flight is distinctive: they course low over the ground with few, but quick, wingbeats and hold their wings in an open V as they constantly tilt the body from side to side. In winter, groups roost together on the ground, sometimes with Short-eared Owls.

Sharp-shinned Hawk, *Accipiter striatus.* Princ. distrib. Texas: Uncommon to locally common M and WR throughout the state, nesting rarely in widely scattered localities, especially in remote wooded areas; however, an increasing number are colonizing urban areas. They rarely venture into wide open country, and they generally avoid areas where Cooper's Hawks—one of their major predators as well as a competitor for food— are nesting.

FEEDING BEHAVIOR. The proportion of birds in their diet probably exceeds that of any other member of the family. In winter they frequently visit bird feeders. They have also been observed capturing bats around cave entrances.

Like Cooper's Hawks, Sharpshins maneuver through trees and bushes with astonishing skill, adeptly using their long tail as a rudder. They are unsurpassed masters of the sneak attack. From a concealed perch they dash out abruptly at a passing bird and relentlessly chase it high in the canopy or even through dense foliage. The average size of their prey is smaller than the prey of Cooper's Hawk.

COURTSHIP BEHAVIOR. Their courtship is believed to resemble closely the behavior of Cooper's Hawk. The few detailed descriptions include high circling, with much calling, and a sky dance.

NESTING. Eggs: usually incubated by the female. The male does almost all hunting from the time the eggs are incubated until the young are in the early nestling stage. Curiously, he usually beheads prey before bringing it to the nest. (See Shackelford et al., 1996, for details of nests in east Texas.)

VOICE. A shrill *keek keek keek* and a thin whine or squeal.

OTHER BEHAVIORS. Sharp-shinned Hawks stubbornly defend their nests against almost all intruders, including humans. During their fast, vigorous flights they alternate quick flaps with glides. They often migrate in large numbers.

Cooper's Hawk, *Accipiter cooperii.* Princ. distrib. Texas: Rare to uncommon SR in much of the e. and c. part of the state; uncommon M and sometimes common WR throughout Texas. They prefer wooded areas adjacent to farmland and thus benefit when Texas forests are broken up by clearings. They are increasingly becoming urban nesters.

Rarely do Cooper's Hawks and Sharp-shinned Hawks nest in the same woods, most likely because the former species drives out the latter.

FEEDING BEHAVIOR. From a concealed perch, they suddenly dash out and fly low over the ground in relentless pursuit of a bird or mammal. They kill their prey by squeezing it with their sharp talons, relaxing their grip, then repeating the process until the animal is dead. There are reports of Cooper's Hawks that have carried animals to water and drowned them.

The tenacity with which they pursue elusive prey is legendary. They follow sparrows from bush to bush in chases that may last up to 45 minutes. Sometimes they pursue ground-dwelling animals on foot.

Individual birds are usually highly selective in what they choose to capture and eat. Some birds take mainly, if not entirely, chickens (hence the colloquial name "chicken hawk"). Paradoxically, other individuals that nest near coops never bother chickens. Prey preferences like these may have resulted from early successes or failures when hunting a particular prey species. Cooper's Hawks (as well as other raptors) have been observed locating Northern Bobwhites by sound.

COURTSHIP BEHAVIOR. In the sky dance, the male, with wings held over his back, flies through a wide arc with slow, rhythmical flaps that suggest a Common Nighthawk. In the high circling display, after both birds ascend to a considerable height, the male pursues the female through several soars and dives. Dives are followed by a slow-motion chase with wings held high and wingbeats exaggerated.

The male also performs a bowing display: he assumes a horizontal standing position, then makes 3–10 quick bowing movements, possibly to convey his submissiveness to the larger, dominant female.

NESTING. Eggs: generally incubated by the female. The male brings prey to the nest and gives it to the female, who feeds the nestlings. When food is scarce, young birds in the nest commonly kill and eat their siblings; it is usually the larger female nestlings who eat their smaller brothers.

VOICE. A loud, deep, and rapid chatter: *ka ka ka ka ka ka kek.*

OTHER BEHAVIORS. On the one hand, Cooper's Hawks are exquisite killing machines whose predatory behavior surpasses in intensity and persistence that of most raptors. On the other hand, some behaviors caricature human beings who rhapsodize on the beauties of nature.

For example, upon awakening early on a spring day, the two birds sing together—the so-called morning duet. The male begins by singing a few notes in his high-pitched voice. He is soon joined by his lower-voiced mate, and together they engage in a duet made up of a variety of unique and intriguing notes. At the conclusion of the performance they copulate.

Another example: While incubating her eggs, the female seems enraptured by the sun. She watches it intensely as it rises in the morning and sets in the evening. She shows an unusual interest in all events and a heightened sensitivity to all movements, including slight ones like passing insects, which she carefully follows with her eyes.

As is typical of accipiters, flight is swift and controlled. The bird flaps four or five times, glides for a few seconds, then repeats the process. It uses its long tail as a rudder when maneuvering through dense vegetation.

Anecdotal evidence suggests that Cooper's Hawks are less aggressive toward humans than are other accipiters, but at the nest they do not hesitate to dive at almost all intruders. Apparently they are solitary other than during the breeding season.

Gray Hawk, *Asturina nitida.* Princ. distrib. Texas: Rare PR in extreme s. Texas, in pastureland, riparian woodlands, and open wooded areas, typically around large stands of willows and cottonwoods that are adjacent to water.

FEEDING BEHAVIOR. From a concealed perch they dart out to capture lizards (probably their preferred food), snakes, small mammals, and insects. Unlike most hawks, they pluck lizards and birds from branches while flying swiftly among trees.

VOICE. A variety of calls, many high, slurred, and flutelike, including a musical piping and a scream (*kree e e e*).

OTHER BEHAVIORS. Their swift, graceful, and agile flight appears more like that of a falcon or accipiter than a buteo. (An older name for this bird, "Mexican goshawk," is said by some to refer to its accipiter behavior, by others to its plumage, which resembles a goshawk.)

Common Black-Hawk, *Buteogallus anthracinus.* Princ. distrib. Texas: Rare M, SR, and winter visitor in parts of w. and s. Texas, nesting in scattered localities. They usually occur near water or moist woodlands.

FEEDING BEHAVIOR. These hawks are singular among Texas raptors in their predatory behavior: they stalk crabs by wading on sandbars and mudflats (hence their older name "crab hawk"). This behavior, observed in Mexican birds, probably occurs only infrequently, if at all, in Texas. Texas birds usually fly out from a low perch, often a branch overlooking the water, and chase down small prey that include crayfish, fish, frogs, and snakes.

COURTSHIP BEHAVIOR. The male performs a series of acrobatic ascents and dives (including a sky dance and tumbling) while calling and dangling his feet. Evidently dangling feet are a releaser for the female's response. Prior to copulation, the male swoops down to the perched female.

NESTING. Nest: usually in a cottonwood or large mesquite near a river or stream. The female constructs a bulky platform with sticks brought by the male, then lines it with mistletoe, leaves, and other plant material. Sometimes an old nest is utilized. Eggs: usually incubated by the female, who feeds the nestlings prey that her mate brings to the nest.

VOICE. Three or four loud, harsh notes, reminiscent of a night-heron's squawk.

OTHER BEHAVIORS. Superficially, their buoyant flight suggests a Black Vulture: the wings and tail are wide and the birds alternately flap and glide. They sometimes make spectacular dives toward the ground, especially during the nesting season.

These relatively tame, sluggish birds often can be approached rather closely.

Harris's Hawk, *Parabuteo unicinctus.* Princ. distrib. Texas: Common PR in the open areas, semiarid mesquite thickets, and thorny shrubs of s.

Texas; locally uncommon PR in the Trans-Pecos, where they favor mesquite grasslands and desert habitats.

FEEDING BEHAVIOR. Pairs or family groups hunt in early morning or evening, dashing like accipiters (flap-flap-glide) into thick vegetation in pursuit of large insects, rabbits, rodents, herons, and even other raptors. Less frequently they attack from a perch or while soaring.

Cooperative hunting is common in this unusually social species, and food is often shared among the individuals of a group. If the first bird of a hunting pair misses the prey and the second captures it, the first bird may be allowed to feed if there is enough food for both birds. When two or more individuals hunt together, the total catch is not always greater, but apparently more birds benefit. If, as is usually the case, many or all individuals in a hunting group are related, then by feeding cooperatively birds help relatives survive at higher rates and therefore help pass along, to future generations, genes that are shared by the group.

COURTSHIP BEHAVIOR. Several adults circle high in the air, stooping, tail-chasing, and vocalizing loudly. In the sky dance, the male partially folds his wings to decrease wind resistance, a common maneuver in raptors that produces astonishingly fast dives. Possibly the sky dance functions in territorial as well as courtship behavior.

Prior to copulating, the male dives, swings upward toward a perched female, and alights on her back.

NESTING. Nest: In a tree, cactus, or utility pole. It is built by both parents. Leafy twigs are added throughout the season as decoration or perhaps to repel parasites. Leafy material may also serve as a visual signal to other Harris's Hawks that the nest is in use. Eggs: normally incubated by the female, who is brought food by her mate. Both parents feed the nestlings.

Cooperative simultaneous polyandry (a female with several mates that help at the nest) is a very rare form of polygamy. It has been recorded in Arizona Harris's Hawks and should be looked for in Texas birds, where apparently most pairs are monogamous.

Some pairs (or trios) build or refurbish two or more nests in the territory; these nests are sometimes used as feeding platforms for prey just captured. Why birds expend so much energy building supernumerary nests is not clear. Possibly the behavior stimulates or reinforces the pair bond.

VOICE. The most common vocalization in the wild is an extended, harsh scream, given when intruders approach. The food begging call of young birds, a shrill, plaintive *eedjeep,* is apparently given by both young birds and adults to indicate submission.

OTHER BEHAVIORS. In flight they accelerate rapidly and hug the landscape contour. At midday these relatively tame birds sit on a conspicuous perch, and when approached they sluggishly fly to the nearest perch. At times they soar at considerable heights in what does not appear to be predatory behavior.

Falconers report that captive Harris's Hawks have a good disposition and a willingness to take a variety of prey. Even in the wild their aggressive interactions are relatively mild, and they usually tolerate conspecifics that intrude into their territories.

Red-shouldered Hawk, *Buteo lineatus.* Princ. distrib. Texas: Common SR in e. and c. Texas and on the coast, where they nest in woodlands and wet areas, especially river bottoms and swamps; uncommon WR on the coast and in the lower Rio Grande Valley. In winter many move to more open habitats.

FEEDING BEHAVIOR. The foraging behavior of these relatively sluggish birds differs from that of most buteos, as they frequently hunt beneath the canopy. In this habitat they employ a perch-and-wait strategy, sometimes waiting an hour or more before prey, usually a reptile, rodent, bird, or insect, appears. They quickly fly out and pounce on the animal, often carrying it to an old nest or other feeding platform before eating it.

In open areas they feed more in the manner of a Northern Harrier than a typical buteo, flying low and leisurely and locating prey both by sight and sound.

COURTSHIP BEHAVIOR. The high circling display is spectacular: several birds, with their tails fanned out, swoop and dive while calling loudly and persistently. They sometimes rise in spirals to considerable heights. In the sky dance, a bird ascends to a high elevation, then plunges almost vertically for several hundred feet — a display that often provokes neighboring pairs to join in and produce "a real melee of birds and yelling" (Palmer, 1988). At the termination of the sky dance, the male dives and copulates with the female.

NESTING. Nest: It rarely has the bulky character typical of buteo nests, and it is lined with fine materials as well as green leaves, which are replenished throughout incubation. These added materials have been termed symbolic, as they appear to be more decorative than functional, although possibly they repel insects and mites. In one case, green leaves were added to the nest 2 months prior to egg laying. Eggs: usually incubated by the female. She is brought food by her mate, who sits on the nest while she eats. He also brings food that she feeds to the nestlings.

Some Red-shouldered Hawks attach themselves firmly to the same territory for a number of generations, the record apparently being a pair and their progeny that occupied the same nesting area for 45 years.

VOICE. These are highly vocal birds. Their two-syllable *kee yeeer* is heard throughout the year, although more frequently during the breeding season.

OTHER BEHAVIORS. The relatively long tail probably enhances maneuverability in this species, as it does in forest-dwelling accipiters.

Curiously, American Crows, which readily mob owls and many hawks, seem indifferent to Red-shouldered Hawks and rarely, if ever, mob them. In fact, American Crows and Red-shouldered Hawks actually mob owls together, although the hawk is hardly an ally, as it does not hesitate to rob American Crows of food.

Red-shouldered Hawks become tame if not molested. They rarely occur in flocks or migrate in large groups.

Broad-winged Hawk, *Buteo platypterus.* Princ. distrib. Texas: Uncommon PR in the deciduous forests of e. Texas; locally abundant M in e. and c. Texas, especially along the Balcones Escarpment. In the Corpus Christi area, kettles totaling more than 300,000 were seen on a single day in September of 1997 and 1998 (Michael Marsden, pers. comm.).

FEEDING BEHAVIOR. Being the smallest of the North American buteos (hardly the size of an American Crow), Broad-winged Hawks must limit their prey to small animals. They employ the perch-and-wait feeding strategy to capture frogs, snakes, toads, insects, and (less frequently) small mammals and birds. They also make short flights within the forest, moving from branch to branch, and conduct extended forays by flapping and soaring above the canopy. Because they utilize trees and utility poles along woodland roads and forest edges, as well as perches along riverbanks and lakeshores, they are classified as edge feeders.

COURTSHIP BEHAVIOR. After arriving on their territories in spring, both sexes ascend to a high altitude, circle around noisily, then suddenly dart at each other. In the sky dance, a bird (probably the male) flies in widening circles to a height almost out of sight, then descends in long, sweeping movements. In the tumbling display, he dips earthward and checks his descent several feet above the ground.

NESTING. Nest: a small, carelessly constructed platform that both sexes build from sticks and twigs and line with bark, lichens, and other materials. They decorate it with fresh, green leaves (referred to as greenery). After feeding her young, the female collects sprigs and leaves from trees to place on top of the young birds; possibly the function of so much fresh vegetation is to repel insects and mites.

Eggs: incubated almost entirely by the female. At this time her mate brings her food.

VOICE. A plaintive whistle (*kweee e e e*), suggesting to some listeners the song of the Eastern Wood-Pewee, to others the Killdeer. The male's call is pitched about an octave higher than the female's.

OTHER BEHAVIORS. This gentle hawk is usually regarded as the tamest raptor in North America.

Swainson's Hawk, *Buteo swainsoni.* Princ. distrib. Texas: Common SR in the Panhandle and much of w. Texas, s. to the lower Rio Grande Valley; common M throughout the state except in the e. Texas forests. In the mid-1990s more than 10,000 individuals died from pesticides on their wintering grounds in Argentina.

They return from their wintering grounds relatively late, usually after other raptors have begun nesting.

FEEDING BEHAVIOR. These graceful hawks forage in open areas, evidently hunting most successfully from a perch, at least when searching for rodents. They also course low over the ground with their wings raised in an open V, like Northern Harriers and Ferruginous Hawks, but at slower speeds; at times they hover above their quarry before making their rapid descent.

These are not small hawks. They are the same size as Redtails, but they have relatively delicate feet that limit the size of prey they can handle. Accordingly, large insects are an important item in their diet. They eat them in flight, as do kites, by lowering their head and taking the prey from their talons. They consume enormous quantities of crickets and grasshoppers, picking them up quickly from the ground as they walk or run around in a manner that suggests a flock of turkeys.

Like many raptors, they learn to look for live animals and carrion along highways and to follow tractors for insects that are flushed. They also hunt near prairie fires for animals that are escaping the flames.

COURTSHIP BEHAVIOR. Their courtship behavior is characterized by vigorous flapping. In one of their more conspicuous displays, two birds circle high over their territory, flashing their white underwings and calling loudly. In the sky dance, the male spirals upward, then makes steep dives with bent wings. Copulation occurs on a tree or post after the female assumes the solicitation posture.

NESTING. Swainson's Hawks build flimsy, ragged nests. Like Broad-winged Hawks, they decorate them with considerable greenery, especially leafy twigs. Both parents incubate the eggs and feed the nestlings. They are very sensitive to nest disturbance and often desert their nests if they are molested.

VOICE. Call: a shrill, prolonged, plaintive whistle.

OTHER BEHAVIORS. These are the most gregarious of our raptors; they are well known for migrating in enormous flocks that frequently spend the night in fields and pastures.

Although buoyant and graceful in flight, they take off with heavy and awkward wingbeats. Once aloft, they are lifted by thermals to very high altitudes, and from these heights they glide to lower elevations, a process that conserves energy during their long migratory flights.

These tame, gentle, and somewhat sluggish birds seemingly live in harmony with most birds. They compete for territory only with Red-tailed Hawks. When resting or feeding, they usually perch near the ground. Their unwary nature makes them vulnerable to killing by humans.

White-tailed Hawk, *Buteo albicaudatus.* Princ. distrib. Texas: Uncommon and local PR in the coastal grasslands and in extreme s. Texas. They are expanding their range in the state and evidently are adapting rather well to the clearing of vast brushlands in s. Texas (Wauer, 1999).

FEEDING BEHAVIOR. They soar in open areas, especially those not farmed or heavily grazed; here they survey the countryside for small prey such as rodents, ground-dwelling birds, reptiles, amphibians, and large insects. They also consume carrion.

They may hover in the same place for several minutes, and in good wind they typically hover before pouncing. They also employ a perch-and-wait foraging strategy.

White-tailed Hawks are consummate opportunists that readily habituate to human activity. They follow tractors for insects that are flushed, and when they detect smoke from a prairie fire (as far as 10 miles away) they fly to the fire to take advantage of fleeing animals.

COURTSHIP BEHAVIOR. They probably establish territories in south Texas as early as December. Courtship displays are incompletely known. A sky dance was observed in the Chisos Mountains in 1901 by the eminent artist Fuertes. He reported that while calling, a bird dove into a canyon with "amazing velocity" and rose again to repeat the process. Also, there are reports of birds grappling for each other's talons (Palmer, 1999). On the ground, the male pulls at grass blades and weeds in what is probably a courtship display.

NESTING. Both birds participate in building the nest, sometimes taking up to 5 weeks to complete it. Evidently the female does most of the incubation.

VOICE. A repeated, high-pitched *keela keela*, sounding to some listeners like a Laughing Gull. (See Farquhar, 1993, for more details.)

OTHER BEHAVIORS. Territorial displays include high circling by one or both birds. White-tailed Hawks rarely exhibit social behavior.

Zone-tailed Hawk, *Buteo albonotatus.* Princ. distrib. Texas: Uncommon SR in Trans-Pecos Texas and w. Edwards Plateau, in arid habitats along watercourses and near wooded canyons.

FEEDING BEHAVIOR. Because they often soar with Turkey Vultures and closely resemble them in appearance and behavior, Zone-tailed Hawks have been regarded as mimics of these vultures. Like vultures, they rock back and forth on up-tilted wings, lazily soaring to heights of 500 feet or more; yet unlike vultures, they dive for prey.

Presumably prey confuse these wolves in sheep's clothing with harmless vultures and thus are less vigilant and more easily captured. That explanation assumes that lizards and other prey species are capable of distinguishing different raptor species soaring overhead, not always an easy task, even for humans, if birds are flying high.

Interestingly, Zone-tailed Hawks take prey that is smaller than might be expected for a hawk this size. This feeding bias does not seem to be related to foot size or any other anatomical feature.

COURTSHIP BEHAVIOR. Some observers report talon grappling as a courtship behavior; pairs also call while circling high in the air and dive steeply from high altitudes.

NESTING. Apparently only the female incubates the eggs. At first she broods the nestlings and her mate brings food; later they both hunt and bring back food to their young.

VOICE. Loud, shrill screams and whistles.

OTHER BEHAVIORS. Territorial defense behaviors include talon grappling. They vigorously defend their nests against intruders.

Red-tailed Hawk, *Buteo jamaicensis.* Princ. distrib. Texas: Common PR in most of Texas; common M and WR throughout the state.

FEEDING BEHAVIOR. They hunt by soaring, kiting, hovering, or by swooping down from a high perch, taking virtually any vertebrate they can capture; however, they are generally incapable of capturing healthy birds in flight. They consume many rodents (up to 85 percent of the diet) and commonly eat carrion on highways.

Their eyesight—remarkable even for raptors—is capable of focusing on potential prey at 500 feet or more.

COURTSHIP BEHAVIOR. In the high circling display, the pair noisily and conspicuously spiral upward, the male usually above the female. He stoops toward her and they interlock talons. In the sky dance, a bird flies to a high altitude, dives toward the earth, then swoops upward at a steep angle. Courtship feeding is common.

Life-long monogamy is apparently typical for this species.

NESTING. Nest: Generally located where birds can view the surrounding territory. Both sexes build a nest or refurbish an old one, in both cases decorating the nest with greenery. Eggs: incubated by both parents. At first the female remains with the nestlings and her mate brings food; later, they both drop food into the nest, evidently to prompt the young birds to feed independently.

VOICE. A descending, hoarse squeal that tapers off at the end.

OTHER BEHAVIORS. Territories are usually bounded by well-defined geographical features such as roads and waterways. Normally simply perching high on a conspicuous tree or bluff is sufficient to deter invaders into the territory; however, Red-tailed Hawks are notably aggressive toward intruders, and sometimes two birds cartwheel toward the ground with interlocked talons. Apparently they also advertise their territories with whirling: the soaring bird rapidly rotates in a complete circle around a more or less stationary wingtip.

They usually are dominant over their best-known competitor, the Red-shouldered Hawk, and they compete ecologically with owls by occasionally appropriating owl nests. Owls reciprocate, however, by feeding on Red-tailed Hawk nestlings.

Red-tailed Hawks play by dropping and catching objects in midair.

Ferruginous Hawk, *Buteo regalis.* Princ. distrib. Texas: Common WR in the High Plains and Trans-Pecos; rare PR in the Panhandle near the New Mexico border. They occur in prairies and other open habitats. In some parts of the United States (and possibly Texas, as well) populations vary according to the availability of jackrabbits.

FEEDING BEHAVIOR. These large, powerful, but relatively trusting and unsuspicious buteos are capable of capturing jackrabbits and other prey species that are beyond the size limit for most hawks. They hover, kite, fly out from a low perch, course low over the ground, or swoop down from great heights to surprise prey. They also forage on the ground by waiting for rodents to emerge from burrows. The bulk of their diet is made up of small mammals, as well as prairie dogs (Allison et al., 1995).

COURTSHIP BEHAVIOR. A soaring Ferruginous Hawk parachutes by holding its wings above its back and dropping rapidly toward the earth. Sometimes the male dives at the female and the two birds grapple with their talons. High circling, high perching, and hovering are other displays.

NESTING. It is not clear why Ferruginous Hawks are so uncritical about details when building nests. Not only do they utilize virtually all possible nesting sites—ground, trees (the preferred site), bluffs, and man-made structures—but they also construct what seem to be unnecessarily large nests, even for a bird this size. Moreover, they add curious pieces of rubbish to the nest, such as bones and cow dung (in the past, bison bones and bison dung). Both parents incubate and bring food to the nestlings. If the nest is disturbed, the pair will most likely desert it.

VOICE. High-pitched screams (*kree ah*), some of which are harsh or gull-like and resemble calls of the Rough-legged Hawk.

OTHER BEHAVIORS. They hover to advertise their territories, especially if another Ferruginous Hawk intrudes. Although their massiveness makes taking flight a laborious process, once aloft they fly swiftly and with ease.

They occur singly or in pairs, and apparently they do not migrate in flocks, as do many buteos. In some parts of the United States (and possibly Texas) they associate closely with Swainson's Hawks. They are generally unwary and easily shot when perched by highways.

Rough-legged Hawk, *Buteo lagopus.* Princ. distrib. Texas: Uncommon WR in n. parts of Texas, much less common to the s., in grasslands, marshes, and other open areas.

FEEDING BEHAVIOR. Rough-legged Hawks hunt close to the ground, often quartering like a Northern Harrier. When flying this low they must flap and glide to keep aloft. Their relatively weak feet limit their prey to smaller rodents (their main diet), and since rodents tend to forage in early morning and in the evening, Rough-legged Hawks are most active at those times. They occasionally eat carrion.

They sometimes stop suddenly in flight and hover (like American Kestrels), then plunge after their quarry with partially closed wings. They frequently attempt to rob Prairie Falcons of their prey.

VOICE. A thin, downward-slurred, plaintive whistle.

OTHER BEHAVIORS. Rough-legged Hawks often migrate in large flocks. When resting, they select low lookout posts such as fence posts, rather than trees. Wintering birds congregate at night and disperse the following morning to hunt.

These gentle and unsuspicious birds are all too often killed while perching by highways.

Golden Eagle, *Aquila chrysaetos.* Princ. distrib. Texas: Rare to uncommon PR in much of w. Texas; rare to uncommon WR throughout the state, generally in w. parts. They occur in open areas near mountains and hills.

FEEDING BEHAVIOR. Alone or in pairs, and mainly during early morning and early evening, these powerful raptors chase down small mammals (especially jackrabbits) and other vertebrates, including rattlesnakes. They also consume carrion. Exactly how many young lambs and other livestock they kill has been difficult to estimate, as many carcasses found in their nests have been shown to be carrion.

Golden Eagles hunt by soaring, by coursing, or by flying out from a perch. Some pairs hunt cooperatively, the second bird capturing animals that the first bird misses. Curiously, they may not eat for many days, even when the supply of prey species seems constant and adequate. When they eventually do eat, they sometimes gorge themselves, then follow this feast with yet another period of fasting. Sometimes they are unable to take flight after eating and can be chased down on foot.

COURTSHIP BEHAVIOR. In the sky dance, the male spirals upward into the sky and dives with wings half open. This display probably serves to advertise the territory and to reinforce the pair bond. Copulation occurs on the ground or on a perch, with the female calling loudly before, during, and after the event. Apparently most pairs mate for life.

NESTING. The nest — a bulky structure of sticks that the pair builds in January — is well appointed: it is lined with grass, bones, and miscellaneous items (in one case, a cowboy hat!). Aromatic leaves placed among the sticks possibly deter harmful insects. Frequently birds attempt to rob sticks from nearby nests, and sometimes more than one nest is claimed.

Eggs: usually incubated by the female, who is fed by her mate at this time. He also brings food for the nestlings.

The nestlings are not treated with equal care, nor do parents interfere when their young act aggressively toward each other and sometimes kill each other. Although in eagles the killing of one's siblings is assumed to be related to food availability, studies of the closely related Black Eagle, *A. verreauxi,* suggests otherwise. Even when siblings of that species are gorged with food they still attack weaker siblings. Moreover, aggression in nestlings is more common in Golden Eagles (which hunt agile prey) than in Bald Eagles (which mainly scavenge or eat fish), even when food supplies for both species appear adequate.

VOICE. Generally silent, but occasionally high-pitched squeals and whistles.

OTHER BEHAVIORS. Eagles soar in buteo fashion, with wings slightly upturned (in contrast to the Bald Eagle's flattened wings). They also flap and glide with slow and powerful wingbeats. When stooping for prey (and also during play) they attain speeds of 150–200 mph. Because of their weight they seek updrafts when taking flight, and they often alight on elevated perches such as cliffs and tall trees.

In high circling, two or more birds soar high, then dive at one another in mock attacks. The attacked bird turns over in the air and reaches up with its talons. (In Scotland, two eagles were found dead with their talons interlocked, evidently the result of a territorial dispute.) In tumbling, a bird suddenly closes its wings and drops like a rock before checking the fall and shooting upward to repeat the display.

Birds play by dropping sticks while in flight and catching them before they reach the ground. Wintering eagles sometimes roost communally when food abundance is high.

CARACARAS AND FALCONS: FAMILY FALCONIDAE

Except for the caracaras, the members of this family are among the fastest flying of our birds of prey. Falcons have long tails and pointed wings, and many dive from above at incredible speeds to capture birds in flight. Males are generally smaller than females, which is typical of raptors that pursue birds and other prey that flee rapidly. Normally falcons nest on cliffs or in holes and do not add material to their nest.

Crested Caracara, *Caracara cheriway.* Princ. distrib. Texas: Locally common PR in s. Texas, north to the Upper Coast and Blackland Prairies, in pastureland, semidesert, cultivated fields, and other open areas.

FEEDING BEHAVIOR. Crested Caracaras are generalists as well as opportunists. They procure food in more ways than any other raptor, a surprising behavior considering the highly specialized family to which they belong.

They search for carrion (their primary source of food) by flying low over the ground. Like vultures, they have long legs and relatively blunt claws that equip them for walking around carcasses.

Early in the morning they begin patrolling highways for road kills, usually before the arrival of their major competitors, the vultures. When competing for the same carcass, the more aggressive individuals drive away the vultures, often so violently that the vultures regurgitate their stomach contents — food that is quickly eaten by the caracaras. In coastal areas Crested Caracaras harass gulls and pelicans to the point that they also disgorge their stomach contents.

Crested Caracaras eat any vertebrate they can chase down, but generally the only birds they are capable of capturing are disabled ones. When searching for insects they walk slowly and deliberately, flipping over debris or scratching the ground like chickens.

COURTSHIP BEHAVIOR. Some observers report that the rattle-call-with-head-toss display is a characteristic courtship display. Both birds stand near each other and croak as they rest their crown on the upper back, the crop protruding like a yellow golf ball. Prior to copulation, they preen each other and engage in mutual courtship feeding.

NESTING. They nest as early as mid-January, even as far north as Houston and San Antonio. This is the only member of the falcon family that builds a nest. Nest: in south Texas, usually placed in Macartney rose, *Rosa bracteata* (Dickinson and Arnold, 1996). It is a deep, bulky cup of sticks, briers, weed stalks, and broomweed, to which the pair sometimes add pieces of mammal hide, dry dung, and trash. Old nests are sometimes utilized. Interestingly, nest construction sometimes begins 2 months prior to egg laying. Most of the incubation is done by the female, but both parents feed the nestlings.

VOICE. As a rule, they vocalize infrequently, and generally during the breeding season. However, they also may produce their cackling call throughout the year, usually from treetops, in the morning and early evening.

The call remotely suggests *cara cara* but is more like *kwi kwik kwik kwik keer,* with the head thrown back over the shoulders on the last syllable.

OTHER BEHAVIORS. Crested Caracaras show little social behavior. They fly low over the ground or just below treetop level, with strong, deep wingbeats that alternate with gliding. They sometimes twist and turn like Northern Harriers or soar like vultures. Pairs often perch in conspicuous places.

They run swiftly, especially when chasing prey. With their long legs and flat claws they are better adapted for walking and running than any other North American raptor, including the Burrowing Owl. Under most circumstances, however, they seem to move sluggishly. Threat behaviors include the head-throwback display. Spectacular aerial maneuvers during the breeding season may represent territorial encounters with intruding males.

American Kestrel, *Falco sparverius.* Princ. distrib. Texas: Common M and WR throughout the state; common SR in n. High Plains, uncommon in other parts of the state. They frequent fencerows, pastures, highways, and other open habitats, as well as urban areas.

FEEDING BEHAVIOR. These seemingly indefatigable birds hunt most persistently in the morning and late afternoon. They dart out from a perch or fly rapidly over open areas in pursuit of insects, rodents, and birds. They often hover—always against the wind—before plunging after their quarry. The principal releaser for a predatory attack is prey movement (for example, running or scampering) rather than what the animal looks like.

In urban areas, American Kestrels take advantage of the abundant supply of House Sparrows, but they also capture other birds, including Mourning Doves and even hummingbirds. Among insects, grasshoppers and crickets are favorite items. American Kestrels regurgitate as pellets the indigestible parts of their prey.

Apparently they require very little water and are capable of surviving for months on the body fluids of their prey.

Unlike many raptors, both males and females eat basically the same size and type of prey. Interestingly, some individuals consistently select a preferred type of food, such as rodents, to the exclusion of other types that are just as easy to capture. Previous success in capturing a particular prey item apparently predisposes a bird to select that prey type habitually; moreover, this bias undoubtedly becomes more entrenched as repeated captures cause the bird to refine its search image for the prey type.

American Kestrels cache food in clumps of grass, natural cavities, and other places, sometimes storing more than a dozen mice before they begin eating them.

COURTSHIP BEHAVIOR. The male calls loudly while flying above the perch, quivering his wingtips as he does so. Both birds perch near each other and bow while the female calls. Courtship feeding occurs in some desert populations and should be looked for in Texas. Promiscuous mating sometimes occurs before birds form their monogamous bond.

Apparently mate selection is by the female, but copulation, which is performed on a perch or platform, is initiated by either sex. Mated American Kestrels copulate frequently and repeatedly outside the fertile period. This behavior is sometimes explained by the mate assessment hypothesis, which maintains that birds assess the genetic potential of other birds by frequently copulating with them.

Although pair bonds appear to be temporarily broken when the sexes occupy different territories in winter, most pairs apparently reunite in spring.

NESTING. Nest: in woodpecker cavities, holes in cliffs, cavities in buildings, and other protected areas. They compete with squirrels and other hole-nesters, but American Kestrels are usually the victors (although European Starlings pose a problem in some areas). Like most falcons, they add very little (if any) material to the nest.

Eggs: incubated by both parents. Young birds remain with their parents a month or more after hatching. Later, juveniles from several families form small flocks.

VOICE. A high-pitched, anxious *ki ki ki ki* or *kily kily kily kily.*

OTHER BEHAVIORS. American Kestrels are among the few birds that defend territories in winter. These territories are generally in agricultural and grassland habitats, and males and females defend different territories. The larger (10–15 percent heavier) females exclude males from the more open sites, which are the preferred areas for hunting (possibly because of reduced risk of predation by bird-eating hawks). When females are experimentally removed, neighboring males immediately occupy the vacant territories.

Displays reported from captive birds include the curtsy, bow, tail spreading, and tail pumping (Mueller, 1971).

Merlin, *Falco columbarius.* Princ. distrib. Texas: Rare to uncommon M in most parts of the state, in woodland openings, marshes, lakeshores, urban areas, and other open areas where there are trees nearby.

FEEDING BEHAVIOR. These are fast-flying raptors that easily overtake most birds they pursue. They dart out from perches or course low over open areas, rising and falling as they search for prey, often maintaining an altitude of less than 10 feet above ground. Less frequently they stoop or hover for prey. Sometimes Merlins follow Northern Harriers and capture animals that the Northern Harriers flush from hiding.

Up to 90 percent of their diet is made up of birds, some as large as curlews, flickers, and teal. Competition between members of a pair is minimal for the larger, less common animals, because the female, which is about 33 percent larger than the male, takes larger prey than the male. However, both sexes eat prey that are plentiful, such as dragonflies and other insects. Merlins cache food in trees and on the ground.

VOICE. Normally silent during migration and winter, but at the nest they sometimes give a *ki ki kii* call or utter other harsh notes.

OTHER BEHAVIORS. Their flight is diagnostic: they take off suddenly, sometimes almost vertically, and quickly attain a fast, steady, rowing flight that suggests a pigeon. (One explanation for their former name, "pigeon hawk," is their pigeonlike flight behavior; another is their fondness for preying on pigeons.) They also fly low over the ground, then bound up to alight on a perch, a maneuver typical of shrikes.

To their detriment they are exceptionally curious about, and relatively unafraid of, humans. At times they badger crows, gulls, and other larger birds in what appears to be a form of diversion or play.

Merlins are generally solitary outside the breeding season. Unlike American Kestrels they do not bob their tails or perch on high electric wires.

Aplomado Falcon, *Falco femoralis.* Princ. distrib. Texas: Reintroduced along the coast, where they breed and will probably become established. They frequently hunt in pairs, capturing snakes, lizards, mice, birds, and insects; they often hover before stooping for their prey. While feeding on prey, reintroduced Texas birds readily chase away raptors, including Peregrine Falcons and Barn Owls (Sean Grimland, pers. comm.). They alight on bare ground as well as watch for prey from perches such as fence posts and utility poles. Their flight is graceful and slower than the Peregrine Falcon's.

Peregrine Falcon, *Falco peregrinus.* Princ. distrib. Texas: Rare to uncommon M throughout most of Texas; common M and WR on the coast, where they are seen in open areas. Local SR in Chisos and Guadalupe mountains, and uncommon WR in extreme s. Texas.

FEEDING BEHAVIOR. These imposing raptors are also known as duck hawks because of their singular method of felling waterfowl. They stoop from high altitudes, at speeds probably more in the range of 100 mph than the 200 mph velocity often quoted, and audibly strike the duck with their powerful feet. They pluck prey out of the air or knock it out of flight. In the latter case they retrieve the bird before it reaches the ground.

Peregrine Falcons also employ a clever foraging strategy known as contouring. On the coast they fly rapidly behind sand dunes and other irregularities (contours) in the terrain, then suddenly appear and take their victims by surprise. More often, however, they hunt from a stationary perch, such as a boat mast, from which they dash out for gulls, terns, shorebirds, and waterfowl.

Many Peregrines fixate on a particular species, such as the Blue Jay, and hunt that species to the exclusion of other species unless resources are scarce (or if they are feeding their young). Reportedly, when domestic pigeons (a favorite prey item) are singled out by a Peregrine Falcon, Mourning Doves feeding nearby are aware of the hawk's indifference to them and do not flee.

Peregrines sometimes follow beaters—animals that flush prey that the Peregrines then capture. Northern Harriers and humans are common beaters. When a pair hunt cooperatively, the larger female usually dives first and takes the larger prey. She also eats first when both are at the same carcass.

These raptors were observed preying on Mexican free-tailed bats (*Tadarida brasiliensis*) at Ney Cave, Medina County, Texas, dashing into a bewildering mass of bats and emerging with a bat in their claws.

They do not ingest whole prey, as do owls; however, like owls and other raptors, they regurgitate pellets, usually one every day, that contain fur, claws, and other indigestible animal parts.

COURTSHIP BEHAVIOR. The male's elaborate courtship displays include an undulating sky dance (steep dives and pullouts) as well as turning upside down during vertical drops. Courtship displays are also performed on ledges, and courtship feeding is common. About 20–30 percent of the females change partners annually.

NESTING. They do not build a nest but instead prepare a scrape (a shallow depression) in rocks or debris on a cliff ledge. Generally the female incubates and broods, and her mate brings her food when she is on the nest.

VOICE. Usually silent, except at the nest; here they respond to intruders with shrill whines that are sometimes rendered as *weetchew weetchew*. They also utter a *kak kak kak kak* sound and a note that suggests wailing.

OTHER BEHAVIORS. Their quick wingbeats suggest the flight of a pigeon, but they also soar on flat wings. They sometimes dive at herons in flight and force them to land. This may be a form of play, as they also chase hawks that are outside their territory.

The male advertises his territory with several displays, including high circling and high perching. In disputes over territory, males lock talons and fall toward the ground, sometimes with fatal consequences for both birds.

Peregrine Falcons capture more food than they eat, and they regularly cache food throughout the year.

Prairie Falcon, *Falco mexicanus.* Princ. distrib. Texas: Rare to uncommon M in w. part of the state, generally on prairies and in open habitats in mountainous regions. Migrating birds frequently pass through cities and towns.

FEEDING BEHAVIOR. They fly rapidly, 10–30 feet above the ground, and perform skillful maneuvers when flushing and capturing their quarry. (Although sometimes the slightly larger Peregrine Falcons also forage at low levels, they usually fly high, then suddenly stoop to strike at prey.) In another foraging tactic Prairie Falcons sit on a perch, sometimes for long periods of time, and wait for ground prey to emerge from burrows or hiding places. Less frequently they hover and then stoop from high altitudes, or employ the accipiter method of capturing prey: flying above vegetation, which serves as a screen, then dashing into a flock of birds.

They kill by gripping the animal with their talons, unlike Peregrine Falcons, which typically strike their prey and let it fall to the ground. Sometimes Prairie Falcons use Northern Harriers as beaters to flush prey; at other times they actually attack the beaters and force them to release their prey. Prairie Falcons may carry their prey to a tree or other feeding site before eating it.

VOICE. A series of high-pitched whistles and cackles (*kee kee kee*).

OTHER BEHAVIORS. At high altitudes, their maneuvering capabilities are actually superior to the Peregrine Falcon's because of their lighter wing-loading; also, their flight speed almost equals the Peregrine's. Their flight is graceful and consists of short, rapid, and powerful wing strokes that alternate with glides.

These highly maneuverable, almost acrobatic birds seem to flaunt their aerial skills, diving at herons, owls, and other large birds or dropping objects, including dried cow dung, in the air and then catching them. Many of these behaviors appear to be play.

GALLINACEOUS BIRDS: ORDER GALLIFORMES

CURRASOWS AND GUANS: FAMILY CRACIDAE

Plain Chachalaca, *Ortalis vetula.* Princ. distrib. Texas: Uncommon to locally common PR in extreme s. Texas, in dense thickets and forested areas by rivers, resacas, ponds, and other wetland habitats, as well as in well-wooded urban areas.

FEEDING BEHAVIOR. Chachalacas forage both on the ground and in trees for berries, buds, leaves, and seeds; apparently Texas birds are herbivorous and/or frugivorous and rarely if ever consume arthropods or other animal matter (Christiansen et al., 1978). However, they have learned to eat junk food in public parks.

COURTSHIP BEHAVIOR. Following a strong vocal contest from the tops of trees, males descend to the ground and strut among the females.

NESTING. Apparently Texas birds do not bring nesting materials to the nests they appropriate, which include nests of Yellow-billed Cuckoos, Groove-billed Anis, and Curve-billed Thrashers (Marion and Fleetwood, 1978). Eggs: incubated by the female. Both parents feed the nestlings regurgitated food, a unique behavior, as this is the only gallinaceous species that feeds its young.

The precocial chicks leave the nest within hours after hatching. There are unverified reports of young birds clinging to their mother's legs when she carries them to the ground during the day and when she returns them to the tree at night.

VOICE. Unchallenged as one of our noisiest birds, especially during the breeding season. Call: delivered from a perch in a tall tree, a loud, resonant, penetrating *cha cha lac.* Birds sing mainly at dawn and at dusk and before thunderstorms. One bird initiates the call, but as neighbors join in, the event quickly cascades into a disorganized, frenzied chorus. They also utter a *kak kak kak* when alarmed. The male's deeper voice has been attributed to his looped trachea.

OTHER BEHAVIORS. These sociable birds form small flocks that spend a large part of the day and night in trees, where they bounce with agility from limb to limb.

They climb well, partly because their hind toe (hallux) attaches to the lower leg at the same level as the other three toes, an arrangement typical of most arboreal birds that helps them grasp tree limbs. Thus, even though Chachalacas walk and trot gracefully, they are less well equipped for running than, for example, turkeys, quail, plovers, and cranes, which have the hallux elevated and out of the way. They fly swiftly but heavily, generally above the treetops and for short distances. For a tree-dwelling species, they are surprisingly fond of dust baths.

These are generally wary birds except in parks and refuges, where they readily habituate to people. Birds from eggs that are incubated by hens become very tame pets.

PARTRIDGES, GROUSE, TURKEYS, AND OLD WORLD QUAIL: FAMILY PHASIANIDAE

Primarily ground-dwellers, these birds take flight explosively but are more likely to escape danger by running.

Members of this family differ in the development of the trailing-edge notch in their primaries, a feature that increases performance in vertical and slow flight but reduces efficiency in level flight. Grouse (including our prairie-chickens), which typically fly from ground to canopy or from branch to branch, have short, broad wings with large trailing-edge notches, as well as light-colored flight muscles (also an adaptation for fast, short flights). Pheasants, quail, and turkeys have wide wings and deep notches and typically fly longer distances from one foraging site to another.

Several species, including prairie-chickens and turkeys, spread their tail feathers when displaying. This behavior, as well as strutting and foot stamping, are probably ritualized intention movements.

Ring-necked Pheasant, *Phasianus colchicus.* Princ. distrib. Texas: Common PR in the High Plains, in fields, margins of playa lakes, roadsides, brushy meadows, and fencerows. Our birds are hybrids of several Asian subspecies that were introduced into North America during the last century.

FEEDING BEHAVIOR. Omnivorous opportunists that consume grains, buds, roots, and other plant parts, as well as insects, spiders, and other invertebrates. They expose food by scratching on the ground and digging into the ground with their bill.

COURTSHIP BEHAVIOR. In defending his territory (called a crowing territory), the male stands on a conspicuous perch, slopes his body toward the opponent, raises his ear tufts, inflates his wattle, and spreads his tail. This posture emphasizes the head ornaments, especially wattle size and length of ear tufts, which are important features for communication between males. The cock then drums with his wings and produces a loud *kock kack* that is followed by a whirring of the wings.

During his courtship display, the wattles on his face become swollen and serve as a signal to male competitors. He struts around a female in a half circle, tilts his back and tail feathers toward her, and droops the wing nearest her. Females select a mate on the basis of traits that include length of tail, black points in the wattle, and length of ear tufts.

Males mate only if they successfully compete for territories; the other males remain as satellites outside the territory. Males that mate often have several hens, each with her own nesting subterritory within his territory.

NESTING. Eggs: sometimes laid in the nest of another bird. The female may feign injury when threatened at the nest. The downy young are able to feed immediately but follow the hen for 10–12 weeks, probably for protection and to learn good foraging sites. Typically the male does not provide parental care.

VOICE. *See* "Courtship behavior." Call: hoarse croaks when flushed; the female produces whistling notes when disturbed.

OTHER BEHAVIORS. When alarmed, pheasants flutter away explosively with a metallic whirring of their wings but then abruptly drop into concealing vegetation, an escape behavior clearly adapted to open grasslands. When cornered against a wall they take flight almost vertically. The trade-off for such immediate bursts of strength is that these birds have little endurance and can eventually be chased down on foot.

Their preferred way to escape danger is by running, which they do with the tail cocked at a 45-degree angle.

In winter, pheasants form flocks that maintain a strong dominance hierarchy. Males and females segregate into separate flocks, those of the females being larger.

Greater Prairie-Chicken, *Tympanuchus cupido.* Princ. distrib. Texas: The few remaining birds that make up the Texas subspecies Attwater's Prairie-Chicken (*T. c. attwateri*) are restricted to tall-grass prairies in Galveston and Colorado counties. Apparently they are extinct on the Coastal Bend, as no birds were located in Refugio County in 2000 (Michael Marsden, pers. comm.).

Their feeding, courtship, nesting, and other behaviors are essentially as in the Lesser Prairie-Chicken, except that the Lesser eats more acorns. The chicks are said to run to their mother when she gives the *brirrb brirrb* call, but freeze when she gives her shrill warning call.

VOICE. Their booming sound is slightly louder and lower in pitch than the Lesser Prairie-Chicken's.

Lesser Prairie-Chicken, *Tympanuchus pallidicinctus.* Princ. distrib. Texas: Locally uncommon PR in the Northern High Plains and w. Southern High Plains. Unlike the Greater Prairie-Chicken, which favors grasslands, this species inhabits sandy rangeland that supports sand shinnery oak (*Quercus havardii*) and sand sagebrush (*Artemisia filifolia*), with grass being less important.

FEEDING BEHAVIOR. They generally forage during the morning and early evening. Their diet is typically gallinaceous: seeds (acorns, in particular), grain, leaves, and insects (especially in late summer). They sometimes travel several miles to feed on waste grain.

COURTSHIP BEHAVIOR. The booming area (called a lek) is a patch of ground that is either bare or covered with very short grass. It is usually less than an acre in size. Here, at dawn as well as at dusk, 8–20 males engage in their remarkable courtship display. With their heads lowered, their tails elevated, the air sacs on their necks inflated, and the feather tufts on their heads erect, the birds stamp their feet, turn in circles, lunge at each other, briefly spar, move around excitedly, and intermittently leap into the air. When engaged in these spirited interactions they inflate their neck pouches to produce a hollow cooing sound (booming) that can be heard up to a mile.

Females watch from the periphery; at the conclusion they leave with one or two of the males, which do most, if not all, of the mating.

Although inflating the neck pouch probably evolved as a unique component of this courtship display, other components may have been derived, through ritualization, from reflexes (feather erection), intention movements (foot stamping, wing drooping), or fixed action patterns (if lowering the head can be interpreted as a ritualized eating or drinking movement).

Peak gobbling intensity and breeding in w. Texas occurs 2–3 weeks earlier than in Oklahoma or Colorado. Lesser Prairie-Chickens also display in October and November, although less vigorously, for the same reasons that many birds briefly resume singing at that time. The testes, which diminish in size following the reproductive period, undergo a modest recrudescence in fall that produces slightly elevated testosterone levels.

NESTING. Eggs: incubated by the hen, who covers them with vegetation prior to incubation. The chicks can feed almost immediately after hatching, but they follow the hen for about 3 months after that.

VOICE. *See* "Courtship behavior." Both sexes produce chickenlike clucks.

OTHER BEHAVIORS. Their flight is strong but infrequent and rarely protracted.

Wild Turkey, *Meleagris gallopavo.* Princ. distrib. Texas: Common in e., c., and s. parts of Texas, locally common (through reintroductions) in other parts of the state. They prefer understory in wooded habitats next to open clearings.

FEEDING BEHAVIOR. Wild Turkeys walk around slowly and deliberately, vigorously scratching in the leaf litter to locate seeds, acorns (a favorite), insects, and spiders. They take essentially any food item they can manage and are quick enough and strong enough to subdue difficult prey such as lizards and snakes. They also climb trees for berries.

Their robust bodies equip them to consume many food items unavailable to most galliform species. For example, the hardness of seeds is apparently no problem for their gizzard, which was experimentally shown to flatten objects that require 400 pounds per square inch to crush.

Turkeys forage mainly in the morning and evening and generally restrict their movements to an area about 5 square miles around their roosting sites.

COURTSHIP BEHAVIOR. The gobbler begins his courtship behavior in February or March. The display, triggered by warm weather and lengthening days, is a caricature of panache. As the wattles on his face become increasingly engorged with blood and his snood becomes extended, he raises and spreads his tail, droops his wings, puffs out his body feathers, and struts before the one or more females that his gobbling has attracted to the site. While doing this he gobbles and makes rattling sounds with the quills of his drooping wings.

Females prefer males with longer snoods, possibly because the fleshy parts indicate the health, condition, and hormonal milieu of a potential mate. The potency of the snood as a signal device is also evident when submissive males retract their snoods after confronting a dominant gobbler. The female head appears to be a powerful releaser for reproductive behavior, as presenting the detached head of a female to a gobbler incites him to copulate.

The male does not have a territory per se; he defends the area where he and the female happen to be at the moment. The spur on his lower leg is used in fighting, but it rarely inflicts injury. His breast sponge functions as a reservoir of energy during the breeding season.

Males are promiscuous and do not invest in family responsibilities, which probably explains why one copulation is sufficient for the hen to lay 12–15 fertile eggs. At the completion of mating the female builds a nest, lays the eggs, incubates them, and cares for the young.

NESTING. Nest: a shallow depression lined with leaves and grass (Thogmartin, 1999). The female sometimes performs a distraction display when disturbed at the nest. Some hens lay eggs in other birds' nests. Although young birds can feed themselves, they accompany the hen for several weeks after hatching and at night sleep couched under her body, wings, and tail.

VOICE. When settling down for the night, turkeys utter a soft *kyow* (made on other occasions as well). When alarmed, they produce an abrupt *putt putt,* which prompts nearby turkeys to dash for shelter.

The male gobbles in response to other males' gobbles as well as to human imitations; they also respond to the calls of American Crows and owls.

OTHER BEHAVIORS. Their flight is swift and strong, but they seldom fly more than a quarter of a mile and are incapable of flying more than about a mile in a stretch. When alarmed, females generally fly away, but males are more likely to run for cover. Before taking flight, turkeys flex their knees and tibiotarsal joints, a movement that may have become ritualized into the male's strut display.

Wild Turkeys are highly gregarious. At night they roost together in tall trees, preferring perches in leafy parts to perches on exposed, bare limbs. They also can swim.

Toward the end of summer they take dust baths in oval depressions scraped out in dirt roads and similar places. The absence of head feathering in turkeys probably functions not only as a signal during courtship but also for thermoregulation in hot weather.

NEW WORLD QUAIL: FAMILY ODONTOPHORIDAE

These birds have chunky, rounded bodies and crests or head plumes. When they are not nesting, they stay in small flocks called coveys. They typically live in grassy or brushy habitats.

Scaled Quail, *Callipepla squamata.* Princ. distrib. Texas: Uncommon to locally common PR in the w., n.w., and s.w. part of the state, in arid grasslands, brush, and juniper slopes, often miles from water. Their numbers are declining in most parts of their range.

FEEDING BEHAVIOR. They forage for seeds from a variety of plant species and eat more weed seeds than any of our other quail; also, in the spring and summer they probably eat more insects than our other species of quail.

COURTSHIP BEHAVIOR. In spring, the male flies up to a fence post, tree, or other conspicuous perch and from this vantage point utters a hoarse *kuck yer* to advertise his territory and to attract females. In the frontal display, he raises his head, lifts and lowers his crest, and droops his wings; in his lateral display he raises his crest and spreads the feathers of his flank.

Scaled Quail sometimes hybridize with Northern Bobwhites and Gambel's Quail.

NESTING. Nest: little more than a shallow depression that apparently only the female prepares. She lines it with grass and leaves. Eggs: incubated mainly by the hen. The chicks, which are able to feed themselves, are cared for by both parents. Often the male perches on a limb or mound and acts as a sentinel as his mate and chicks feed below.

VOICE. *See* "Courtship behavior." One common call is a low, nasal *pecos pecos.* In this species, four calls have been related to aggregation and contact, three to sexual attraction, three to alarm or distress, and one to threat-attack encounters. (See Anderson, 1978, for a detailed account of vocalizations in the Scaled Quail.)

OTHER BEHAVIORS. Other than during the early part of the nesting season (when they are alone or in pairs), they generally move in coveys of about 30 individuals; however, unlike Northern Bobwhites, which rarely have more than 30 individuals in a covey, sometimes as many as 150 Scaled Quail flock together.

Their gait is swift and powerful. When in coveys they generally avoid danger by running, but individuals and pairs tend to fly away. Scaled Quail can be trusting; they readily adapt to human presence, especially to farms and ranches if food and water are available.

Anting has been reported for the species: a female was observed stroking her wings, body, and tail with ants she was holding in her bill.

Gambel's Quail, *Callipepla gambelii.* Princ. distrib. Texas: Common PR in extreme w. Texas, in desert habitats, especially near tamarisk-lined streams and water holes. Among our quail, they inhabit the most arid of habitats, yet they are actually more dependent on nearby water than Scaled Quail. Breeding depends on the availability of water, and in very dry years many birds simply do not nest. In El Paso they frequent arroyos and brushy habitats well within residential districts, where they respond to food and water placed in gardens.

FEEDING BEHAVIOR. They forage on the ground for plants and plant parts, ascending into low trees and bushes for leaves, buds, and berries. Apparently they eat fewer insects (at least as adults) than other quail. How much of this insectivorous behavior reflects prey preference rather than insect availability is not known.

COURTSHIP BEHAVIOR. The male struts around the female while bowing his head. Fights between competing males are frequent at this time.

NESTING. The pair spend considerable time locating the ideal nesting site. The female makes a shallow depression and lines it with soft plant materials, two females sometimes sharing the same nest. Eggs: incubated by the hen. Although young birds are soon able to feed by themselves, for several months their parents lead them to good food sources.

VOICE. Call: From a high perch, the male utters a clear, descending note. When traveling from place to place, both sexes make a piglike *oik oik*. Birds isolated visually from the flock give a location call, *chi CA go go*.

OTHER BEHAVIORS. They live in coveys (up to about 50) except in early spring. Unlike Northern Bobwhites and Montezuma Quail, they rarely freeze when approached but instead run to the nearest cover.

A sentinel usually watches over a foraging flock. At night coveys roost in bushes or low trees.

Northern Bobwhite, *Colinus virginianus.* Princ. distrib. Texas: Uncommon to common (or locally abundant) PR throughout the state except in far w. Texas. Many individuals are from releases of birds raised in hatcheries. They occupy a wide variety of habitats but favor farmlands, brushy pastures, roadsides, and open country. Where this species and the Scaled Quail overlap in range in s. Texas, they apparently compete very little for the same habitats.

FEEDING BEHAVIOR. They walk along with their head down, searching for seeds, insects, leaves, berries, acorns, and other plants and animals. Occasionally they move up into shrubs to forage among the leaves and stems.

COURTSHIP BEHAVIOR. The male bows to the female and turns his head to the side to expose his facial markings. He droops his wings so that the tips touch the ground, then pushes his elbows forward so that his

FIGURE 4. Male Northern Bobwhite tidbitting: offering food to the female. (H. W. Williams et al., 1968, *Auk* 85: 464–476. With permission.)

wings make a feathered wall. Next he fluffs his body feathers and rushes at the female. He also fans his tail, fluffs up his feathers, and walks slowly around her. Sometimes he presents her with an insect (tidbitting) while making a special call (Fig. 4).

Males in coveys occasionally engage in vigorous fights.

NESTING. Site probably selected by both sexes. Nest: a shallow depression that both parents line with leaves and grass, to which they add an arch woven from grass and weeds. Both parents incubate the eggs and accompany the chicks, which are able to feed themselves immediately after hatching. The parents sometimes perform a distraction display when threatened.

If the eggs of bobwhites and other precocial species were to hatch asynchronously, as would be expected since an egg is laid each day, the last chick to hatch would be left unattended, because by that time the parents would have already departed with the older chicks. In some species, parents compensate for asynchronous hatching by partially incubating the first eggs (thereby slowing their development) and completing incubation of the clutch only after the last egg is laid. Other species differentially incubate the eggs by giving more attention to the last ones laid.

Northern Bobwhites deal with asynchronous hatching in a different manner. The youngest embryos, which are in the most recently laid eggs, produce nonvocal clicking sounds that pass through their shells and into the shells of older embryos. These sounds, by means of a physiological process not well understood, slow embryonic development in the older embryos, with the result that all eggs hatch at approximately the same time.

VOICE. Call: unarguably onomatopoetic. Careful listeners claim that at close range even the *b* and the *w* are audible in the male's penetrating, desperate-sounding *bob white!* The female answers with an *a loya a hee.* These are only two of the many calls in the species' repertoire; others include a catlike caterwauling heard throughout the year.

OTHER BEHAVIORS. Coveys remain together throughout autumn and winter, and at this time coveys wander widely, sometimes into cities. Most confine themselves to a feeding territory that they defend against other coveys, and groups reject newcomers when the number of individuals reaches 30 or so. At night birds roost together in a circle, their heads pointing outward and their tails pointed inward, supposedly to conserve heat.

Coveys scatter explosively when alarmed and sometimes do not reassemble until the next morning. The maximum distance these quail can fly without stopping appears to be about a quarter of a mile.

Montezuma Quail, *Cyrtonyx montezumae.* Princ. distrib. Texas: Locally uncommon PR in several w. Texas mountain ranges (most common in Davis Mountains) and s.w. Edwards Plateau. They frequent wooded mountain slopes, canyons, and openings in coniferous forests, usually with oaks and grasses, commonly bunchgrass.

FEEDING BEHAVIOR. These striking quail are superbly adapted to the rough terrain of Trans-Pecos Texas. They walk slowly and noiselessly, carefully searching for insects, seeds, bulbs, berries, and small fruits. They use their powerful feet and sharp claws (much stouter than in our other quail) to dig into the rocky soil.

COURTSHIP BEHAVIOR. The breeding season in Arizona is timed to the summer rains, but apparently this relationship has not been documented in Texas birds.

NESTING. Nest: on the ground in tall grass. It is a well-constructed, shallow depression that the female makes and lines with grass. She covers it with a grass dome that has an entrance at the front. Eggs: most likely incubated only by the hen. She incubates them a little longer (up to 26 days) than do most quail.

Although the chicks are able to feed by themselves almost immediately after hatching, they follow their parents to feeding areas for several months. Parents perform a distraction display when their young are threatened.

VOICE. Call: The male advertises his territory with a quavering whinny, suggesting an Eastern Screech-Owl; this sound carries remarkably far to be so soft.

OTHER BEHAVIORS. This is our tamest quail. When threatened, they crouch and remain motionless (hence the colloquial name "fool quail"), then fly away noisily, only to drop quickly to the ground. They fly and run less than our other quail. Alarmed birds spread their crests laterally, giving the appearance of "a mushroom sliced in half vertically" (Oberholser and Kincaid, 1974).

Autumn coveys, which most likely are family units, rarely contain more than 10–12 birds. These coveys occupy relatively small feeding areas; indeed, Arizona birds wander less than 100 yards a day. At night, Montezuma Quail roost on the ground.

RAILS, CRANES, AND ALLIES: ORDER GRUIFORMES

RAILS, GALLINULES, AND COOTS: FAMILY RALLIDAE

These are birds of the marshes, swamps, and other shallow-water habitats. The expression "thin as a rail" refers to the ability of rails to compress their bodies laterally, an advantage when moving through dense marsh vegetation. Coots and gallinules are more ducklike.

Most members of the family are omnivores that eat roots, leaves, aquatic invertebrates, and sometimes frogs and fish. Many of their displays are predictably vocal because vision is so restricted in marshes; on the other hand, others incorporate specific postures and movements, as well as courtship feeding and mutual preening. Several species constantly lift the tail when they walk and in doing so expose white feathers under the tail, quite likely to initiate courtship, announce the bird's territory, or maintain contact between members of a pair.

Yellow Rail, *Coturnicops noveboracensis.* Princ. distrib. Texas: Rare M in e. half of Texas; locally uncommon WR in coastal marshes, probably the highest wintering density being in the rice fields of Calhoun County and northward (Michael Marsden, pers. comm.). They inhabit freshwater and brackish marshes, occasionally showing up in unexpected places like city parks. They forage in mud and water for crayfish, snails, insects, and other invertebrates, as well as seeds. They appear indifferent to people, even though in marshes they are not easy to locate. When approached too closely, they generally freeze before running.

VOICE. Call: a series of five clicking notes (*kuks*), with a pause between the second and third note; the call has been compared to striking two stones together.

Black Rail, *Laterallus jamaicensis.* Princ. distrib. Texas: Rare M in e. half of Texas, and rare and local PR on the c. Coast. They inhabit coastal tidal marshes and inland grassy marshes, where they feed on insects and other invertebrates, as well as on seeds. They are very secretive and usually escape detection by running through the vegetation.

VOICE. The male utters a *kik kik kik,* and the female, a cuckoolike *croo croo croo.* Calls are rarely heard away from the breeding areas.

Clapper Rail, *Rallus longirostris.* Princ. distrib. Texas: Common PR on the coast, much more frequently encountered in salt than in brackish or freshwater marshes.

FEEDING BEHAVIOR. They probe or pick in mud and water for crustaceans, insects, small fish, frogs, and other aquatic animals, taking advantage of muddy areas that are exposed at low tide.

COURTSHIP BEHAVIOR. Holding his tail erect, the male approaches the female, points his bill down, and swings his head from side to side; he approaches her with his bill open and his neck outstretched. Males also engage in courtship feeding.

NESTING. Nest: a well-shaped and sturdy cup of grasses and sedges that is built by both sexes. They weave a canopy of vegetation over the top, a structure that both conceals the eggs and contains them during flooding. A ramp of grasses and other materials leads to the nest. Both parents incubate the eggs. They begin incubating before the last egg is laid; consequently, eggs hatch asynchronously. The downy young leave the nest soon after hatching and sometimes occupy a separate nest, where they are brooded by their parents. Possibly this behavior reduces the risks of predation or parasite infestation.

VOICE. Although they are most vocal at dawn and dusk, at any time of day or night a loud noise usually precipitates a chorus. Call: a harsh, staccato *kak kak kak kak* that diminishes in pitch and volume.

OTHER BEHAVIORS. They swim and even dive, but rarely for very long periods of time; in severe floods they will drown if they cannot find a floating log or other secure structure. They walk carefully and deliberately, bobbing the head and twitching the tail to reveal the white undertail coverts that probably serve to maintain contact between individuals. When alarmed, they are more likely to run swiftly through the dense vegetation than to fly. They become quite tame if not molested.

King Rail, *Rallus elegans.* Princ. distrib. Texas: Common PR on the coast; locally common PR in e. half of Texas, favoring freshwater or brackish marshes over salt marshes. The species has been considered a freshwater race of the Clapper Rail on the basis of presumed (but not universally accepted) hybridization between the two species.

FEEDING BEHAVIOR. They forage in shallow water near dense cover, taking insects, crustaceans, and other invertebrates, sometimes immersing their head into the water. They carry larger prey items to solid ground where they can more easily dismember them.

COURTSHIP BEHAVIOR. The male raises his tail to expose the white undertail coverts and in this posture walks near the female. After the two birds have paired, the female approaches the male, who responds by raising his tail, pointing his bill downward, and swinging his head from side to side. Apparently, courtship feeding is common.

NESTING. The male appears to do most of the nest building. He constructs a solid platform of sedges, grasses, and other plants, and adds a woven canopy over the top and a ramp leading to the entrance. He also builds other platforms nearby, the so-called brood nests, where both parents care for young rails that have recently hatched. Both parents incubate the eggs and feed the nestlings, which forage independently after about 3 weeks. When disturbed at the nest, parents perform a distraction display.

VOICE. Call: a dry *chuck chuck chuck,* likened to a person clucking a horse, and a *chupe chupe chupe* used to advertise the territory.

OTHER BEHAVIORS. This is our largest rail. Birds can be quite casual around humans, and sometimes they forage in full view as they make their way along the edge of a marsh.

Virginia Rail, *Rallus limicola.* Princ. distrib. Texas: Uncommon M throughout most of Texas. Local and uncommon WR on the coast, and local, uncommon SR in several scattered localities. They show a definite preference for fresh and brackish marshes, but on the coast, wintering birds also frequent salt marshes.

FEEDING BEHAVIOR. They probe the mud or shallow water for insects, crayfish, and other invertebrates, sometimes quietly stalking animals before capturing them with a quick thrust of the bill. A smaller part of their diet consists of seeds gathered while climbing up in vegetation (to which they cling tenaciously) or when visiting fields after harvest.

They often associate with Soras, but the two species probably do not compete for resources, as Soras prefer seeds, and Virginia Rails eat small, aquatic animals.

VOICE. Call: a metallic *kid ik kid ik kid ik,* suggesting a hammer bouncing off an anvil, as well as a variety of other calls.

OTHER BEHAVIORS. When flushed, Virginia Rails normally escape by running instead of flying, but they will swim and dive to avoid predators. They migrate in large flocks but otherwise seem to associate with other individuals only opportunistically, such as when drying wetlands concentrate large numbers of aquatic invertebrates.

Sora, *Porzana carolina.* Princ. distrib. Texas: Common M in most parts of Texas, common WR on the coast, and uncommon local SR in scattered localities. Although favoring marshes and wet grass, they occur in almost any wet area where vegetation is present.

FEEDING BEHAVIOR. They walk daintily across the mud, bobbing their head, flicking their tail up and down, and sometimes probing the mud with their bill. They pick seeds from the ground, water, and aquatic plants and occasionally capture insects and other invertebrates. Their long toes allow them to forage on lily pads and other floating structures. (*See* Virginia Rail.)

VOICE. Call: a plaintive, quaillike *er wee,* which rises in pitch; also a whinny and several peeping calls. Soras are quick to respond vocally to noises such as loud clapping of the hands.

OTHER BEHAVIORS. When alarmed, they cock their tail and expose the white undertail coverts. They usually escape by running, but they can also swim; when migrating, they fly with strong wingbeats. These are curious birds that will approach an observer who sits quietly by the marsh and waits for them.

After the breeding season, Soras congregate in enormous numbers to feed on rice or grain. Because they migrate at low altitudes at night, they frequently fly into radio towers and tall buildings.

Purple Gallinule, *Porphyrula martinica.* Princ. distrib. Texas: Uncommon and local M and SR in e. half of the state, in freshwater marshes, swamps, ponds, and other densely vegetated wetlands that have still or slow-moving water.

FEEDING BEHAVIOR. Purple Gallinules swim, walk, and climb through vegetation in search of insects, fish, frogs, snails, and other prey. They also take eggs and young birds, and they eat seeds and fruits of both aquatic and terrestrial plants, sometimes climbing up stalks in grain fields to forage on seed heads.

COURTSHIP BEHAVIOR. Apparently unrecorded.

NESTING. Nest: a platform, often floating, that both parents construct from grasses, sedges, and other marsh vegetation. It is concealed under a canopy and has a ramp leading down to the water. Eggs: incubated by both parents. Soon after the chicks hatch, the parents sometimes move them to a second nest built specifically for brooding young.

VOICE. Call: in flight, henlike cackles; also, a shrill, extended laugh, *hitty hitty hitty hitup.*

OTHER BEHAVIORS. Conspicuous and noisy, and constantly flicking their tails, Purple Gallinules move about busily on muddy shores or clamber laboriously through shoreline vegetation. Their long toes enable them to walk effortlessly on lily pads and other floating vegetation. They associate with American Coots and Common Moorhens.

When disturbed, they readily take flight, dangling their feet and beating their wings feebly. They usually fly for a short distance and alight on branches before finding concealment in the foliage.

They constantly nod the head when swimming. Under some circumstances they become fearless to the point of taking food from park visitors' hands (and in one case, lips!).

Common Moorhen, *Gallinula chloropus.* Princ. distrib. Texas: Uncommon SR throughout the state (except w. Texas); common to locally abundant PR in s. Texas. They favor freshwater marshes, highly vegetated lakes, and other still or slow-moving waters. Common Moorhens require more marsh growth than American Coots but do not need the floating vegetation that Purple Gallinules apparently require.

FEEDING BEHAVIOR. They eat stems, leaves, and seeds of aquatic plants, as well as fruits and berries of terrestrial plants. Less frequently they forage for insects, spiders, tadpoles, and other invertebrates, as well as for bird eggs and carrion. They dive for rootlets, stems, submerged leaves, and other plants.

COURTSHIP BEHAVIOR. On land, the male chases a potential mate; when she stops, he bows deeply and the two preen each other. In another display, the male lowers his head, partially raises his wings, and elevates his tail to expose the white undertail coverts. He also bends his head downward and inward toward his feet.

NESTING. Common Moorhens sometimes nest in small colonies. Both sexes build a solid platform of cattails and other plants and usually provide the structure with a ramp (also built from plants) that leads down to the water. They build additional platforms nearby on which they rest when they brood their young. Eggs: incubated by both parents. Young birds swim very soon after they hatch; they also climb adeptly, thanks to wing spurs. They are fed by both parents as well as by helpers (older siblings from earlier broods who also help their parents defend the territory). In about 3 weeks the young birds feed independently. Younger female moorhens lay eggs later in the season and have smaller clutches.

VOICE. Call: many chickenlike notes, constantly uttered, as though the bird must maintain uninterrupted vocal contact with its family. Other calls include loud, grating, and complaining sounds. The male's courtship call is a loud *ticket ticket ticket.*

OTHER BEHAVIORS. They swim buoyantly, sometimes tipping up like a duck, sometimes diving. When swimming they bob their head, and when walking they jerk their tail nervously, exposing the white under the tail that probably serves to maintain contact with other individuals. This white patch is also visible in the front-down, tail-up posture in swimming birds. Moorhens walk with ease over lily pads and maneuver skillfully through the densest shoreline vegetation.

They fly heavily and awkwardly. When taking off they run along the surface of the water to gain momentum, but often, before becoming airborne, they seem to reconsider the effort involved and suddenly drop back down to the water.

These are not especially shy birds, and in urban habitats they often forage in full view, frequently with American Coots. However, they are not as bold or as gregarious as American Coots, and they rarely associate in large groups.

American Coot, *Fulica americana.* Princ. distrib. Texas: Uncommon to locally common SR throughout most of Texas; locally abundant WR in some parts of the state, in fresh, brackish, and salt wetlands as long as there is sufficient open water.

FEEDING BEHAVIOR. American Coots usually forage in small groups. They dive (as deep as 25 feet) for submerged vegetation, pick food from the surface of the water, or upend in shallow water; they also commonly graze on land adjacent to lakes and ponds, in particular golf courses and public parks. American Coots are primarily vegetarian and eat stems, leaves, roots, and other plant parts, especially underwater algae. They also take insects, worms, crayfish, and other invertebrates, as well as an occasional fish or tadpole.

COURTSHIP BEHAVIOR. The male's courtship displays include paddling toward the female while flapping his wings. He also arches his wings, spreads and elevates his tail to expose the white under the tail, and swims with his head and neck lowered (Fig. 5). The female reciprocates with basically the same postures and movements.

NESTING. American Coots build one or more nests, essentially floating platforms made from cattails and other marsh plants. The eggs, which are incubated by both parents, hatch asynchronously. Females sometimes lay eggs in the nests of other American Coots.

A. NORMAL POSTURE

B. PATROLLING

C. CHARGING

D. PAIRED DISPLAY

E. SWANNING

F. WARNING

G. BOWING AND NIBBLING

H. BRACING

I. ARCHING

J. SPLATTERING

K. BEGGING

FIGURE 5. Displays of the American Coot. (G. W. Gullion, 1952, *Wilson Bulletin* 64: 83–95. With permission.)

At night parents brood their young on a platform constructed by the male. Young birds are excellent swimmers.

VOICE. The name "mud hen" is applied to them probably as much for their constant, henlike complaining (various clucks, grunts, and similar noises) as for their chubby physical appearance. Their numerous varied vocalizations are often accompanied by considerable splashing.

OTHER BEHAVIORS. They are highly territorial. During territorial disputes, males rear up and attack each other with their sharp-clawed feet. The white frontal shields help paired birds recognize each other.

American Coots are the most aquatic members of the family Rallidae. They are better adapted for diving than dabblers and many other ducks, as their feet are set back on the body and their toes bear lateral flaps. Like grebes and diving ducks, they must run across the water when taking flight. It is said that when severely frightened they dive and cling to underwater vegetation, sometimes until they drown.

CRANES: FAMILY GRUIDAE

The physical appearance of cranes is arresting. They have long legs and long necks and walk in a slow, dignified manner. They locate seeds, grain, roots, and small animals by probing in dirt or mud or by picking at the surface. When engaged in their well-known dancing displays, they jump into the air with considerable calling and flapping of the wings.

Cranes are capable of projecting trumpeting notes for several miles, thanks to their extraordinarily long trachea (windpipe), so long that only part of it fits in the neck. The part that does not fit loops down into the chest cavity and becomes permanently encased in the sternum (breastbone).

Sandhill Crane, *Grus canadensis.* Princ. distrib. Texas: Uncommon to locally abundant M and WR in w. and s. Texas, chiefly in prairies and other vast, open areas.

FEEDING BEHAVIOR. Omnivorous. They consume grain, berries, and seeds, as well as any animal they can subdue, including lizards, frogs, snakes, and young birds. Wintering Texas birds apparently consume very few animals (Ballard and Thompson, 2000). Sandhill Cranes forage on land or in shallow water by walking with long, deliberate strides and locating food items by sight. They also use their long bill to probe and dig into the ground.

COURTSHIP BEHAVIOR. Often communal, courtship behavior is sometimes observed in Texas outside the mating season. In the dancing display, two birds leap as high as 8 feet in the air, flap their wings, bow, and call loudly.

Sandhill Cranes have a long-term pair bond, and most pairs probably mate for life.

VOICE. Call: perhaps best described as bugling and almost always given in flight, a guttural *ger ooooah*. It has enormous carrying capacity.

OTHER BEHAVIORS. In the High Plains, just before sunrise, thousands of wintering Sandhill Cranes disrupt the inexorable silence that normally dominates this part of Texas by rising noisily from their evening roosting site—often a shallow lake—and flying to neighboring cultivated fields. They fly with powerful wingbeats in vee formation, and they soar at very high altitudes.

Their plumage acquires a rusty coloration when they dig into soil colored by ferric oxide and spread this soil over their feathers as they preen.

Unlike Whooping Cranes, they seem to require water only at night.

Whooping Crane, *Grus americana.* Princ. distrib. Texas: Common WR in Aransas National Wildlife Refuge and Matagorda Island (see Wauer, 1999).

FEEDING BEHAVIOR. Whooping Cranes walk through marshes and shallow waters with their characteristic long and deliberate stride. Here they search for insects, crabs, frogs, snakes, small fish, and other animals, as well as berries, roots, and acorns. They locate food items by sight or by probing the soft substrate.

Pairs or family groups defend feeding territories in winter, returning to the territories each year. Generally young cranes (called colts) stay with their parents throughout the winter.

COURTSHIP BEHAVIOR. Pairs usually mate for life. During their courtship dance they repeatedly leap into the air, point their bill to the sky, and flap their wings. They also toss tufts of grass into the air, bow, and make trumpeting sounds (the whooping call).

VOICE. Their sonorous trumpeting sounds can be heard for several miles; the call is rendered as *kerloo ker lee oo.*

OTHER BEHAVIORS. When alarmed, these wary birds fly only a short distance, with slow wingbeats that have a powerful jerk on the upbeat. They also flap, soar, glide, and swim (usually only in young birds). When traveling long distances they fly in a line or assume a vee formation. They stand in water when roosting.

Whooping Cranes are not as gregarious on their wintering grounds as Sandhill Cranes, but newly paired birds often establish a winter territory next to their parents.

SHOREBIRDS, GULLS, AUKS, AND ALLIES: ORDER CHARADRIIFORMES

LAPWINGS AND PLOVERS: FAMILY CHARADRIIDAE

Foraging behavior in most plovers is diagnostic: they generally run a few steps, pause to pick up food items from the ground or shallow water, then run a few more steps. Some species shuffle one foot over the sand or mud, apparently to disturb small invertebrates or possibly to lure them to the surface by simulating raindrops.

Several species perform injury-feigning displays when intruders threaten their eggs or young. Migrants are more restless in spring (Oberholser and Kincaid, 1974) and take less time to travel between their breeding grounds in spring than to their wintering grounds in fall.

Black-bellied Plover, *Pluvialis squatarola.* Princ. distrib. Texas: Common to locally abundant M on the coast, less common inland. Uncommon WR on the coast. Whereas most coastal birds inhabit beaches and salt marshes, inland migrants favor lakeshores, mudflats, plowed fields, and moist meadows. Black-bellied Plovers generally do not coexist with American Golden-Plovers on the dry, short-grass prairies that are the latter's preferred habitat, and only rarely do large flocks of migrants inhabit bare fields.

FEEDING BEHAVIOR. The foraging strategy of this well-proportioned bird (the largest North American plover) is to run a few quick strides, then stop abruptly and assume an alert, dignified posture. One observer, however, reported behavior similar to sandpiper foraging: "Most birds foraged in water 1–3 cm deep, and they did so consistently by walking slowly, in the manner of a foraging sandpiper, and picking at the surface at short intervals" (Paulson, 1990). Although they typically forage singly, they usually fly and roost in flocks.

In saline habitats they search for marine worms, insects, and mollusks; inland they take beetles, grasshoppers, other insects, and, less frequently, seeds. They have been noted taking fewer steps after successfully capturing prey than after failing.

VOICE. A rich, melodious, and easily imitated *tee u eee* or *whee u eee,* given in flight.

OTHER BEHAVIORS. They generally migrate in smaller flocks than golden plovers, although on the coast large flocks are sometimes observed. They associate with Ruddy Turnstones and Sanderlings. Their swift, powerful flight ranks them among the fastest of the shorebirds. They swim readily, but rarely for long distances.

In one threat display seen in spring migrants, birds walk stiffly toward each other; in another, they peck at the ground and throw pebbles over their shoulder.

American Golden-Plover, *Pluvialis dominica.* Princ. distrib. Texas: Common M throughout most of Texas in spring, but rare in fall because their clockwise migratory pathway bypasses Texas and directs them along the Eastern Seaboard en route to South America. On returning in spring they pass through Texas and the Midwest. Migrants favor short-grass prairies, golf courses, plowed fields, and burned pastures.

FEEDING BEHAVIOR. When searching for insects and other invertebrates they run swiftly for a few steps, bobbing the head as they move, then halt abruptly.

VOICE. Their calls are numerous and are harsher than the Black-bellied's. One is a quavering, descending *quee e e a.*

OTHER BEHAVIORS. Their strong, swift, and graceful flight has been clocked at over 100 mph, making this species the fastest flying North American wader. They fly in compact flocks that quickly change shape and direction; upon alighting, they momentarily hold their wings above the body. On land they can hop for a considerable distance on one leg.

Snowy Plover, *Charadrius alexandrinus.* Princ. distrib. Texas: Uncommon PR on the coast; locally uncommon M throughout Texas. They breed rarely on the coast and in scattered localities throughout w. Texas. Coastal birds prefer barren beaches and sandy flats away from the surf; inland birds occur on freshwater, alkaline, and saline lakes.

FEEDING BEHAVIOR. These spirited little birds feed in small flocks, usually employing the run-and-peck method of foraging. Sometimes, like sandpipers, they run back and forth with each advance and retreat of the waves. In still water they sometimes vibrate one leg (termed pattering) to agitate prey. Their diet includes marine worms, insects, dead fish and other animals.

COURTSHIP BEHAVIOR. In the nest-scrape display, the male raises a wing, ruffles his throat feathers, and points his bill toward the scrape.

NESTING. The pair utilize one of several scrapes that they dig in the sand and line with shells, twigs, leaves, or grass. Eggs: incubated by both parents; normally both parents also tend to the young during their first days of life. Presumably, all Snowy Plovers that nest east of the Rockies are monogamous, whereas male birds in California may have more than one mate.

VOICE. Although generally silent, they make a low-pitched, flutelike *poo-weet;* in flight, a soft, rapid trill.

OTHER BEHAVIORS. These are relatively tame birds. When approached too closely, they usually crouch or run, and they take flight only if necessary. They fly in both loose and compact flocks and follow a direct or a zigzag course.

Wilson's Plover, *Charadrius wilsonia.* Princ. distrib. Texas: Locally common SR and M on the coast, in open sand flats and areas between dunes.

FEEDING BEHAVIOR. Although Wilson's Plovers eat basically the same food items as other small plovers (crabs, shrimp, mollusks, insects, and other invertebrates), their larger bill permits them to take slightly larger prey. They forage more slowly than plovers of comparable size, and they often continue feeding after dusk.

COURTSHIP BEHAVIOR. The male prepares several scrapes in front of the female. While she crouches, he circles around her with his head lowered, his wings drooped, and his tail spread, pattering all the while with his feet.

NESTING. Nest: selected by the female from one of several scrapes prepared by the male, who lines it with bits of vegetation. Both parents incubate the eggs, and both tend to the young.

When threatened at the nest, parents engage in extraordinary distraction displays. The male, after running around anxiously, suddenly sits down in the sand and begins scraping a new nest hollow (apparently displacement behavior). His more theatrical mate feigns incapacitation by dropping helplessly to the ground and rolling over on her side. In this defeated posture she wails loudly and desperately gasps for air.

VOICE. Although not particularly vocal, they frequently make a musical *wheet* in flight or while on the ground; they also produce a series of higher-pitched whistled notes.

OTHER BEHAVIORS. Wilson's Plovers form compact flocks that fly with remarkable swiftness and agility, performing stunning maneuvers near the ground or on water. Even then, they are slightly less gregarious than most small plovers. Usually they can be approached with ease. If threatened when away from the nest, they are more likely to run than fly. Males erect their throat feathers when threatened by another male.

Semipalmated Plover, *Charadrius semipalmatus.* Princ. distrib. Texas: Common M and WR on the coast, migrants becoming less common to the w. They favor tidal flats, muddy shores, and other open habitats, usually avoiding areas overgrown with vegetation.

FEEDING BEHAVIOR. These smart-looking birds feed in typical plover fashion, gleaning the damp soil for crustaceans, marine worms, insects, and other invertebrates.

VOICE. A plaintive two-note *chee wee,* the second note slightly higher; also (although rarely heard in Texas), a series of notes that accelerates rapidly.

OTHER BEHAVIORS. Like many shorebirds, they fly in compact flocks low over the ground, twisting and turning in unison. After alighting, they usually disperse rather than remain together. When alarmed, and perhaps also when curious, they jerk or bob their head.

At sunset, birds fly to an isolated beach or flat where they settle on the sand or mud, sink the head snugly between their shoulders, and roost for the night.

Piping Plover, *Charadrius melodus.* Princ. distrib. Texas: Uncommon M in e. Texas, wintering locally on the coast on tidal flats and sandy beaches.

FEEDING BEHAVIOR. Piping Plovers forage in the run-and-halt manner characteristic of plovers, but they seem more deliberate in their movements, suggesting to some observers the foraging tactics of American Robins. They eat marine worms, crustaceans, and other invertebrates that at low tide are exposed on wet sand or flats. They also feed at night; in fact, nocturnal foraging is generally regarded as the basic foraging strategy in Piping Plovers.

VOICE. A soft, melodious whistle, *peep peep peep lo,* the last note dropping in pitch.

OTHER BEHAVIORS. They migrate singly or in small flocks. Their flight is swift and agile, and they generally twist and turn even more than Sanderlings (Drake et al., 2001).

Killdeer, *Charadrius vociferus.* Princ. distrib. Texas: Common PR throughout Texas, in virtually all open and semiopen habitats.

FEEDING BEHAVIOR. Killdeers forage on lakeshores, lawns, and other open areas. They move in a cautious and deliberate manner, taking a few long strides, stopping with the head held high to survey the surroundings, then abruptly jabbing the beak at insects. Sometimes they follow tractors for insects that are stirred up.

COURTSHIP BEHAVIOR. In one display, the male flies erratically over the territory, flapping his wings slowly and calling repeatedly. In another, performed on the ground, he makes ritualized nest-scraping movements.

NESTING. Nest: a shallow scrape in a substrate of stones, gravel, or dirt. Killdeers nest in almost any open area (occasionally even on flat, pebbled rooftops), and not necessarily near water, as long as they have an extended view of their surroundings. Eggs: incubated by both parents. To cool eggs in hot weather, birds soak their breasts and abdominal feathers with water, as well as stand over the eggs to shade them. If their eggs are disarranged, they rearrange them, like a pie, with the pointed end of the eggs toward the middle. This compact arrangement allows the parents to cover the eggs better.

Although the chicks are able to feed themselves almost immediately after hatching, they remain with the parents for several weeks.

When their eggs or young are approached, parents engage in an injury-feigning bout of the first order: they relentlessly scream at the intruder while flopping around helplessly as though partially paralyzed, a truly believable performance. Ethics aside, few persons would regret provoking this unforgettable show at least once during a birding career. Interestingly, Killdeers respond differently to nonpredatory animals such as cows and horses, which could step on their eggs — they fly in the animal's face.

Monogamy is probably the rule for this species, as pairs often return each year to the same territory.

VOICE. Their several calls, all more or less with the same penetrating quality, include a piercing *kill dee, kill dee, kill dee;* an emphatic *teea eee! teea eee! teea eee!;* and a ringing *kill dee, tee, tee, tee.* Killdeers depend on volume and repetition to dissuade intruders, and indeed, half an hour of their strident calling usually encourages one to seek other birding sites.

OTHER BEHAVIORS. Flight is swift, wavering, and rarely prolonged. When suspicious or alarmed, these alert and high-strung birds nervously jerk their head and tail up and down.

If they need to defecate while foraging in shallow water, they go ashore, then return to the water to resume feeding.

Mountain Plover, *Charadrius montanus.* Princ. distrib. Texas: Rare, sometimes uncommon, local M throughout the state, nesting rarely in w. Texas. They inhabit upland, short-grass prairies and overgrazed grasslands, including bare ground, plowed fields, and prairie-dog towns. They spend the winter on coastal prairies, alkaline flats, and other open areas, often far from water. These misnamed birds generally avoid mountains. Their favorite habitats are disturbed prairies, plowed fields, or semidesert instead of tall-grass prairies, making their behavior more in line with that of most plovers, which show clear preferences for bare-ground habitats.

FEEDING BEHAVIOR. They engage in the stop-and-go foraging tactics typical of plovers and generally forage for grasshoppers, crickets, and other insects.

VOICE. Call: a sharp *kipp*. As they feed, Mountain Plovers utter plaintive, musical whistles that sometimes develop into sounds that are harsh and high in pitch, suggesting to some listeners frogs croaking.

OTHER BEHAVIORS. They are generally tame and approachable. When threatened, they first attempt to escape by running across the prairie, but if this response fails, they quickly crouch on the ground and blend into the surroundings. They fly low and swiftly, usually for short distances by flapping and gliding on down-curved wings.

OYSTERCATCHERS: FAMILY HAEMATOPODIDAE

American Oystercatcher, *Haematopus palliatus.* Princ. distrib. Texas: Locally uncommon PR on the coast.

FEEDING BEHAVIOR. These strictly coastal birds walk along the shore probing for marine worms, mussels, crabs, and other invertebrates, but sometimes they wade out into the water up to their bellies. Their chisel-shaped bill leaves conspicuous, oblong pocks in the sand at low tide. They not only feed at exposed oyster beds but also bring mussels to shore, where they open and eat them. In both cases, opening these hard shells is a learned behavior (described under "Nesting").

COURTSHIP BEHAVIOR. In the piping display, two birds, with their bills pointed toward the ground, walk or run next to each other and call loudly. Sometimes birds from adjacent territories join in to form a piping tournament.

Prior to copulating, the male calls softly and repeatedly as he lowers his tail, retracts his head, and stealthily approaches the female.

NESTING. Nest: a shallow scrape on a sandy or gravelly area. Both parents incubate the eggs and brood the nestlings, feeding them until they are about 2 months old.

In a frequently cited experiment, a large wooden egg was placed next to an oystercatcher's much smaller egg when the parents were away from the nest. Upon returning, the parents ignored their own egg and began incubating the larger wooden egg, even when it was 20 times the size of the normal egg. The larger egg, called a super sign stimulus or super releaser, evidently provided a stronger stimulus for incubating. That it was artificial was unimportant.

The closely related Eurasian Oystercatchers exhibit a culturally transmitted foraging behavior quite remarkable among birds. Parents teach offspring how to open mussel shells in one of two ways, depending on how they themselves were taught. They teach them to be stabbers, which sneak up on open mussels and quickly ram their bill into the soft meat, or to be hammerers, which shatter the shell with powerful blows aimed at weak points in the shell. Presumably the reason some young birds remain with their parents so long (up to a year) is so they can refine the feeding technique their parents taught them.

Oystercatchers also steal from each other. Stealing is particularly prevalent in young birds, probably because they are not yet capable of efficiently obtaining their own food.

VOICE. Call: a piercing *wheep, wheep, wheeup* or a loud *pic, pic, pic,* the latter commonly heard from birds taking flight. In spring one sometimes hears their song, a musical trill reminiscent of a plover.

OTHER BEHAVIORS. Oystercatchers are shy and wary. They feed alone or in small, exclusive groups. In winter they are weakly territorial. Their flight is strong, swift, and graceful. They only occasionally swim and dive.

The piping display is also used to maintain the pair bond and to advertise the territory.

STILTS AND AVOCETS: FAMILY RECURVIROSTRIDAE

Black-necked Stilt, *Himantopus mexicanus.* Princ. distrib. Texas: Uncommon to locally common SR throughout the state, except the forested areas of e. Texas; locally common PR on the coast. They frequent mudflats, grassy marshes, pools, shallow lakes, and similar wetland habitats, favoring fresh water over brackish water. (The related American Avocets tolerate a broader range of habitats.)

FEEDING BEHAVIOR. These singularly elegant and striking birds walk in shallow water or on muddy shores in search of insects, crustaceans, and seeds, gracefully balancing themselves on their extraordinarily long legs and occasionally wading up to their belly. Sometimes they plunge their head into the water to capture invertebrates or an occasional small fish; at other times they quickly grab passing insects.

COURTSHIP BEHAVIOR. The displaying male pecks the ground or water, preens his breast, repeats this process several times, then wades around the female while flicking water with his bill.

NESTING. In loose colonies. Nest: open or partially concealed in a depression or on a mound, often paved with mussel shells. Both parents incubate and care for the young in an extraordinary manner. To cool the eggs and chicks and to increase nest humidity, they carry water to the nest in their breast and abdominal feathers, sometimes making as many as 100 trips a day.

They exhibit remarkable defensive behaviors for thwarting intruders in their nesting areas. Besides calling loudly and furiously as they fly above the intruder, they bob vigorously, strike the surface of the water with their breasts, feign injury, and engage in mock incubation.

VOICE. When alarmed or disturbed they utter a persistent, penetrating *pep, pep, pep,* ascending in pitch to the point of resembling frenzied yelps. Other calls are quite ternlike.

OTHER BEHAVIORS. Both immatures and adults swim and dive, but they do so less frequently than American Avocets. Young birds use their wings to swim.

Black-necked Stilts regularly rest with their tarsi (lower part of the leg) on the ground.

American Avocet, *Recurvirostra americana.* Princ. distrib. Texas: Locally common SR in w. third of the state, uncommon SR on the Coastal Bend, and common WR on the coast. They occupy almost any open body of water, but inland they favor salt or alkaline lakes.

FEEDING BEHAVIOR. One of the rewards of seeking out shorebirds is to come across a group of foraging avocets walking slowly and deliberately through the shallow water, shoulder to shoulder, while moving their heads from side to side. They locate insects, crustaceans, and small fish both by sight and by touch, sometimes stirring up insects from the bottom, other times skimming the water's surface for both live and dead insects. They also forage on mud or sand flats for plants as well as animals, occasionally reaching up to snatch an insect out of the air. Their webbed feet permit them to swim efficiently, and when swimming they occasionally tip up like a dabbling duck to feed on submerged aquatic organisms.

COURTSHIP BEHAVIOR. The male's displays include crouching, bowing, wading, and — with wings spread ceremoniously — dancing. The female's solicitation posture is noteworthy: she stiffens her body and extends her bill far forward.

NESTING. Usually in loose colonies and sometimes alongside Black-necked Stilts. The nesting territories they defend are not always close to their feeding territories. Nest: hidden among tufts of vegetation, and lined with materials that include dry grass, weeds, and mud chips. Eggs: incubated by both parents. Young avocets can swim within a few hours after hatching.

When ground predators, including humans, approach a colony, several adults fly directly at the intruder. In the tightrope display, birds approach intruders while tipping their extended wings from side to side. American Avocets also engage in a distraction display.

It is not known why brooding behavior is so strong in this species. It is apparent even in young birds. The chick of a captive European Avocet tried to brood its nest mates when it was only 2 days old.

VOICE. In the breeding season, American Avocets are distressingly noisy, repeatedly showing alarm or protest with their yelping *wheeps* (also described as *wheek* or *kleep*); they also utter a softer *whuck* or *whick* that probably maintains contact between individuals.

OTHER BEHAVIORS. Upon alighting, they raise their wings high over their back. They swim in an upright position, with their tail elevated.

JACANAS: FAMILY JACANIDAE

Northern Jacana, *Jacana spinosa.* Princ. distrib. Texas: Rare, local, and irregular visitor in s. Texas, occasionally to the Upper Coast.

Jacanas' long, slender toes allow them to walk and run on mats of floating vegetation. They seem especially fond of lily pads, which they constantly glean for insects. After landing, they momentarily hold their wings high before folding them.

SANDPIPERS, PHALAROPES, AND ALLIES: FAMILY SCOLOPACIDAE

Most of our sandpipers are small to medium-sized birds that busily locate food by probing mud, sand, or shallow water. In flight, they twist and turn in highly coordinated flocks. The smaller species are referred to as peep sandpipers, or simply peeps.

Greater Yellowlegs, *Tringa melanoleuca.* Princ. distrib. Texas: Common M throughout Texas, and common WR on the coast and in scattered inland localities. They are found on mudflats, riverbanks, tidal flats, salt-marsh shores, and flooded pastures, yet they generally avoid open beaches on the coast.

FEEDING BEHAVIOR. They take insects and other animals from shallow water. They peck at the surface (instead of probing deeply), skim the surface by gracefully swinging the head back and forth in an arc, or plow the surface with their lower mandible. At times they wade out into the deep water, dash after minnows or other small fish, and even swim. They often forage in small flocks.

See Short-billed Dowitcher regarding nocturnal feeding.

VOICE. Three to five loud, ringing notes, *kyeu kyeu kyeu;* in flight, a *whee oodle.*

OTHER BEHAVIORS. Greater Yellowlegs frequently bob their body up and down as they walk, most likely to maintain contact with other birds or to estimate distances. They appear to be more alert, more wary, and more easily alarmed than Lesser Yellowlegs. They fly with strong, swift wingbeats.

Lesser Yellowlegs, *Tringa flavipes.* Princ. distrib. Texas: Common M throughout Texas; common WR on the coast. They occupy basically the same habitats as the Greater Yellowlegs but are more partial to freshwater shores than to tidal flats and tend to favor smaller ponds.

FEEDING BEHAVIOR. Although they occasionally wade into deep water (even up to their breasts), they typically forage in shallow water; here they scour the surface for aquatic insects, crustaceans, and other invertebrates. They do not swing their bill back and forth as frequently as the Greater Yellowlegs, and they seldom probe the mud for food.

VOICE. A soft, flat *cu cu,* as well as a series of *kip*s, which probably serve to maintain contact between individuals.

OTHER BEHAVIORS. Lesser Yellowlegs seem slightly more gregarious than their larger counterpart and often join flocks that include other species. They are more easily approached than Greater Yellowlegs, and at times they actually appear relatively tame. Both yellowlegs periodically interrupt their graceful stride to abruptly raise and lower their head. Some individuals defend a territory in winter.

Solitary Sandpiper, *Tringa solitaria.* Princ. distrib. Texas: Locally common M throughout the state, occurring singly or in small groups in freshwater pools and lakes. They generally avoid saltwater habitats.

FEEDING BEHAVIOR. These dainty and graceful sandpipers actively forage in shallow pools and along streamsides, constantly nodding their head as they probe the mud or water for insects and other invertebrates. Occasionally they snatch insects from the air. When feeding, they sometimes pause, suspend the lead foot in the water, then shake it vigorously, evidently to stir up insects from the bottom. In deeper water they wade more slowly than do either of the yellowlegs.

VOICE. Similar to the Spotted Sandpiper's but higher, a shrill *peet wheet wheet.*

OTHER BEHAVIORS. Solitary Sandpipers teeter like Spotted Sandpipers, but not so relentlessly. When flushed, they take flight almost vertically, possibly an adaptation to living in waters at the edge of woodlands. Their flight is light and swallowlike, with deep, deliberate strokes; also like swallows, they follow a zigzag path. Upon landing they briefly extend their wings high above their back, then gracefully fold them.

Solitary Sandpipers occur singly or in pairs. Normally they are unwary and can be easily approached. They swim readily and will dive to avoid predators.

Willet, *Catoptrophorus semipalmatus.* Princ. distrib. Texas: Common PR on the coast on sandy beaches, mudflats, and tidal estuaries; elsewhere, rare to uncommon M.

FEEDING BEHAVIOR. Willets forage slowly and deliberately, probing the mud or water with their bill as they search for fiddler crabs, mollusks, fish, and other small animals. They also pick up items from the shore or from the surface of the water.

COURTSHIP BEHAVIOR. The male noisily flutters over the nesting area in an aerial display that calls attention to his bold wing pattern.

NESTING. In a salt marsh or sand dune. Nest: a simple, shallow scrape or a cuplike nest made of grass. Eggs: incubated by both parents. The male makes a low bow when he relieves his mate at the nest.

Soon after hatching, the parents lead their young to a marshy area, where the chicks immediately begin feeding. When they are about 3 weeks old they are abandoned by their mother and left to their father's care.

VOICE. They are extraordinarily garrulous during the nesting season, producing a shrill but not unmusical *pilla willa willet,* or a *whee wee wee wee.*

OTHER BEHAVIORS. Willets maintain separate feeding and nesting territories and show a strong fidelity to their previous year's territories (including their winter feeding territories). The male frequently flashes his prominent white wing pattern during the breeding season, no doubt to advertise his territory as well as to maintain contact with his mate or with conspecifics. He also hovers on quivering wings in what appears to be another territorial display.

They wade in water up to their belly and even swim. They also perch on trees, fences, and occasionally buildings.

Spotted Sandpiper, *Actitis macularia.* Princ. distrib. Texas: Common M and local WR throughout Texas in rivers, sewage ponds, ditches, and other aquatic habitats, especially where there are rocks and debris. On the coast they occur in rocky bays and on rock jetties.

FEEDING BEHAVIOR. These industrious sandpipers constantly teeter as they walk along the shore, on rocks, and on logs next to the water. They pick up insects, crustaceans, and other invertebrates from the surface of the water or from the ground, and sometimes they reach out to snap up insects that fly past. In one interesting, catlike foraging tactic they crouch when they see a crab or ground insect, then slowly stalk it before rushing for the capture.

VOICE. An anxious *peet weet*, followed by a series of similar notes; also, a one-syllable *peet.*

OTHER BEHAVIORS. The details of their reproductive behavior are not treated here, but it may be pointed out that the larger and more aggressive females sometimes mate with more than one male during a single nesting season.

These are typically solitary birds, but they sometimes roost in loose flocks. Although normally they teeter along shorelines, they also swim, dive, and walk on the bottom of streams. Their foot structure allows them to perch easily, which they readily do on limbs, cattails, wires, and similar structures. Young birds swim well and teeter even on the first day of life.

Their flight is diagnostic: they fly low and rapidly over the water with stiff, arched wings that produce a rapid series of flutters and glides. It has been proposed that their manner of diving from a low flight over water, followed by swimming underwater, is a normal antipredator behavior for them, making them, in a sense, inverted flying fish. In winter they are territorial and defend a feeding area.

Upland Sandpiper, *Bartramia longicauda.* Princ. distrib. Texas: Common spring M in e. Texas and fairly common fall M in s. Texas; less common to the w., where they are a rare, local SR. They favor pastures, grassy fields, and coastal prairies and are observed much less frequently on large bodies of water, mudflats, sandy shores, and other typical sandpiper habitats.

FEEDING BEHAVIOR. When making their way through the grass in search of prey, they appear more like plovers than sandpipers. They hold their small, dovelike head high, run for a few paces, then stop abruptly. Indeed, they were called upland *plovers* in the past. They eat insects, spiders, and occasionally grains.

VOICE. A rich, mellow, three-note *whoo hee lee,* rather melancholy in quality, and frequently heard on late summer nights. On their breeding grounds they make other sounds, including one described as an "enchantingly" weird, long drawn-out whistle, *whooooleeeeee, wheeeloooo* (Oberholser and Kincaid, 1974).

OTHER BEHAVIORS. Upland Sandpipers fly swiftly and with strong wingbeats. Their flight is graceful and buoyant, thanks to wings that are long relative to the bird's weight. They perch on fence posts, rocks, and utility poles, and upon landing they momentarily hold their wings extended. Migrating birds are more wary in fall than in spring.

Whimbrel, *Numenius phaeopus.* Princ. distrib. Texas: Locally common M in e. Texas, more common on the coast, in salt-marsh pools, beach grass around sandy shores, and mudflats. During migration, large numbers congregate at preferred sites, making them appear locally abundant even when they are absent in adjacent areas.

FEEDING BEHAVIOR. They feed singly or in small groups, picking up insects, crustaceans, spiders, and other invertebrates; they sometimes probe wet sand or mud, although rarely as deeply as their long bill would indicate. They walk and run rapidly, pecking and probing when they stop.

VOICE. A loud, clear *pip pip pip;* in flight, a long series of liquid notes.

OTHER BEHAVIORS. In migration, Whimbrels fly overhead in vee formation; locally, they fly low over the ground or water with deliberate, steady wingbeats, sometimes gliding for a considerable distance. Often large numbers roost together on isolated sandbars or marsh islands. Sometimes they rest in a curious posture by placing their tarsi (lower part of the leg) flat on the ground.

Long-billed Curlew, *Numenius americanus.* Princ. distrib. Texas: Locally common SR on the High Plains as well as on the coast, where they are

FIGURE 6. Long-billed Curlew ejecting undigested part of prey. (From a photograph by Rod Rylander.)

also a common WR; locally common M throughout most of the state (except e. Texas forests). Although inhabiting mudflats and shores, they definitely favor short-grass prairies, closely grazed wet pastures, and other grasslands.

FEEDING BEHAVIOR. These are the largest of all North American sandpipers. They walk with a graceful stride and employ their extraordinarily long bill to capture insects, crustaceans, and other invertebrates. They probe just beneath the soil's surface or poke the bill into small burrows, cattle hoofprints, and other places where small animals hide. The long bill provides them with a broad feeding area immediately in front of their path and therefore a good supply of insects that can be snapped up with a quick turn of the head. As do many shorebirds, they regurgitate indigestible matter as pellets (Fig. 6).

COURTSHIP BEHAVIOR. Birds flutter high into the air and glide down while calling loudly. On the ground, their displays include ritualized nest-scraping movements.

NESTING. In moist or dry grasslands, often in loose colonies. Nest: a scrape in the ground that is lined with grass. Two females occasionally lay eggs in the same nest. Both parents incubate the eggs. Although males join together to mob approaching intruders vigorously, incubating birds hold their ground so tenaciously that an intruder can approach them very closely. To conceal their presence, incubating birds stretch their head out on the grass.

VOICE. A clear, plaintive *curr lee* (from which the name is derived) that rises in pitch, and a rapid *kli li li li*. In spring they utter a melodious, liquid *curleeeeeeuuu*.

OTHER BEHAVIORS. They roost together at night and disperse the next day to feed singly or in small groups. These are generally wary birds, but when undisturbed, as in Rockport and other coastal areas, they become quite tame. They take flight erratically but quickly straighten their path and flap steadily and deliberately. On landing, they momentarily hold their wings high above their back.

Hudsonian Godwit, *Limosa haemastica.* Princ. distrib. Texas: Uncommon to common spring M on the coast, in marshes, mudflats, playa lakes, and other wetlands; on the coast, spring migrants favor wet meadows and pastures (Michael Marsden, pers. comm.).

FEEDING BEHAVIOR. They forage deliberately and aggressively, rapidly thrusting their bill into the mud to locate mollusks, crustaceans, and other aquatic animals. They sometimes wade in water so deep that their head is submerged.

VOICE. Generally silent when migrating, although occasionally they utter a low *qua qua qua*. On their breeding grounds they produce the sound for which they were named, *god wit,* or perhaps more accurately *ta it.*

OTHER BEHAVIORS. Their flight is strong and swift. They swim readily, and when not feeding, they walk with the neck drawn into the body.

Marbled Godwit, *Limosa fedoa.* Princ. distrib. Texas: Common M and WR on the coast, in tidal flats, marshes, and bay beaches.

FEEDING BEHAVIOR. Marbled Godwits search the shores and mudflats for insects, mollusks, crustaceans, and other aquatic or marine organisms, often totally burying their bill into the mud as they probe. The female's slightly longer bill suggests that the sexes minimize competition by extracting food items from different levels in the mud. These godwits also wade into the water, often up to their head.

VOICE. In flight, a low *ter whit god wit god wit,* with the accent on the second syllable—a sound suggesting laughter to some listeners. Marbled Godwits are noisier than Hudsonian Godwits, but both species are generally quiet in winter.

OTHER BEHAVIORS. They are among the last shorebirds to leave the coastal pools when the tide comes in. At night large flocks roost together in wet meadows or shallow water. Their flight is steady and rapid.

Ruddy Turnstone, *Arenaria interpres.* Princ. distrib. Texas: Common M and WR on the coast. Summer individuals are probably nonbreeding first-year birds that remain on their wintering grounds until the spring of their second year. In Texas, turnstones favor tidal flats, sandy beaches, rocky shores, and jetties.

FEEDING BEHAVIOR. In keeping with their name, these versatile foragers flip over stones as they busily search for marine invertebrates. They also examine the underside of virtually any object on the beach, including shells and bits of shoreline debris. They have been observed rooting like a pig under clumps of seaweed, turning over large objects by shoving them with their breasts, and digging holes in the sand that are as large as their body. They eat carrion and food scraps on the shore as well as garbage at landfills, and they have been known to take bread from a person's hand. Their impact on nesting terns and other birds (whose eggs they readily eat) is probably significant in Texas but has not been quantified.

VOICE. A metallic, staccato *tuk a tuk* or *cut i cut.*

OTHER BEHAVIORS. Although turnstones migrate in large flocks that include Black-bellied Plovers and other shorebirds, they prefer to be alone or in small, loose groups. Flushed birds fly with rapid, powerful wingbeats, coursing in a straight line low over sand or water. They swim well and perch easily on posts and other structures.

These feisty little birds give no quarter when defending their feeding territories. They hunch their back into a menacing posture, lower their head, and rush at intruders. They defend feeding territories (small areas along the beach) only when resources are patchily distributed and clumped.

Red Knot, *Calidris canutus.* Princ. distrib. Texas: Uncommon to locally common M on the coast and barrier islands, frequenting mudflats and especially sandy beaches at low tide.

FEEDING BEHAVIOR. Winter flocks and some migratory flocks can be large, and many include dowitchers and other shorebirds. Most migrants in Texas forage in small, dense flocks, probing the mud with the touch-sensitive bill for buried mollusks and other small marine animals.

VOICE. A low, soft *knut;* also a *tlu tlu* and a mellow *wah quoit.*

OTHER BEHAVIORS. Knots seem a bit more gregarious than other sandpipers. They rarely leave their flock, whether foraging, twisting and turning in flight, or simply resting shoulder to shoulder on the beach.

Sanderling, *Calidris alba.* Princ. distrib. Texas: Common to abundant M, SR (nonbreeding) and WR on the coast, where they forage almost exclusively on sandy beaches.

FEEDING BEHAVIOR. Certainly everyone who visits our sandy beaches notices these alert, seemingly indefatigable birds chasing and dodging waves in what appears to be a Sisyphean task. The receding waves, however, expose an abundant supply of sand fleas, crabs, mollusks, and other small invertebrates whose momentary vulnerability Sanderlings quickly take advantage of. The pocks formed in the sand by the bird's partially opened bill are often arranged in lines or arcs.

Sanderlings also avail themselves of junk food discarded on the beach, as well as dead fish and other carrion, and occasionally seaweed. They vigorously defend feeding territories by fluffing up their feathers (giving them a threatening, hump-backed appearance) and chasing approaching plovers and sandpipers so aggressively that intruders often take flight.

VOICE. A sharp, high-pitched *twick* when flushed; however, apart from that sound and a soft twittering while feeding, Sanderlings are generally silent.

OTHER BEHAVIORS. Flock dynamics are interesting. Birds sometimes rest together in groups of 100 or more, and in this regard are clearly gregarious (as is typical of most sandpipers); however, they generally feed alone or in small groups. A company of Sanderlings moving along the beach changes membership quickly, suggesting that individual birds have no need for specific foraging partners. Moreover, even when Sanderlings forage with other sandpipers, one's impression is that the other sandpipers are seeking company with the Sanderlings, not the reverse.

The minimum number of birds that must depart before the entire flock takes flight depends on flock size. Only a few birds need depart to incite a small flock to leave together, but for large flocks, approximately 80 or more birds must leave before the entire group will take flight.

Most individuals return faithfully to the same wintering grounds. Here they defend feeding territories that differ in size depending on prey density. Territory size is smaller when prey densities are high, possibly because high prey densities attract so many intruders that a large territory cannot be defended.

Semipalmated Sandpiper, *Calidris pusilla.* Princ. distrib. Texas: Locally common M throughout Texas, on beaches, mudflats, and other typical shorebird habitats.

FEEDING BEHAVIOR. With the abundant energy that characterizes most peeps, Semipalmated Sandpipers walk briskly or run quickly and erratically along the water's edge, probing the wet mud or picking prey items from the surface. They sometimes follow the advancing and receding waves, like Sanderlings, and occasionally they investigate floating vegetation. They feed both during the day and at night, consuming mainly insects, crustaceans, small mollusks, and other invertebrates.

VOICE. A hoarse, high-pitched *cherk,* uttered in flight; also, an abrupt *ki i ip,* produced when they are flushed. Occasionally in late spring the cicadalike courtship song is heard.

OTHER BEHAVIORS. They fly swiftly and with rapid wingbeats. Flocks twist and turn in unison, suddenly separate into two or more groups, then, just as quickly, reunite. At high tide they rest or sleep on the sand with other peeps, huddling together with their bill tucked in their back feathers. They frequently stand on one foot and occasionally hop away without lowering the retracted leg.

Western Sandpiper, *Calidris mauri.* Princ. distrib. Texas: Common M and WR on the coast, less common inland, frequenting shores, mudflats, and beaches.

FEEDING BEHAVIOR. Basically the same as the Semipalmated Sandpiper's, but they feed more often in shallow water, sometimes with their head immersed. They rarely venture out into deep water.

VOICE. A squeaky *chree eep,* which is coarser and more querulous than the higher-pitched Semipalmated's. Feeding birds constantly palaver with soft whistles.

OTHER BEHAVIORS. Like Semipalmated Sandpipers, they perform highly coordinated aerial maneuvers. Most behaviors of migrating and wintering birds are essentially like those of the Semipalmated Sandpiper.

Least Sandpiper, *Calidris minutilla.* Princ. distrib. Texas: Common to abundant M throughout most of Texas, WR mainly in e. half of the state, most commonly on the coast. Inland migrants are observed on river bars, rain pools, wet pastures, and other wetland habitats. Most coastal birds prefer bay or marsh flats to sandy beaches on the surfside.

FEEDING BEHAVIOR. These are notably gregarious birds. Their manner of feeding is similar to the Semipalmated Sandpiper's, but they pick more than probe and favor muddy substrates over sandy beaches. They generally feed more slowly than Semipalmated Sandpipers.

COURTSHIP BEHAVIOR. During their courtship display, performed by migrants in late spring, males circle on quivering wings while uttering clear, musical trills.

VOICE. A thin, reedy *kree eeep,* much higher and more drawn out than the Semipalmated Sandpiper's call (almost "mousy" — Michael Marsden, pers. comm.).

OTHER BEHAVIORS. Migrating and wintering birds closely resemble Semipalmated Sandpipers in their behavior, but Least Sandpipers appear tamer and sometimes crouch rather than fly when approached. When flushed, they suggest Common Snipes or Pectoral Sandpipers: they utter a grating *scree ee eee gree eet gree eet* and zigzag off with extremely rapid wingbeats.

White-rumped Sandpiper, *Calidris fuscicollis.* Princ. distrib. Texas: Uncommon to common spring M throughout most of Texas, much less common to the w. They frequent many wetland habitats, including mudflats, sandbars, sewage ponds, and shallow, grassy pools.

FEEDING BEHAVIOR. On the coast they feed at low tide, often probing deeply and methodically in one place. They sometimes wade out waist-deep and immerse their head as they search for marine worms, insects, and other invertebrates, as well as for vegetal material.

VOICE. A squeaky, abrupt, batlike *tzeep,* likened to the sound made by striking two marbles together.

OTHER BEHAVIORS. Compared with the other peeps, they move about slowly and deliberately. They are relatively tame, and when disturbed they generally fly only a few feet away and resume feeding. Their flight is swift and undulating.

At high tide, migrants join large, mixed flocks of shorebirds to rest on higher ground.

Baird's Sandpiper, *Calidris bairdii.* Princ. distrib. Texas: Uncommon to common M throughout Texas. They are seen just beyond the wave-washed part of the shore, or in damp mud by coastal ponds and marshes. In other parts of North America, Baird's Sandpipers have been recorded at higher elevations (including mountains) than have the other peeps.

FEEDING BEHAVIOR. These are very active birds. They run swiftly for 8–12 feet or more, then abruptly stop to pick up crustaceans and other invertebrates (and sometimes seaweed) from the mud or sand. They are much less inclined to probe in water or wet mud than most peeps.

VOICE. A soft *kreep,* uttered in flight.

OTHER BEHAVIORS. In flight Baird's Sandpipers join mixed flocks of peeps, but upon landing they abruptly leave the others.

Pectoral Sandpiper, *Calidris melanotos.* Princ. distrib. Texas: Uncommon to common M throughout Texas in most aquatic or semiaquatic habitats, but clearly favoring wet pastures, marshes, and other damp, grassy areas, making it one of the sandpipers dubiously termed grasspipers. Coastal migrants are more likely to occur away from the beach, especially in fresh or brackish marshes.

The first fall migrants in Texas are probably males, as males leave their Arctic nesting sites while their mates are still incubating their eggs.

FEEDING BEHAVIOR. Pectoral sandpipers eat insects, spiders, and occasionally seeds. They pick up items from the ground or probe rapidly in wet mud or shallow water. Sometimes 30 or more birds feed together in loose flocks.

VOICE. *See* "Other behaviors."

OTHER BEHAVIORS. Their profile is distinctive: they extend their neck, hold their head high, and walk slowly (Fig. 7). They often stop and momentarily stand without moving.

Their behavior suggests miniature snipes, also birds of grassy areas. ("Grass-snipe" was a former name for the Pectoral Sandpiper.) When alarmed, they fly off on a zigzag course while uttering a harsh, grating *krriek krriek krriek krriek* (similar to the sound of the Western Sandpiper). Also like snipes, they arrest their flight by suddenly dropping down into the grass.

When a group of birds is approached, they first freeze, or crouch, then flee one at a time. Otherwise they fly in compact and synchronized flocks. Their flight is strong, swift, and direct.

FIGURE 7. Typical postures of Pectoral Sandpipers. Alert (a–c); normal feeding (d); displacement feeding (e); wing-away (f); supplanting another individual (g). (W. J. Hamilton, Jr., 1959, *Condor* 61: 161–179. With permission.)

Dunlin, *Calidris alpina.* Princ. distrib. Texas: Locally common M throughout the state, except on the coast, where they are an abundant M and common WR. They frequent muddy pools, tidal flats, sewage ponds, and flooded fields.

FEEDING BEHAVIOR. Dunlins forage industriously, both day and night, often in company with Sanderlings and other peeps. They eat mainly insects and invertebrates that live in the mud and sand. They probe very rapidly (2–3 times per second), recalling the sewing-machine movements of dowitchers.

VOICE. A nasal, grating *czuup* when flushed; in flight, a plaintive and melodious *purrra.*

OTHER BEHAVIORS. Although generally tame and sluggish, they sometimes can be quite restless. Their swift and highly coordinated aerial flights are outstanding: compact flocks fly low over the water or sand and twist and turn with astonishing coordination.

Dunlins are highly gregarious. Some enormous migrating flocks have been likened to swarming insects.

Stilt Sandpiper, *Calidris himantopus.* Princ. distrib. Texas: Common M in e. and n.w. parts of the state, locally common elsewhere; uncommon WR on the coast. They occur in marshes and on mudflats and sandy beaches, but they generally favor quiet, shallow pools and lagoons.

FEEDING BEHAVIOR. Aptly referred to as a yellowlegs that feeds like a dowitcher, they even associate with the two species of yellowlegs. Stilt Sandpipers forage together in such closely bunched groups that their bodies almost touch. They walk slowly and methodically in belly-deep water, submerging their head for several seconds while rapidly probing like a sewing machine. They detect invertebrates in the bottom ooze with sensitive touch receptors in their bill. They also sweep their bill from side to side in the bottom mud or pick insects from the water's surface.

VOICE. A soft, coarse *whru whru,* suggesting the call of the Lesser Yellowlegs.

OTHER BEHAVIORS. These are generally unwary birds and can be approached rather closely. They fly in compact flocks and lift their wings slightly when they land.

Buff-breasted Sandpiper, *Tryngites subruficollis.* Princ. distrib. Texas: Locally uncommon M in e. part of the state, more common in fall (Texas Ornithological Society, 1995). However, they are more common in spring on the Coastal Bend (Michael Marsden, pers. comm.). They prefer short-grass prairies and plowed fields to wet mudflats.

FEEDING BEHAVIOR. Their foraging behavior suggests a plover: they run a few steps, pause motionless to look around, run a bit farther, then stop to pick up insects and other invertebrates. Frequently they pause momentarily and raise one or both wings high above their back, perhaps to maintain contact between individuals.

VOICE. They are generally quiet. Their sharp *teek teek,* sounding like striking two stones together, is only infrequently heard in Texas. They also utter a low, rolling trill. (*See also* "Other behaviors.")

OTHER BEHAVIORS. These are gregarious birds. They are relatively tame and normally attempt to avoid intruders by running (and sometimes hiding) instead of taking flight. When in the company of other sandpipers, they are usually the last to fly, and even then they often fly just a few feet before alighting. Before taking flight they sometimes freeze for a few seconds with their head held erect. Their gait has been characterized as high-stepping.

Birds that are suddenly disturbed fly away like snipes, zigzagging off while uttering a low *pr r r ree eek* that sounds somewhat like a Pectoral Sandpiper (which also flees in this manner). Flocks flying low over the ground are truly impressive as they twist and turn in synchronized aerial maneuvers.

Besides exhibiting ploverlike behaviors, the Buff-breasted Sandpiper has a round head, long neck, and upright posture that suggest a plover; indeed, migrants seem to prefer the company of plovers, in particular American Golden-Plovers, as well as the ploverlike Upland Sandpiper.

Short-billed Dowitcher, *Limnodromus griseus.* Princ. distrib. Texas: Common M and WR on the coast, uncommon inland; locally common WR on the coast. They favor salt- and freshwater mudflats, marshes, and other wetlands.

Both species of dowitchers occur in fresh- and saltwater habitats, but the Long-billed species favors freshwater habitats, and the Short-billed salt and brackish areas. Otherwise, migrating and wintering birds appear to behave identically.

FEEDING BEHAVIOR. Dowitchers bunch together in shallow water, where they probe deeply and deliberately with their long bill, often pushing their head entirely underwater. The rapid, vertical motion of the bill (almost universally compared to the action of a sewing machine) usually identifies these birds at a distance (although Stilt Sandpipers feed in a similar manner). They move forward slowly while probing, occasionally stopping momentarily to concentrate in one place. Sometimes they come to shore to feed in wet mud. They eat insects, crustaceans, and other invertebrates, as well as seeds and grasses.

Comparative studies of the retina (specifically, the rods and cones) show that dowitchers are better adapted for nocturnal vision than the Greater Yellowlegs, even though all three species feed at night.

VOICE. The Short-billed Dowitcher's mellow *tu tu tu* contrasts sharply with the Long-billed Dowitcher's thin, high-pitched *keeek*.

OTHER BEHAVIORS. These relatively tame birds can be approached closely when they are feeding. They fly swiftly, with strong and rapid wing-beats, and swim readily and easily.

Long-billed Dowitcher, *Limnodromus scolopaceus.* Princ. distrib. Texas: Uncommon to common M and WR throughout Texas, abundant in some areas on the coast, where it is common in winter. *See* Short-billed Dowitcher.

Common Snipe, *Gallinago gallinago.* Princ. distrib. Texas: Common M throughout Texas, and common WR in e. part of the state, frequenting bogs, wet fields, pastures, and grassy shores.

FEEDING BEHAVIOR. These shy, secretive birds prefer dense grass and grassy shorelines to open water and mudflats. Curiously, they also investigate cow patties for potential food items, especially those that are half-dried. They usually feed alone and are most active at dawn and dusk. On the coast in winter, however, they are more gregarious, and 50 or more individuals sometimes forage together.

They probe soft mud with their flexible, touch-sensitive bill. The upper mandible can curve to grasp buried invertebrates (including earthworms). Prey is swallowed with the help of spines on the tongue and serrations on the bill; these structures direct food backward. Indigestible material is regurgitated in the form of pellets. It is said that snipes drink unusually large quantities of water every day.

VOICE. Generally only the flight call is heard in Texas, but the call accompanying the territorial display can sometimes be heard in wintering and migrating birds. As the birds dive, they spread their tail feathers so that air passing between them produces an eerie, bleating sound, rendered as *woo woo woo woo woo.*

OTHER BEHAVIORS. When approached in their grassy hiding places, snipes usually freeze, but if approached too closely, they take flight abruptly and explosively, zigzagging off swiftly and excitedly while uttering a rasping, nasal *skaip*, like two rocks being scraped together. Oberholser and Kincaid (1974) graphically describe their descent: "a bird drops vertically from the air—wings lifted and bill pointed up—and drifts slowly into the grass."

Snipes are capable of swimming, diving, and standing on fences or in trees, but these behaviors are rarely observed in Texas.

American Woodcock, *Scolopax minor.* Princ. distrib. Texas: Locally uncommon WR in e. Texas, where they are also a rare SR. They occur in moist woods, thick swamps, and especially damp areas near rivers.

FEEDING BEHAVIOR. During the day these solitary, unsociable birds confine themselves to specific resting and feeding areas in the forests. At night they venture out into open fields, swamp edges, and other exposed areas, where they locate and extract earthworms from the damp earth with their long, flexible, and highly sensitive bill. They also forage on dry ground, turning over leaves to expose insects, snails, and other invertebrates. It has been estimated that they eat their weight in earthworms every 24 hours.

Some observers suggest that the rocking motion these birds make while standing on the ground incites earthworms and other invertebrates to move so they may be detected visually, and that the forward position of the birds' ears (in front of their eyes) enables them to hear insects on the ground.

VOICE. A nasal *peent* that suggests the call of the Common Nighthawk. They are generally silent except during the breeding season.

OTHER BEHAVIORS. It is thought that because the eyes are set far back on the head, woodcocks are able to survey what is above and behind them while probing in the ground. When flushed, they ascend almost vertically, zigzag skillfully through the branches in a weak and irregular flight pattern, then drop quickly to cover. There are unsubstantiated reports that females, when flushed, carry their young with their legs and feet, a behavior that has been verified only in the closely related European Woodcock.

A twittering sound produced in flight has been attributed to air passing between the outer three primaries.

Wilson's Phalarope, *Phalaropus tricolor.* Princ. distrib. Texas: Common to abundant M throughout most of Texas, nesting locally in the Panhandle

(Seyffert, 1985). They inhabit both freshwater and saltwater marshes, as well as shallow lakes, especially alkaline playa lakes.

Wilson's Phalaropes are aquatic, as are all phalaropes, but they are also the most terrestrial of the three species (as evidenced by the absence of lobes on their toes).

FEEDING BEHAVIOR. Besides walking and probing like most small sandpipers, phalaropes swim jerkily and in tight circles up to 60 revolutions per minute, bobbing their head as they stab for prey at each turn. These movements apparently stir up food and possibly bring items to the surface. Flies are an important item in their diet, and many passing flies are captured on the wing.

Wilson's Phalaropes take advantage of food stirred up by avocets, ducks, and other larger birds. (See Williams, 1953, for an account of the behavior of south Texas birds.)

COURTSHIP BEHAVIOR. The highly unusual reproductive behavior of phalaropes deserves mention. The female is larger and more colorful than the male. She defends her mate against other females, and after laying her eggs, she abandons her mate and sometimes seeks another. Her mate remains in the area to incubate the eggs and care for the nestlings. The extraordinarily high level of testosterone in the female's ovaries (sometimes surpassing that in the male's testes) probably accounts for her brighter plumage and more aggressive behavior.

VOICE. A soft honking sound, rendered as *wunrk;* also, a number of shrill, nasal notes.

OTHER BEHAVIORS. All phalaropes swim buoyantly, apparently because of air trapped in their breast feathers. Compared with the other two phalaropes found in North America, Wilson's spends less time swimming and more time walking and wading. It seems better adapted for feeding on shore or in shallow water, as it has a longer bill, neck, and legs.

Their swift, light, and graceful flight has been compared to that of the Lesser Yellowlegs. Both immature and adult birds play by picking up and dropping objects.

Red-necked Phalarope, *Phalaropus lobatus.* Princ. distrib. Texas: Uncommon to locally common M in Trans-Pecos Texas, more common in fall. Their behavior is basically as in Wilson's Phalarope. Call: a sharp *pe eet,* usually repeated.

SKUAS, GULLS, TERNS, AND SKIMMERS: FAMILY LARIDAE

Most gulls are noisy, highly social, and aggressive. The majority live in saltwater habitats and feed on fish and marine invertebrates. They forage while swimming, walking, or flying. Several species are extremely opportunistic and unhesitatingly consume carrion, garbage, and junk food left on the beach.

Most terns are smaller than gulls and more graceful in flight. They do not swim as readily, and they usually dive from the air to capture animals from the surface of the water.

Courtship feeding is common in terns and seems to be a good indicator of the male's ability to provide fish for the young. Female Common Terns assess the quality of potential mates by comparing the number of fish their suitors bring them; they usually select the male that brings the most fish. Quite likely many other terns also assess mate quality in this way.

Gull behavior played an important role in some pioneering studies by early ethologists. Many decades ago Nikolaas Tinbergen exhaustively analyzed displays in gulls and formulated concepts that even today contribute immensely to our understanding of animal behavior.

Laughing Gull, *Larus atricilla.* Princ. distrib. Texas: Abundant PR on the coast in virtually all saline habitats; inland they occur on lakes and rivers and at waste disposal sites. This is the only gull that regularly nests in Texas.

FEEDING BEHAVIOR. Noisy, aggressive, and persistent, these high-profile birds forage in almost every way possible, capturing and consuming virtually any animal they can handle. They dive, swim, wade, plunge into the water, and even pursue flying insects in the manner of swallows. They follow shrimp boats for offal and tractors for insects, and because of their fondness for animal fats and oils, they readily take greasy junk food thrown to them (including bread, provided it is buttered). After rains they visit fields for exposed earthworms.

Laughing Gulls steal prey from pelicans by reaching into the pelican's large pouch, and they themselves are victimized by frigatebirds that chase them and force them to drop food they are carrying.

COURTSHIP BEHAVIOR. The male courts the female with head tosses and by directing prolonged calls toward her.

NESTING. Colonies sometimes contain thousands of birds that nest in coastal vegetation. Nest: a scrape dug by both parents and lined with grasses. Both parents incubate the eggs. They feed their nestlings half-digested regurgitant.

VOICE. A penetrating, high-pitched laugh that predictably is rendered by some authors as *ha ha ha ha ha,* but the last notes descend into a wail that hardly suggests mirth. This call is often delivered ceremoniously from a conspicuous perch. In flight, birds utter a *kee ah, kee ah.*

OTHER BEHAVIORS. Laughing Gulls fly with light, strong, and graceful wingbeats, but they also soar on thermals like hawks and vultures. They swim buoyantly and often. Like most gulls, they spend considerable time loafing or strolling on the beach, which they seem to do with poise and self-assurance.

Franklin's Gull, *Larus pipixcan.* Princ. distrib. Texas: Uncommon to common M throughout most of Texas, in marshes, prairies, flooded pastures, and other open areas.

FEEDING BEHAVIOR. These so-called prairie gulls (they nest in the marshes of the Northern Great Plains) forage for insects, grubs, cutworms, and similar prey. They follow tractors for insects that are flushed, hawk gracefully and buoyantly for dragonflies, and hover over water before dropping down for small fish or aquatic invertebrates. In pastures flocks leapfrog over one another, the individuals in back flying forward over those in front to assume the lead position. These gulls also search for invertebrates while walking around on beaches or wading in shallow water. Like most gulls, they feed while swimming, but they rarely dive for food.

VOICE. A soft *kruk kruk kruk;* in flight they utter a shrill *wee kah wee kah wee kah.*

OTHER BEHAVIORS. Migrating flocks sometimes reach well into the hundreds. For no apparent reason, other than perhaps to play, birds ascend as a group high into the air, dive suddenly, then immediately repeat the process. They swim buoyantly, and when flushed they rise rapidly from the water. On land they walk with side-to-side body movements.

Compared with most gulls, Franklin's Gulls are notably gentle and non-aggressive. They live peacefully with grebes, ducks, terns, and other waterbirds, and on their breeding grounds they occasionally even care for their neighbor's young.

Bonaparte's Gull, *Larus philadelphia.* Princ. distrib. Texas: Uncommon to common WR on the coast and on larger inland reservoirs; locally common M throughout the state.

FEEDING BEHAVIOR. Like other gulls, they eat fish, crustaceans, marine worms, and other marine or aquatic animals, but they seldom visit landfills. Insects make up a large component of their diet, and it is typical for these gulls to dip down to the water for insects trapped on the surface.

VOICE. A strident, nasal, high-pitched, ternlike *cheeeer.* The most common sounds heard in Texas are weak, conversational whistles they make while feeding.

OTHER BEHAVIORS. These are not typical gulls. Their plump bodies suggest a pigeon, and their delicate, buoyant flight and their habit of pointing the bill downward in flight suggest a tern. They are equally as graceful when alighting and taking off from the water. Pough (1951) describes their flight eloquently: "With each bird taking a few deep strokes, then gliding with dangling legs to the surface to flutter for a moment in one place before bounding off again, a feeding flock has something of the appearance of a gathering of white butterflies."

Ring-billed Gull, *Larus delawarensis.* Princ. distrib. Texas: Common M throughout the state; common to abundant WR on the coast and on inland reservoirs, urban parks, landfills, and sewage treatment ponds.

FEEDING BEHAVIOR. Besides the usual foods consumed by most gulls (aquatic and marine invertebrates, fish, garbage, and insects caught on the wing, to mention a few), these birds also kill and eat rodents. Sometimes they misjudge their capabilities: there are records of Ring-billed Gulls choking to death on ground squirrels that they could only partially swallow.

Like many gulls, they follow boats for refuse as well as tractors for insects. They walk along the shore at wave's edge searching for any food item they can find or capture. They are not shy around people and readily avail themselves of scraps scattered around fast-food restaurants.

VOICE. Basically the same *kree kree kreee* as the Herring Gull's call, but higher in pitch. Birds sometimes vocalize when feeding, but otherwise they are usually quiet.

OTHER BEHAVIORS. In winter and migration they generally occur alone or in loose flocks. They are agile, graceful, and buoyant on the wing. In water they swim well (and frequently), and they take flight easily.

Herring Gull, *Larus argentatus.* Princ. distrib. Texas: Common M and WR on the coast, less common inland, in virtually all coastal habitats where food is available.

FEEDING BEHAVIOR. They forage in the same way and for essentially the same foods as the Ring-billed Gull, but they plunge more frequently into the water. Their large size equips them to prey on larger birds, fish, and mammals than most gulls are capable of subduing. They also consume garbage, carrion, and certain plants. They have been observed opening crabs, mollusks, and other hard-shelled items by carrying them high in the air and dropping them on rocks — possibly a learned behavior passed down culturally. Herring Gulls defend a feeding area on the beach or in the intertidal zone.

VOICE. A loud, harsh *kak kak kak kak* as well as a trumpeting *kyow.* Call notes are not identical among members of a colony. In an instance cited by the pioneering ethologist Nikolaas Tinbergen, birds in a colony ignored the alarm call of one individual, apparently because they learned that the bird frequently cried "Wolf."

OTHER BEHAVIORS. For many years Herring Gulls have been the subject of numerous behavioral studies, including pivotal studies of fixed action patterns (FAPs) and releasers (*see* Introduction). In one study, the red spot on the parent's bill was shown to be the releaser for the FAP in which a young gull pecks at the spot, causing the parent to regurgitate food that the young gull quickly devours. Because chicks do not peck when the parent's spot is painted over, but do peck at spots painted on pieces of cardboard, clearly the releaser for the FAP is the spot on the bill rather than another part of the parent's body.

These are gregarious birds. Large flocks circle together, calling excitedly and sometimes rising almost vertically into the air. At times they move inland a short distance to bathe and rest on freshwater ponds and lakes.

Glaucous Gull, *Larus hyperboreus.* Princ. distrib. Texas: Rare M and WR on the coast.

FEEDING BEHAVIOR. Glaucous Gulls are omnivorous and eat a wide range of food items, including berries, bird eggs, seaweed, carrion, and garbage. Because of their very large size they are able to capture on the wing prey that are too large for other gulls to overpower, including ducks and even other gulls. After alighting with their prey, they secure it with one foot and tear into the flesh or devour it whole, feathers and all. In general temperament and to some extent in appearance they are clearly the most raptorlike of our gulls.

VOICE. A harsh, nasal, and prolonged call, suggesting a Common Raven, as well as several more typical gull sounds.

Gull-billed Tern, *Sterna nilotica.* Princ. distrib. Texas: Locally common PR on the coast, in salt marshes and plowed fields. (An earlier name was "marsh tern.")

FEEDING BEHAVIOR. Gull-billed Terns resemble gulls not only in their large bill, broad wings, and stocky frame but also in their behavior. They only infrequently dive for fish in typical tern fashion; more often they dip down and take invertebrates and small fish from the water's surface, as might be expected for a marsh dweller. Insects, especially grasshoppers, are an important component of their diet. They are captured in flight or picked up from the ground.

Gull-billed Terns locate insects by walking in fields or following tractors. They also hover over grass and marsh fires to take advantage of fleeing insects and other animals (including small mammals).

COURTSHIP BEHAVIOR. The male performs a number of displays that include feeding the female and pointing his bill upward.

NESTING. Nest: a scrape or shallow depression in an open area on beaches and islands, usually in the vicinity of other terns. The parents add plant material and debris. Both parents incubate and care for the nestlings, and both aggressively dive at intruders, intensifying the assault by flying overhead and defecating on the person or animal.

An apparent increase of nests on gravel rooftops may be due to the destruction of suitable nesting sites in coastal areas (Smalley et al., 1991).

VOICE. A harsh, raspy *keywek keywek,* vaguely suggesting laughter; also, a distinctive, throaty *za za za.*

OTHER BEHAVIORS. Their flight is steady, powerful, and not especially buoyant, but they stoop and maneuver as precisely as a swallow and are capable of performing impressive aerial acrobatics. Gull-billed Terns walk more adeptly than most terns, thanks to their relatively long legs. They seldom swim or dive. Threat displays include an oblique posture and a head toss.

Caspian Tern, *Sterna caspia.* Princ. distrib. Texas: Common PR on the coast, in protected bays, rivers, and lagoons instead of the open sea, where Royal Terns so often occur.

FEEDING BEHAVIOR. Caspian and Royal Terns are often considered together because of their similar size and plumage. Both species eat insects and other invertebrates, as well as fish, bird eggs, and young birds; Caspians, however, often focus on locations where a single species of fish is concentrated and then feed on those fish to the exclusion of those nearby.

Their large size (they are the largest terns in the world) allows them to forage for large fish that they detect from relatively long distances away; thus, they enjoy the option of flying high over the water in search of prey.

Royal Terns fly low over the water in search of flying fish, or dip down to the surface for floating animals.

Both species plunge into the water after fish, and both momentarily hover before plunging. Caspian Terns also forage like gulls by picking animals from the water while swimming; Royal Terns seldom alight on the water, perhaps because they have weaker feet and are poor swimmers.

Like the larger gulls, both species readily steal from other birds, suggesting that any gull or tern is temperamentally suited to steal prey if it is able to.

COURTSHIP BEHAVIOR. The male flies over a colony of terns with a fish in his mouth. A female follows him, and after they alight, he feeds her the fish.

NESTING. Nest: a shallow depression on bare ground. Eggs: incubated by both parents. The young of both this species and the Royal Tern are dependent on their parents for an unusually long time (up to 8 months), often throughout winter. This prolonged adolescence may be related to the fact that neither species breeds until the third or fourth year. By remaining with their parents for a long time, young birds may be learning how to forage more efficiently.

VOICE. A deep, harsh *kraaa ka ka kaaa*.

OTHER BEHAVIORS. These shy and wary birds are the least sociable of our terns. They usually occur singly or in small groups, in contrast to Royal Terns, which sometimes form enormous flocks.

Caspian Terns fly with strong, gull-like wingbeats. Like gulls, they soar to great heights for no apparent reason.

Royal Tern, *Sterna maxima.* Princ. distrib. Texas: Common PR on the coast, less likely than the Caspian Tern to be seen inland, and more likely to occur on the open sea. Worldwide, this species tends toward warmer waters than the Caspian Tern.

FEEDING BEHAVIOR. *See* Caspian Tern, which shares many feeding behaviors with this species. Royal Terns sometimes feed at night.

COURTSHIP BEHAVIOR. Two or more birds spiral high in the air. On the ground, after both birds bow and strut in circles, the male feeds a fish or crab to his mate.

NESTING. In colonies on low islands and comparable habitats. Nest: a scrape on the bare ground. Eggs: incubated by both parents. Both parents feed the young, and for an exceptionally long time (6 months).

Soon after hatching, young terns leave the nest and join a creche of other young terns that usually includes young Sandwich Terns. When parents return to the creche to feed their young, parents and young identify each other by voice.

Creching, which also occurs in flamingos, penguins, ostriches, and other birds, frees both parents to seek food. Why this apparently efficient reproductive strategy occurs in so few of our species (in Texas, only this species and the Sandwich Tern) is unknown. *See* Caspian Tern.

VOICE. A high-pitched, shrill bleat, rendered as *kee eer;* also, a whistle like a plover.

OTHER BEHAVIORS. Royal Terns are highly gregarious; in some areas, up to 10,000 nest together in a dense colony. They associate with Laughing Gulls, evidently not recognizing that these gulls eat their eggs. On shore they often rest with Sandwich Terns, which were their creche mates during the first few weeks of life.

Terres (1980) describes Royal Terns as flying "with strong, unremitting flappings; looks like a slim gull in the air, much like flight of Common Tern but less buoyant."

Sandwich Tern, *Sterna sandvicensis.* Princ. distrib. Texas: Common SR on the coast, less common in winter; along beaches, mudflats, sandbars, and jetties. Like Royal Terns, they rarely visit inland bodies of fresh water, more often occurring far offshore.

FEEDING BEHAVIOR. They often hover at a considerable height before plunging into the water to disappear, then reappear with a fish. Fish are the principal items in their diet, but they also consume numerous marine invertebrates, as well as insects caught on the wing.

COURTSHIP BEHAVIOR. The pair chatter while spiraling high in the air, then glide downward. On the ground they assume a striking posture, with wings drooped and bills pointed up, and turn their heads from side to side. The male usually feeds a fish to the female.

NESTING. In colonies on sandy islands, lagoons, and beaches. Nest: a shallow scrape on the ground or sand prepared by both parents. Both parents incubate and care for the young, which often join a creche, especially if the nest is disturbed (*see* Royal Tern). Maturation is relatively lengthy, taking about 4 months.

VOICE. A loud, grating *kirr rit* or an abrupt *kwit kwit,* less harsh than most terns.

OTHER BEHAVIORS. Their flight ranks as one of the most swift, skillful, and graceful of our terns. The Sandwich Tern has been referred to as the Royal Tern's "sidekick" (Kaufman, 1996) because they almost always associate with the latter species, including when they are nesting. They roost on coastal sand banks with various gulls and terns.

Common Tern, *Sterna hirundo.* Princ. distrib. Texas: Common to uncommon fall M on the coast, less common in spring, on sandbars and beaches.

FEEDING BEHAVIOR. Common Terns fly over water searching for fish. After spotting one, they hover briefly, plunge headfirst into the water, and disappear under the surface before quickly emerging with their quarry. They also dip down to the water's surface for marine invertebrates, and they capture insects on the wing. They often focus their attention on a specific prey item (for example, shrimp or a particular kind of fish), and they readily steal food from other terns.

VOICE. A rasping, descending *kyee arr r r r r* that is harsh and prolonged; also, a series of high-pitched notes suggesting *kek kek kek kek kek,* often heard in hovering birds.

OTHER BEHAVIORS. Strong, deep wingbeats give their flight an extraordinarily graceful appearance. Their unerring sense of direction is legendary though perhaps exaggerated: sailors lost in the fog reportedly have regained their direction by observing the homeward flight of these birds.

Their aggressive dives (mainly on the nesting grounds) are coordinated so as to defecate on intruders at the very moment they pull out of the dive.

Forster's Tern, *Sterna forsteri.* Princ. distrib. Texas: Common PR on the coast and uncommon M throughout most of Texas, frequenting marshes, bays, lakes, and beaches.

FEEDING BEHAVIOR. They forage in the typical tern manner, hovering and then plunging for fish beneath the surface. They also fly low over the water, sometimes skimming it, before gently dipping down to pick up invertebrates from the surface. Over marshes they hawk for insects, in particular dragonflies, capturing most on the wing. They perform all flight movements gracefully and with extraordinary skill.

COURTSHIP BEHAVIOR. Apparently their courtship displays have not been reported.

NESTING. Most pairs nest in loose colonies on grassy islands in salt bays. Nest: compactly woven (unusual for terns) from reeds and grasses, and placed in dense vegetation or on the ground. Both parents incubate the eggs, care for the young, and vigorously defend the nest.

VOICE. A buzzy note recalling the Common Nighthawk's *zreeenp,* as well as a shrill *chick chick chick,* which has been compared to a person clucking a horse.

OTHER BEHAVIORS. "In flight, Forster's moves its wings with a quick, sharp snap instead of with the slower, deeper strokes of the [Common Tern]" (Pough, 1951). They rarely alight on water.

Least Tern, *Sterna antillarum.* Princ. distrib. Texas: Common SR and M on the coast, less common inland. Its numbers are increasing in some localities.

FEEDING BEHAVIOR. These, our smallest terns, fly low over the water looking for invertebrates and small fish. They hover (more regularly than most terns), then plunge after prey in typical tern fashion. They also dip with buoyant movements to pick up food items from the surface or to capture insects on the wing.

COURTSHIP BEHAVIOR. The male, flying upward with a fish in his bill, is relentlessly chased by several terns, including the female; then all the birds glide down. Courtship feeding is common. Rod Rylander (pers. comm.) observed the following behavior at Rockport: "The male, carrying a tiny shrimp in his beak, located himself beside the slightly crouched female. He kept chattering and swinging the shrimp back and forth for about 15 minutes. When he mounted her, she raised her bill and he gave her the shrimp."

NESTING. They nest as isolated pairs or in colonies, on open ground or, in some parts of the United States, gravel roofs. Ironically, their tolerance of human activity contributes to their demise in some regions, as nests on public beaches are frequently destroyed by domestic animals and human activities. Nest: a shallow scrape lined with grass, debris, and pebbles. Eggs: incubated by both parents. During hot weather, parents dip their breast feathers into water, fly to the nest, and shake water on the eggs. Both parents care for the nestlings. In defending their territories, they hover over intruders, then suddenly defecate on them.

VOICE. Often uttered in flight, a harsh, high-pitched *zree eek,* sounding to some listeners like jingling sleigh bells; also, a sharp *kip* and a variety of excited cries from birds hovering over a school of fish. Studies show that birds are able to recognize their mate's call.

OTHER BEHAVIORS. They fly gracefully and buoyantly, with wing-beats deep and quick — the quickest of all terns. Although not particularly sociable, Least Terns occasionally associate with Snowy Plovers and Wilson's Plovers.

Bridled Tern, *Sterna anaethetus.* Princ. distrib. Texas: Common visitor along the Continental Shelf, often after storms. They fly low over the water and take small fish and invertebrates by hovering, then dipping down to the surface. They often take advantage of fish that have been chased to the surface by predatory species, and they sometimes forage at night. They rarely dive.

Their flight is light and buoyant, and occasionally they alight on floating debris.

Sooty Tern, *Sterna fuscata.* Princ. distrib. Texas: Rare and local SR on the coast.

FEEDING BEHAVIOR. They usually hover before swooping down gracefully to pick up prey, often small fish that are jumping out of the water to avoid larger, predatory fish. At sea, fishermen have identified good fishing locations by observing Sooty Terns hovering over schools of tuna.

VOICE. A high-pitched *ker wacki whack,* somewhat nasal in quality.

OTHER BEHAVIORS. Their long wings equip them for extended journeys at sea. Their flight suggests a Common Tern, but their wingbeats are stronger and steadier. They sometimes soar or circle high in the air, often in association with frigatebirds. It is said that their plumage is so absorbent that should they alight on water, they would quickly become waterlogged and drown. Because they have rarely been observed perched, they are presumed to spend most of the day in flight.

Black Tern, *Chlidonias niger.* Princ. distrib. Texas: Common M throughout Texas. Along the coast they are observed in open habitats; inland, primarily on lakes and rivers (in particular those with sandbars). They seem fond of rice fields, perhaps because the fields resemble the marshy habitats farther north where they breed.

FEEDING BEHAVIOR. Loose flocks forage tirelessly over land as well as water. They seldom plunge into the water but instead dip just beneath the surface for fish and invertebrates. They also fly low over grassy or marshy areas looking for frogs, spiders, and other animals and sometimes follow tractors for insects that are flushed. When pursuing insects, they twist and turn erratically like a flycatcher or a Common Nighthawk.

VOICE. In flight, a sharp, metallic *keek keek keek.*

OTHER BEHAVIORS. Their flight is swift, graceful, and buoyant, with relatively slow wingbeats and frequent hovering. Compared with other terns, their movements appear lighter and more erratic because the downbeat is notably stronger than the recovery. Black Terns rest on land but seldom walk around.

Black Skimmer, *Rynchops niger.* Princ. distrib. Texas: Locally common PR on the coast, in lagoons, estuaries, and other sheltered areas that provide calm water.

FEEDING BEHAVIOR. Flying low over the water with strong, rapid wingbeats, these oddly proportioned birds forage alone or in synchronized groups. The longer lower mandible drags in the water to make a furrow, and when the touch-sensitive bill detects a fish, the bird snaps the bill shut to capture it. A mechanism in the neck vertebrae jerks the head backward and downward at the time of contact so as to halt the head briefly and provide a pause for the prey to be snatched up.

Possibly the furrow in the water attracts to the surface animals that the skimmers capture as they return along the same path. For no apparent reason, they skim the sand near their nest.

They feed at dawn, at dusk, or at night, when the water is calmer and fish are near the surface. Like cats (and unlike almost all other birds), their pupils (the dark center of the eye) are vertical slits instead of round openings. This type of pupil permits more complete closure for protection against the glare of the daytime sun.

On rare occasions skimmers scoop up fish and crustaceans while wading in shallow water.

COURTSHIP BEHAVIOR. Not well known. Sometimes two or more males chase a female in a zigzag path. Males offer potential mates a fish, stick, or leaf.

NESTING. In colonies on sandbars and beaches. Nest: little more than a shallow scrape. Their territories are no larger than a narrow space around the nest. Both parents incubate the eggs and feed regurgitant to the nestlings. Curiously, males spend more time incubating the eggs and brooding the nestlings, and females spend more time feeding and defending them.

The upper and lower mandibles are the same length in young birds; if the mandibles were unequal, the young would be unable to pick up the regurgitant deposited on the ground by the parents. Unlike gulls and terns, skimmers respond to intruders with well-developed distraction displays that include belly-flopping. Young birds avoid predators by lying flat on the sand or in a shallow depression they make in the sand.

VOICE. Nasal yaps or barks (*kak kak kak kak*), which are heard day and night. They also make softer sounds (*kow kow kow*).

OTHER BEHAVIORS. Their flight is strong and graceful, but their wingbeats are shallow, necessarily so since they fly just above the water's surface. They rarely swim, never dive, and when they walk on land they do so awkwardly. Before and after breeding, adults regularly lie flat with their chin and bill on the ground.

They are gregarious throughout the year and rest together on the sand with their bills all pointing the same direction.

PIGEONS AND DOVES: ORDER COLUMBIFORMES, FAMILY COLUMBIDAE

When considered worldwide, birds called "pigeons" are not technically distinguishable from birds called "doves." On the other hand, in North America, the former name usually applies to larger species with square or rounded tails and the latter to smaller species with pointed tails.

Most members of this family are plump and have a short neck and small head. Their feathers appear dry and powdery (perhaps because of a reduced or absent oil gland) and are loosely attached, possibly giving the birds a slight edge for escape when predators attack. Most species are pigeon-toed (the toes point inward), and most have a characteristic mincing gait. Almost all bob their head when they walk.

Both parents feed their young regurgitated pigeon milk ("crop milk" is a more appropriate term for this nutritious substance secreted by the crop). Other birds (e.g., penguins) also produce crop milk.

Because incubation of the eggs and maturation of the nestlings are often relatively rapid, many doves and pigeons have time to produce several broods during the nesting season, accounting for the large numbers of pigeons and doves observed toward the end of summer.

Pigeons and doves are used extensively in behavior experiments. In one experiment, when female Ringed Turtle-Doves (*Streptopelia risoria*) were surgically prevented from performing the nest *coo*, they did not produce mature eggs, even when male courtship was normal. Thus, the female's own cooing triggers specific endocrine changes in her own body that are required for ovulation (Cheng, 1992).

In another experiment, the innate nature of the courtship displays in Ringed Turtle-Doves was demonstrated by raising birds in isolation; yet the influence of learning was evidenced when these displays became less awkward with practice.

Rock Dove, *Columba livia.* Princ. distrib. Texas: Common to abundant throughout the state in almost all urban habitats, in particular around bridges, public buildings, and other large structures. In remote areas, small numbers inhabit canyons and rocky bluffs.

Almost all pigeons observed outside captivity are feral, that is, escaped or released domesticated pigeons that successfully reproduce in the wild.

Natural selection seems to be reconstituting in feral pigeons reasonable facsimiles of the original wild birds. For example, the feral pigeon's skeleton more closely resembles the ancestral skeleton (as represented in native European Rock Doves) than it does the skeleton of the domestic pigeon. If natural selection acts on anatomical structures in feral pigeons, it most likely acts on innate behavioral traits as well.

FEEDING BEHAVIOR. Rock Doves show a fondness for seeds, but they also consume many buds and leaves and occasionally insects and other small animals. Large numbers may congregate at preferred feeding sites.

COURTSHIP BEHAVIOR. With neck feathers ruffled and tail feathers spread, the male struts around the female, repeatedly bowing and uttering polysyllabic coos. Between bows he runs short distances with his head held high and his tail feathers spread and dragging the ground.

In the mutual billing display, probably derived from infant feeding behavior, the female puts her bill into her mate's mouth and the two birds rhythmically move their heads up and down. In another display, the pair gently peck at each other's head feathers (nibbling).

Females initiate the majority of copulations, and to prevent females that solicit copulations from misdirecting their efforts, these females are guarded more closely by their partners.

NESTING. Under eaves, bridges, and on ledges. Nest: The female constructs a flimsy nest of sticks, leaves, grasses, and other materials that her mate brings her. When choosing a nest site, the male or the female crouches and repeatedly nods at a location. Eggs: incubated by both parents. Both parents feed crop milk to their young.

VOICE. Call: variable, including *co ah roo, co roo coooo*, and *cock a war.*

OTHER BEHAVIORS. Rock Doves regularly reinforce the pair bond with the driving display, during which the two run in tandem. The female, who is in front, appears to be driven by the male. When its territory is threatened, a bird will bow, spread its tail, ruffle its plumage, and open its mouth.

Feeding or resting individuals are keenly aware of each other's movements. Individuals signal flight intention by crouching before taking flight, and the other birds ignore the departing bird; however, if this flight intention movement is not made, the flock quickly takes flight.

Flock dynamics also work at more subtle levels. Dominant birds consistently force themselves to the center of the flock, where they have more time to forage than their subordinates at the edge, who must periodically interrupt their feeding to look for predators.

If Rock Doves are experimentally frustrated while attempting to get food, they pace in a stereotyped manner reminiscent of penned cats. In conflict situations, they commonly exhibit displacement activities, such as preening or pecking indiscriminately at the ground.

They are capable fliers; they wheel and glide with considerable ease and have sufficient dexterity to alight on water for a moment, swim, drink, then fly back up.

The clapping sound made by birds that are flushed is especially common in sexually active birds. The duration of clapping prior to taking flight indicates approximately how long the bird intends to fly. Pigeons, which are among the fastest flying birds, have been clocked at 94 mph.

Red-billed Pigeon, *Columba flavirostris.* Princ. distrib. Texas: Rare SR and local WR, in s. Texas, in woodlands near water, as well as open areas if trees are nearby. For Red-Billed Pigeons to continue nesting in s. Texas, it will probably be necessary to maintain the existing riparian habitat and to restore the deteriorated forests (Brush, 1998).

FEEDING BEHAVIOR. They consume berries, acorns, and nuts, showing a weakness for mistletoe. Although most birds feed high in the crowns of trees, some individuals have learned (perhaps from watching Mourning Doves and White-winged Doves) to drop down into stubble fields to take advantage of waste grain.

COURTSHIP BEHAVIOR. Apparently unrecorded.

NESTING. Nest: a frail and shallow platform of twigs, grass, rootlets, and other plant materials, probably built by both parents and placed in the fork of a horizontal tree limb. Eggs: incubated by both parents. Both parents care for the nestlings, feeding them crop milk for 2–3 weeks. When threatened at the nest, incubating birds sometimes feign injury by dropping to the ground and fluttering about.

VOICE. Call: a loud and deep *wooooOOOO up cup a coo cup a coo cup a coo.* It is used to advertise the territory. Cooing notes resemble the domestic pigeon's.

OTHER BEHAVIORS. They spend most of the day in trees, seeking "the tallest timber and brush [they] can find" (Oberholser and Kincaid, 1974). Normally they come to ground only for water. Their flight is swift and strong; birds flying overhead suggest a flock of parrots.

Band-tailed Pigeon, *Columba fasciata.* Princ. distrib. Texas: Common SR in Guadalupe, Davis, and Chisos mountains, on mountain slopes and in heavily forested canyons.

FEEDING BEHAVIOR. These handsome birds cling to trees, sometimes hanging upside down, as they search for berries, seeds, young leaves, blossoms, and insects, including many grasshoppers. They are especially fond of acorns, which they consume in large quantities.

COURTSHIP BEHAVIOR. The male perches near his mate (or potential mate), inflates his neck, swings his lowered head slowly from left to right, and coos. While flying, he extends his neck, circles horizontally on spread wings, and produces an excitement call. Following copulation, the male stands stiffly and points his bill upward.

NESTING. In a tree or on the ground. Nest: a platform or shallow cup, with very little cross-weaving, constructed by the female from twigs brought by her mate.

Unlike many North American pigeons and doves, pairs nest in adjacent territories. Concentrations of food resources often influence nest site selection. They also show a high degree of philopatry (site fidelity): about 90 percent of birds banded in Colorado returned the following year to sites less than 35 miles from where they were captured.

Eggs: incubated by both parents, who make soft, guttural cooing sounds when they relieve each other at the nest. Both parents feed crop milk to the nestling for about 3 weeks.

VOICE. Call: a repeated *whoo-hoo* (the male's advertising call). The same (or very similar) sounds are part of his courtship display. The male's excitement cry, given during his courtship flight, is a chirping or wheezing sound.

OTHER BEHAVIORS. Flocks fly long distances to take advantage of concentrations of acorns and other food items. They do this even during the breeding season when most birds avoid all conspecifics except their mates and offspring. In autumn they descend to lower elevations to feed (altitudinal migration).

Disturbed or agitated birds preen behind the wings (displacement behavior). When escaping danger they fly away rapidly and make a clapping sound with their wings. Their strong, direct flight resembles the Rock Dove's.

Eurasian Collared-Dove, *Streptopelia decaocto.* Princ. distrib. Texas: Increasingly common PR locally, especially on the coast but also in most cities inland. They favor urban areas, farmland, and other open habitats with scattered trees, where they usually search for seeds, berries, and insects by walking on the ground. Males display by noisily flying up and gliding down with spread wings and tail, as well as by calling and bowing. Song: a persistent *coo COO cuh.*

White-winged Dove, *Zenaida asiatica.* Princ. distrib. Texas: Common PR in s. Texas and along the Rio Grande, along river bottoms and in mesquite woodlands. They are becoming increasingly numerous in urban areas and are currently expanding their range northward and eastward.

FEEDING BEHAVIOR. White-winged Doves eat berries, acorns, fruit, and waste grain, up to 11 percent of their body weight in dry food per day. They seem to prefer undigested corn and grains in the manure of penned livestock instead of the same food items in livestock feeding troughs. Leatherstem (*Jatropha dioica*) is an important food item in the Trans-Pecos.

They sometimes fly 25 miles or more for food and water. In the Sonoran Desert (and perhaps in the Chihuahuan Desert, as well) they rely heavily on cactus fruit for water. During the breeding season they have been observed eating snails, possibly as a source of salt and lime. This interesting behavior has not been documented in Texas birds.

COURTSHIP BEHAVIOR. Courtship begins early (February in Big Bend National Park). With his body tipped forward, the male rapidly spreads and closes his tail to reveal the prominent black and white tail pattern. In one of his aerial displays, he circles around the female with stiff wings.

NESTING. They breed as isolated pairs or in large colonies. Nest: a flimsy structure built by the female from sticks, twigs, and grass that her mate brings her. Sometimes old nests are used. Both parents incubate the eggs and feed crop milk to the nestlings for the first few days.

When disturbed at the nest, birds rarely perform the distraction display for which Mourning Doves are so well-known.

VOICE. Call: a flutelike *Who cooks for you?* advertises the territory but has other functions as well.

OTHER BEHAVIORS. White-winged Doves are quite gregarious; they feed in large flocks, especially after the breeding season. They often roost together in the very trees where they nest. Compared to Mourning Doves, they fly with fewer zigzags and less whistling of the wings. Like Rock Doves, they make a loud clapping sound with their wings when taking flight.

In south Texas, White-winged Doves and Inca Doves generally displace Mourning Doves and Common Ground-Doves, respectively, at least in urban areas.

Mourning Dove, *Zenaida macroura.* Princ. distrib. Texas: Common PR and locally abundant WR throughout the state in almost all terrestrial habitats. Many individuals have learned to avoid hunters by confining themselves to residential areas during the hunting season.

COURTSHIP BEHAVIOR. The male glides over the female in a spiral pattern; he also struts before her with spread feathers while nodding his head. The pair frequently preen each other.

NESTING. Usually in the fork of a horizontal limb, but sometimes on the ground or in man-made structures. The male leads his mate to several sites. She selects one of them, where she builds a flimsy platform of sticks. Eggs: Clutches of three or four eggs probably represent the addition of eggs by another female to the typical clutch of two. The eggs are incubated by the male most of the day and by the female the remainder of the day and throughout the night.

In their distraction display (injury feigning or broken-wing act), birds threatened at the nest suddenly drop to the ground and flutter off as if partially paralyzed. Although their own eggs are white, apparently egg color is not critical for Mourning Doves to accept an egg. They accept variously colored eggs placed in their nests, but they quickly reject eggs of any color (including their own white eggs) that are cracked or punctured.

For the first 3–4 days, both parents feed crop milk to the nestlings. Because crop milk is critical to fledgling survival, the number of eggs that female doves lay is probably limited by the amount of crop milk that parents can produce in one day.

Many pairs stay together through several successive breeding seasons.

VOICE. Call: a low, soft-pitched *ooh aH oo oo ooo,* that suggests a mourning sound; the call is important for maintaining the pair bond. (An earlier name for this species was "moaning dove.")

OTHER BEHAVIORS. When defending his territory, the male bows, lifts his head, and coos repeatedly. He also vigorously flies at intruding males.

Although solitary nesters, Mourning Doves congregate in large flocks outside the breeding season. Large flocks also form in midsummer. Flight is strong and swift (like a small hawk). When taking flight and also during flight, both sexes produce a noticeable whistling of the wings; when checking their flight, the wing feathers produce "a loud sound much like that of stout cloth tearing" (Oberholser and Kincaid, 1974). They lift their tail when they take flight (a flight intention movement) as well as when they alight.

Penned doves establish a rank order. Dominant males mate with dominant females, and these dominant pairs are the first birds in spring to pair, as well as the most successful in raising their young.

Inca Dove, *Columbina inca.* Princ. distrib. Texas: PR throughout much of the state, especially in cities and towns where rainfall is low to moderate.

FEEDING BEHAVIOR. Inca Doves bob their head as they walk around busily looking for seeds. Around human habitations they readily take advantage of such waste food items as dog food and table scraps.

COURTSHIP BEHAVIOR. Holding his tail vertically (to display his black and white markings), the male struts in front of the female and repeatedly coos and bows (Fig. 8). If a second male intrudes, both males crouch and freeze, then fight violently while making growling sounds. During pauses, the combatants peck on the ground (redirection) or preen above or under the wings (displacement behavior). Some pairs probably mate for life.

NESTING. In a variety of nest sites, including a ledge, a hanging basket, or the horizontal fork of a tree. Nest: The male brings twigs, grasses, and leaves to the female, who uses them to build a small platform that she lines with grass. Eggs: incubated by both parents. For the first few days, both parents feed the nestlings crop milk.

VOICE. Call: heard throughout the hottest summer afternoons, including August, when almost all birds are silent. It is used to advertise the territory. Many listeners describe the call as monotonous and melancholy (*no HOPE, no HOPE*); others hear in it a resigned *too HOT! too HOT! too HOT!* Birds also produce polysyllabic variants of this call, as well as desperate-sounding growls.

OTHER BEHAVIORS. Flight is rapid and jerky. The dry, buzzing flutter (produced by air passing through the outer flight feathers) probably signals a bird's presence to other birds.

Frequently, 50 or more birds flock together in winter. In cold weather a dozen or more perch on each others' backs to form a pyramid two or three tiers deep. They remain like this for about an hour during the day, evidently to maintain body warmth.

Common Ground-Dove, *Columbina passerina.* Princ. distrib. Texas: Primarily in s.e., s.w., and w. parts of the state, generally in open country, cultivated fields, and urban areas. Where Common Ground-Doves and Inca Doves both exist, the Common occupies cultivated areas and the Inca nearby brushy rangeland (Oberholser and Kincaid, 1974). Indeed, in s. Texas, Common Ground-Dove populations increase after pastures are cleared of trees and brush.

FEEDING BEHAVIOR. The diet of these tiny doves includes seeds, grain, and insects. They raise their tail as they forage, possibly a signal to maintain contact with other individuals.

FIGURE 8. Courtship fanning of the tail by a male Inca Dove (top); roosting posture (center); aggressive tail fanning (bottom). (R. F. Johnston, 1960, *Condor* 62: 7–24. With permission.)

COURTSHIP BEHAVIOR. The courting male puffs out his feathers, bows his head, and struts around the female while cooing softly. Even when courting his own mate, he may join other males to engage in bouts of chasing her. In the bow coo display, which precedes copulation, he lowers his body to an angle of about 30 degrees from the horizontal, flicks his wings, and utters a *browr* sound.

Prior to copulation, the female solicits food from her mate, who responds by feeding her regurgitant. Evidently some individuals pair for life.

NESTING. On the ground, in bushes, or in small trees. Nest: a flimsy platform of grass and sticks probably built by both parents. The eggs are incubated by both parents. It is assumed that both parents feed the nestlings crop milk during the first few days of life. Injury-feigning behavior has been reported in incubating birds.

VOICE. Call: a mournful *coo-oo* that rises at the end. It is repeated, with periods of rest, for hours at a time. Curiously, Common Ground-Doves call primarily during the breeding season, whereas Inca Doves vocalize throughout a large part of the year.

OTHER BEHAVIORS. Apparently very little ritualized behavior is necessary for maintaining the pair bond. During his territorial display, the male holds his body parallel to the ground, flicks his wings, and utters a *towah*. Unlike Inca Doves, Common Ground-Doves do not readily respond to artificial threats to their territories, such as tape recordings of their calls.

When walking, they thrust their head forward in the typical dove manner. These exceptionally tame birds usually do not fly away unless approached very closely. When taking flight they produce a slight whistling sound. They feed together in small groups throughout the year, the flocks increasing in size as winter approaches.

White-tipped Dove, *Leptotila verreauxi*. Princ. distrib. Texas: Common PR in riparian woodlands and adjacent cultivated areas of extreme s. Texas, usually inhabiting the tallest trees or thickest understory.

FEEDING BEHAVIOR. They forage on the ground in dense underbrush, less frequently seeking the low branches of hackberry, elm, mesquite, and other trees. They eat mainly seeds and berries.

COURTSHIP BEHAVIOR. The male hunches his shoulders, lowers his head, and runs toward the female, only to stop suddenly and coo. Birds also perform a bowing display.

NESTING. In dense brush or at the fork of an inclined branch. Nest: probably built by both parents, a flimsy structure made from sticks and weed stems. Eggs: presumably incubated by both parents. Probably both parents feed crop milk to the nestlings.

VOICE. Call (usually delivered from a concealed perch): a soft, deep *ooo whoooooooo*, suggesting the sound made by blowing across the mouth of a bottle.

OTHER BEHAVIORS. These ground-dwelling birds are quite wary and are less easily spotted than Mourning Doves or White-winged Doves. When alarmed or agitated they show displacement behaviors, especially exaggerated tail dipping and head nodding (probably derived from the nodding behavior in walking birds). Startled birds are disinclined to take flight immediately, attempting first to escape by walking briskly through the underbrush.

Flight is swift and (unlike White-winged Doves and Red-billed Pigeons) usually routed below the treetops. Their wings produce a whir-ring in flight.

PARAKEETS AND PARROTS: ORDER PSITTACIFORMES, FAMILY PSITTACIDAE

Few birds clamber about in trees as skillfully as parrots. They climb with their bill, hold food with their feet, and balance with their wings. The heavy bill allows them to crack hard seeds, but they also eat fruit and blossoms. Most nest in cavities, and many species mate for life, reinforcing the pair bond with mutual preening and partner feeding.

Most parrots are gregarious other than during the breeding season. They tend to be sedentary except for daily movements from roosting to feeding sites. Their social behavior is complex, with considerable ritualized play and fighting behavior. Many vocalizations are loud, metallic, and screeching. Some parrots have been taught to associate meanings with words, but they do not talk in the human sense.

A number of **Monk Parakeets,** *Myiopsitta monachus,* have escaped from captivity to become established in Texas, where they are common in some cities. Unlike most parrots, which nest in tree holes, they build bulky stick nests for sleeping and for raising their young. They are highly vocal, usually uttering a high-pitched chattering while feeding; in flight they shriek loudly and almost continuously.

The **Green Parakeet,** *Aratinga holochlora,* and **Red-crowned Parrot,** *Amazona viridigenalis,* are established breeders in south Texas. Many populations probably represent a mixture of naturally occurring birds and birds that escaped from captivity. Human disturbance of their natural habitat in northeastern Mexico may influence northward movements of Red-crowned Parrots (Neck, 1986). Both species fly in noisy flocks.

CUCKOOS AND ALLIES: ORDER CUCULIFORMES, FAMILY CUCULIDAE

This unusual family includes the European Common Cuckoo, *Cuculus canorus,* a species known for laying eggs in other birds' nests (brood parasitism), as well as for being represented in cuckoo clocks. Generally, our North American species are not brood parasites, although occasionally Black-billed Cuckoos and to a lesser extent Yellow-billed Cuckoos show a hint of this behavior. Furtive behavior, such as sneaking through foliage, is seen in Black-billed and Yellow-billed Cuckoos and points to the probable evolution of these species from brood parasites that had to approach the nests of potential foster parents surreptitiously.

Probably communal nesting of Groove-billed Anis also represents the general tendency for brood parasitism in this family.

Black-billed Cuckoo, *Coccyzus erythropthalmus.* Princ. distrib. Texas: Locally common M in e. and c. part of the state, where they inhabit thickets, forest edges, and other dense habitats, especially where there are young trees and tall shrubs.

FEEDING BEHAVIOR. These cuckoos appear more secretive than Yellow-billed Cuckoos. They hop silently among the branches of trees and shrubs, gleaning for caterpillars and other insects and occasionally taking food items as unrelated as snails, bird eggs, and fruits.

VOICE. Call: 2–5 low *cu*s that are soft and dovelike; also, a series of rapid *kuck*s.

OTHER BEHAVIORS. Their flight is graceful and rarely prolonged. Some observers suggest that they are more active at night than Yellow-billed Cuckoos.

Yellow-billed Cuckoo, *Coccyzus americanus.* Princ. distrib. Texas: Common M and SR in most parts of Texas, in woodlands, parks, and orchards, often near water; they prefer dense second growth to deep forests.

FEEDING BEHAVIOR. They move sluggishly through the foliage gleaning for insects, especially the hairy tent caterpillars that many birds avoid (325 reported from one bird's stomach). They temporarily invade areas following outbreaks of tent caterpillars or gypsy moths. They also eat other insects and occasionally take lizards, berries, and the eggs of other birds.

COURTSHIP BEHAVIOR. Courtship feeding is common: the male alights on the female and places food in her bill. Other courtship displays include bobbing the head, lowering the wings, spreading the tail, and swaying from side to side.

NESTING. Nest-building behavior is not well developed, perhaps because these birds belong to a family where brood parasitism is common. Nest: a flimsy platform in a tree or shrub that is carelessly constructed by both parents from twigs and grass. It is often not completed until after the first egg is laid.

Generally females lay more eggs in years when caterpillars and other insects are abundant. These are the same years that they more frequently deposit eggs in other birds' nests. Both parents incubate the eggs and feed regurgitated insects to the nestlings.

Because the eggs hatch asynchronously, the chicks clamber out of the nest on different days. In some cases the male tends to the first fledglings and the female to the later ones.

VOICE. Call: a series of hollow, guttural *kuk kuk kuk kuk*s, diminishing in pitch and tempo and ending in *cow cow cow*. Many people maintain that they call more frequently prior to rainstorms (hence the colloquial name "rain crow"); usually dismissed as folklore, the belief has been neither verified nor shown to be false. A less frequent, but still common, vocalization is a series of melancholy, dovelike, down-slurred *coo coo coo*s.

OTHER BEHAVIORS. They fly gracefully and directly, usually disappearing quietly into the foliage of another tree. Their eyes are situated laterally, allowing them to focus on objects either in front of or behind the head.

Greater Roadrunner, *Geococcyx californianus.* Princ. distrib. Texas: Uncommon to locally common PR throughout most of the state, especially in much of w. and s. Texas; they favor open regions with scattered brush.

FEEDING BEHAVIOR. Their legendary speed and agility make it easy for them to capture insects, scorpions, centipedes, and other invertebrates. They also take numerous vertebrates, especially lizards, mice, and birds and their young. Larger prey are usually beaten on the ground until limp, then swallowed headfirst. In one case, a Greater Roadrunner choked to death on a horned lizard (Holte and Houck, 2000). The skillful maneuvers they employ to kill rattlesnakes (apparently without ever being bitten) have been documented for Texas birds with extraordinary photographs

(Meinzer, 1993). Greater Roadrunners also eat fruits and berries from plants such as tasajillo. They forage singly or as mated pairs.

COURTSHIP BEHAVIOR. The male produces a loud, dovelike *kowoo kowoo kowoo,* which at a distance suggests a lost puppy. He usually does this early in the morning. With each deliberate, melancholy *kowoo,* he thrusts his head toward the ground as if he were retching. He also parades with his head held high and his wings and tail drooped while producing a *pop* sound with his wings.

The male often presents the female with a lizard or other food item, fol-lowing — instead of prior to — copulation; afterward he sometimes tries to retrieve and eat the item (Wyman Meinzer, pers. comm.). Interestingly, sometimes courtship feeding becomes less consistent after the first batch of young fledge. In California, by the second round of breeding, some males do not award a lizard or other food prize to the female after mating (J. Cornett, pers. comm.). Greater Roadrunners apparently mate for life, and pairs remain together throughout the year.

NESTING. In dense shrubs or cacti, usually near the ground. Nest: con-structed of sticks, grass, feathers, snake skins, dry cow manure, and other items. Both parents incubate the eggs, and both bring food to their young, even after the young can catch their own (at about 3 weeks of age). When threatened at the nest, parents may perform a distraction display. They "squat low on the ground and grovel with their beaks almost flat on the ground" (Wyman Meinzer, pers. comm.), or they feign a broken leg in-stead of a broken wing, the limb utilized by most birds that feign injury (J. Cornett, pers. comm.).

VOICE. *See* "Courtship behavior." Birds commonly produce a clatter-ing by repeatedly snapping their bill together. Females clack at a higher pitch than males.

OTHER BEHAVIORS. Greater Roadrunners run 15 mph or faster in advance of approaching vehicles, bursting into short flights only if ap-proached too closely. They raise their crest frequently, apparently when confronting situations that threaten or excite them.

Females and nonincubating males lower their body temperature at night to conserve energy. In early morning they sun themselves by perch-ing in the top of a tree or shrub, spreading their wings, and fluffing their feathers to expose the underlying black skin to the sun's warmth.

Groove-billed Ani, *Crotophaga sulcirostris.* Princ. distrib. Texas: Common SR in extreme s. Texas, uncommon to the n. and w., occasionally nesting elsewhere in scattered localities; they inhabit open areas, such as pastures, if there are dense thickets nearby.

FEEDING BEHAVIOR. They hop and run about in grassy areas, where they capture grasshoppers, beetles, and other large insects, frequently those stirred up by cattle. They also chase and capture lizards and other animals and climb into shrubs for berries. They have been observed picking external parasites from the backs of cattle.

COURTSHIP BEHAVIOR. Besides courtship feeding, birds engage in reverse mounting, which occurs about half as frequently as normal mounting. Its function is not known.

NESTING. Anis engage in cooperative breeding. In a low tree or shrub, one to four pairs build together a bulky structure of sticks and other plant materials, in which all the females in the group deposit their eggs. Additionally, all parents incubate the eggs (normally, the dominant male at night) and all care for the nestlings. Sometimes helpers (nonbreeding birds) assist in raising the young, recalling comparable behavior in birds like the Mexican Jay.

However effective this method of raising young may be, it is not without occasional glitches that point to this species' evolution from a brood parasite. Often the cooperating females attempt to throw each other's eggs out of the nest, just as Old World cuckoos destroy or eject eggs from the nests they parasitize.

VOICE. Call: a rapid, flickerlike *PLEE koh PLEE koh PLEE koh PLEE koh;* in flight, a variety of clucking sounds. They also produce numerous other sounds (chucks, quacks, whines, etc.), each apparently with its own function.

OTHER BEHAVIORS. Gregarious throughout the year, they rest together on limbs, sometimes spreading out their wings. At night 30 or more may roost together in trees.

Ecologically, anis occupy a position between our two arboreal cuckoos and the cursorial Greater Roadrunner. Their flight is weak: they launch from a bush, flap a few times, then sail a few feet before flapping again.

OWLS: ORDER STRIGIFORMES

Most owls feed nocturnally, locating prey both by sight and by sound. Their large eyes detect objects in dim light, but their ears — unequal in size and positioned asymmetrically — equip them to detect prey in total darkness. Also, the facial disk probably enhances hearing by directing sounds toward the ears.

Because owls cannot focus on close objects, they must back away from food items to bring them into focus before pouncing.

Owls fly silently, thanks to soft wing feathers that have a serrated leading edge. Noiseless flight probably helps owls hunt, not only by permitting them to approach animals silently but also by making it easier to hear sounds produced by prey.

Indigestible parts of animals are regurgitated as pellets of fur and bones, which accumulate beneath roosting areas.

Like many dull-colored crepuscular and nocturnal species (for example, whip-poor-wills), owls communicate more with vocal than with visual signals; accordingly, their vocalizations are varied and distinctive.

Most owls do not build a nest; they usually lay their eggs in a cavity or an old nest. Since the female in many species begins incubating after the first egg is laid, eggs hatch on consecutive days, and therefore the nestlings are of unequal ages.

BARN OWLS: FAMILY TYTONIDAE

Barn Owl, *Tyto alba.* Princ. distrib. Texas: Locally common PR in most of Texas, frequenting farms, sparsely wooded lowlands, and other habitats containing open areas for foraging and sites suitable for nesting.

FEEDING BEHAVIOR. Chiefly nocturnal, they fly low over open ground hunting for mice, rats, and other mammals but rarely for birds. They also hunt from perches. Barn Owls can be lured by making squeaking sounds that suggest prey.

COURTSHIP BEHAVIOR. During the male's display flight he claps his wings loudly. He also feeds the female during courtship. Pairs apparently mate for life.

NESTING. In caves, hollow trees, holes dug in banks, and other cavities (including nest boxes); the nest consists of little more than rearranged debris. The male brings food to his mate when she is incubating, and both parents feed the nestlings. Clutch size depends on prey availability.

Eggs hatch asynchronously, sometimes over an interval of about 14 days. This extreme hatching schedule results in a marked disparity in the sizes of the fledglings.

VOICE. Call: a lengthy, rasping screech; when flying overhead, a harsh, gasping *eee SEE eeek*. The flight call is unusual in that most owls rarely, if ever, vocalize in flight. When agitated, Barn Owls snap their bill loudly and rapidly.

OTHER BEHAVIORS. They greet intruders by weaving the body back and forth and nervously bobbing the head. They sleep soundly during the day and are not easily aroused by humans. As a rule, Barn Owls are not gregarious, even though, as an exception, as many as 50 birds have been reported roosting together in winter.

Their flight is slow and mothlike, except when they pounce on their quarry.

TYPICAL OWLS: FAMILY STRIGIDAE

Flammulated Owl, *Otus flammeolus.* Princ. distrib. Texas: Locally uncommon SR in mountains of Trans-Pecos Texas, in both deciduous and coniferous forests. They are probably more common than assumed, as their call is soft and low-pitched, and their retiring habits make them difficult to locate during the day.

FEEDING BEHAVIOR. They generally hunt at dawn and after dusk, flying out from a perch to capture moths, beetles, and crickets and taking other animals from the air or ground. They also hover briefly next to foliage and grab insects with their talons.

VOICE. Call: a single, soft, low-pitched *boot,* or a double *boo BOOT,* uttered by the male mainly on moonlit nights; his mate occasionally answers with a quavering note. They readily approach imitations of their call.

OTHER BEHAVIORS. They are capable of maneuvering skillfully among tree branches. During the day they conceal themselves in dense foliage close to the trunk of a tree. Apparently the rusty and gray morphs exhibit identical behaviors.

Western Screech-Owl, *Otus kennicottii.* Princ. distrib. Texas: Locally uncommon PR in w. part of the state. This species was considered a subspecies of the Eastern Screech-Owl until the 1980s, but their different calls evidently prevent interbreeding and thus establish them as separate species. The behavior of the Western Screech-Owl is essentially the same as the Eastern Screech-Owl's, except that a distraction display has been observed in the Western species but not the Eastern.

VOICE. Call: a soft, monotonous, and evenly pitched *hoo hoo hoo hoo hoo,* which increases in speed toward the end, the so-called bouncing-ball cadence.

Eastern Screech-Owl, *Otus asio.* Princ. distrib. Texas: Common SR throughout Texas except n.w. and far w. parts, in virtually any wooded habitat, including urban areas (where they are frequently common). The presence of suitable nesting holes seems critical to their distribution.

FEEDING BEHAVIOR. From a perch they visually locate insects, mammals, and other animals; they then swoop down to the ground or to a shrub to capture their prey, sometimes chasing it by running or hopping on the ground. These birds also locate prey by sound. Some individuals have been observed catching small fish.

COURTSHIP BEHAVIOR. The courting male bows, raises his wings, clicks his bill, blinks, then approaches the female. He also engages in courtship feeding, but not without prior ceremony. After considerable hopping and bowing, he offers her food by laying it in front of her. Often the two preen each other's plumage, as well as call in duet.

NESTING. In abandoned woodpecker holes and other natural cavities, but also nest boxes. The male brings food to his incubating mate. Both parents feed the nestlings, and both protect the nest by diving at intruders, including humans.

Interestingly, in central Texas, parents bring blind snakes (*Leptotyphlops*) to the nest for reasons other than to feed the nestlings. These small, wormlike snakes are released in the nest's debris, where they burrow out of sight and eat insects and mites. Owlets in nests with these snakes grow faster—and have a lower mortality—than those raised without snakes.

Also, certain ants in the nest maintain a mutually beneficial relationship with these owls by feeding on debris in the nest. They help the owls by swarming at intruders (but they do not bother the owls).

VOICE. Call: a mournful, tremulous whinny that descends in pitch. This blood-curdling call contrasts sharply with the gentle hooting of many owls. Eastern Screech-Owls also utter a long, unbroken trill. They readily approach whistled imitations of their call, as well as squeaking sounds.

Birds sometimes duet with a bounce song that differs slightly between males and females and from one individual to another. This song advertises the presence of a mated pair, identifies a mate, and signals the sex of a prospective mate. When agitated, birds indignantly snap their bill.

OTHER BEHAVIORS. They fly through the lower tree canopy with uniformly rapid wingbeats. Although screech owls have daylight vision, they are among the most strictly nocturnal of our owls, spending almost all day secluded in dense vegetation or in tree holes.

When approached during the day, they freeze by stretching their body high, pressing their wings tightly against their body, and appearing like a vertical stump on a tree limb. They can be lifted from this position without resistance, and formerly many were captured in this posture and kept as pets.

Great Horned Owl, *Bubo virginianus.* Princ. distrib. Texas: Common PR in most parts of Texas; they are less common in the e. Texas forests. They occur in almost all habitats, although they favor areas with trees.

FEEDING BEHAVIOR. These remarkable predators are capable of capturing prey as large as Canada Geese, domestic turkeys, Red-tailed Hawks, jackrabbits, skunks, and porcupines and animals as swift and evasive as bats. Their plumage and nests frequently reek of dead skunks. One reason they are so successful as a species is that they are ecological generalists and can adapt to different habitats and diets, probably more so than any of our other owls.

They spot their quarry from a high perch, then swoop down to grasp it with their powerful talons. Their keen hearing and eyesight allow them to hunt throughout the night as well as at dusk. In lean times they fly at midday to pastures to hunt for insects and small mammals.

They carry their prey to a feeding roost, such as a stump or old nest, and dismember the animal before eating it. In winter they cache uneaten prey and later thaw the frozen carcass by sitting on top of it.

COURTSHIP BEHAVIOR. They do not breed until 2 years of age. Breeding begins in late December and is initiated with courtship feeding and courtship flights. The pair bond is periodically reinforced with bobbing, calling, and clicking sounds.

NESTING. Nest: on a ledge or on the ground, usually with very little if any new material added. The eggs hatch asynchronously and are incubated by the female. Both parents bring food to the nestlings, but sometimes leave so many carcass remains in the nest that the nestlings die from the unsanitary conditions.

Great Horned Owls readily and effectively defend the nest by diving at intruders, often injuring them with their sharp claws.

VOICE. Call: a series of five low, soft, and highly resonant notes, generally following the cadence *hoo hoo HOO hoo hoo,* uttered as the bird leans forward, lifts its tail, and vibrates its white throat feathers. They respond to imitations of their calls.

OTHER BEHAVIORS. During the day Great Horned Owls roost high in trees, perching next to the trunk; when settled like this, they are usually difficult to disturb. They fly silently and often glide for considerable distances. They are frequently mobbed by American Crows, jays, and other birds.

Ferruginous Pygmy-Owl, *Glaucidium brasilianum.* Princ. distrib. Texas: Locally common PR in extreme s. Texas, in live oak, Texas ebony (*Pithecellobium*), and mesquite habitats.

FEEDING BEHAVIOR. They usually forage at dawn and dusk, darting out silently from an elevated perch to capture insects and other invertebrates, as well as small vertebrates.

VOICE. Call: deliberate, monotonous, and evenly spaced whistles (*puks*), sometimes more than 60 in an uninterrupted series. With each note the bird throws its head back and jerks its tail.

OTHER BEHAVIORS. These are remarkably bold and aggressive birds, considering their small size. When perched, they repeatedly cock their long tail and momentarily hold it in that position.

They generally roost in caves on hillsides, but their roosting sites in Texas have not been reported.

Elf Owl, *Micrathene whitneyi.* Princ. distrib. Texas: Locally common SR in the Trans-Pecos and extreme s. Texas, in desert scrub, mesquite woodlands, wooded canyons, and hillsides. They seem to be benefiting from the revegetation of abandoned agricultural fields in s. Texas (Gamel, 1997).

FEEDING BEHAVIOR. They forage primarily at dusk and at dawn, swooping down from a perch, hovering briefly, then capturing insects and other invertebrates (rarely small vertebrates) with their talons. They also catch insects on the wing, in the manner of flycatchers; or they may hover next to shrubs and trees and flush insects from the foliage. They occasionally are seen around campfires hawking for moths. They are said to remove the stinging parts of scorpions before swallowing them.

VOICE. Extraordinarily loud, considering that this is the smallest owl in the world. Call: an excited *whi whi whi whi whi whi whi,* as well as yips and barks that suggest a puppy. Complex vocalizations probably represent a pair calling in duet.

OTHER BEHAVIORS. During the day they roost inside woodpecker holes.

Burrowing Owl, *Athene cunicularia.* Princ. distrib. Texas: Uncommon to common PR in w. part of Texas, and rare WR in s.e. part, frequenting open prairies with very short grass and patches of bare ground; they occur regularly in prairie-dog towns.

FEEDING BEHAVIOR. Except during the breeding season, when they hunt during the day to meet the nutritional demands of their young, Burrowing Owls are generally active just after sunset and before sunrise. They dart out from a bush or fly low over the ground and capture prey by grasping it with their talons.

Their diet consists of insects and other invertebrates, as well as small vertebrates such as mice, ground squirrels, and lizards, typically captured by a rapid swoop from a low perch. They also run along the ground searching for prey, follow dogs or horses for stirred-up insects, and occasionally hawk for flying insects. They hover as high as 20 feet above ground in what appears to be another method of foraging.

COURTSHIP BEHAVIOR. In spring males arrive unpaired at the burrows they occupied the previous season (Martin, 1973). The male and female stand on the rim, display with their necks and bills, stretch their wings and legs, and call softly. Later they engage in ritualistic feeding. Since females do not show a strong bond to the previous year's burrow, it is doubtful that birds remain paired for life.

Although females are heavier (perhaps due in part to egg weight) and males larger for all linear dimensions, size was shown to be unrelated to mate selection in west Texas birds (Plumpton and Lutz, 1994).

NESTING. Normally in loose colonies. Nest: at the end of a burrow 10 feet or longer that is dug by prairie dogs, ground squirrels, or other animals. Burrowing Owls are physically capable of digging their own burrows in most substrates, and they modify burrows extensively. Often they line the entrance and nest chamber with cow manure. The male brings food to the female, who remains inside the burrow to incubate the eggs. Both parents feed the fledglings. Sometimes parents move the young to a new burrow a few weeks after hatching, as burrows tend to become highly infested with fleas. The family unit does not break up until fall.

It is generally regarded as folklore that Burrowing Owls, rattlesnakes, and prairie dogs inhabit the same burrow, perhaps because common sense would rule out such a cooperative living arrangement. Moreover, when disturbed in their burrows, young owls mimic a rattlesnake's rattle almost perfectly, and this behavior could explain the origin of the myth. On the other hand, it is doubtful that many prairie-dog burrows have been examined carefully enough to exclude this possibility.

VOICE. Call: a staccato, chattering *kak kak kak kak,* when alarmed, and sometimes given in flight (flight calls are unusual for owls); in the evening, especially, they produce a series of mournful notes that are almost cuckoolike in quality.

OTHER BEHAVIORS. They are active during the day, when they can be seen perched on fence posts, bobbing up and down. When they stand at the entrance of their burrow (apparently a favorite loafing site) they frequently nod, bow deeply, or turn their head almost completely around to scan the sky and surrounding area. They also bow before they leave an area and after they land.

The male responds to intruders by bending forward and bobbing, holding his wings back, and displaying the white patches of the throat and brow. This display can be evoked by a person who makes a clicking sound while mimicking the display with his arms and legs (Sam Braut, pers. comm.).

Flight intention appears to be signaled by a back-turning movement. Flight is labored and usually only for short distances. In the dim light of early evening a Burrowing Owl, with its erratic flight path and intermittent wing fluttering, suggests an enormous bat.

Spotted Owl, *Strix occidentalis.* Princ. distrib. Texas: Rare PR in Guadalupe and Davis mountains, in wooded mountains and canyons, especially in tall trees adjacent to cliffs.

FEEDING BEHAVIOR. Spotted Owls (the western counterpart of the Barred Owl) hunt in forested areas for small mammals, birds, reptiles, and large insects. They swoop down from a perch and grasp the animal with their talons, and they take prey from trees. They have been seen snatching bats out of the air. Except during the nesting season they generally hunt only at night. They regularly cache food for later consumption.

VOICE. Usually heard at dawn and dusk in late winter and early spring. Call: a deep, hooting *whoo whoo hoo hoo.* Like the White-winged Dove's call, it has been rendered as *Who cooks for you?* The call carries a mile or more with the "full-throated explosive effect of a baying hound" (Terres, 1980).

These owls also utter a variety of barking sounds, as well as a musical *weeeeeeee*. They respond readily to squeaking, imitations of their calls, human voices, and other noises, and therefore are generally easy to detect.

OTHER BEHAVIORS. Although Spotted Owls remain well concealed on their daytime roosts, once located, these gentle birds can be closely approached. (On the other hand, they aggressively attack human intruders who attempt to molest their young.) They roost in shaded rocky gorges and wooded canyons, in particular on the north sides of bluffs and slopes where temperatures tend to be lower. Their thick plumage makes them intolerant of even warm temperatures.

Spotted Owls are agile in flight and fly with quick wingbeats that alternate with glides.

Barred Owl, *Strix varia.* Princ. distrib. Texas: Uncommon to common PR in e. two-thirds of the state, in wooded river bottoms, swamps, and dense forests that have few or no clearings.

FEEDING BEHAVIOR. They usually hunt at night, but because their eyesight is effective in bright light, they occasionally forage during the day as well. They course low through the forest, with slow, quiet wingbeats (often gliding), or they fly out from a perch, occasionally hovering before dropping down to grab prey with their strong talons. They take mammals as large as foxes, opossums, and rabbits, numerous other smaller vertebrates, and occasionally large insects and fish (caught as the owl wades in water). Often they carry their prey to a feeding perch near the nest.

COURTSHIP BEHAVIOR. The pair perch side by side, bob and bow with half-spread wings, and call loudly, often in duet. Also, the male engages in courtship feeding.

NESTING. Eggs: laid in a natural cavity or an old hawk nest (often used in alternate years by hawks or other owls). The male brings food to the incubating female and to the fledglings.

VOICE. Call: usually eight accented hoots, rendered (like the White-winged Dove and Spotted Owl) as *Who cooks for you? Who cooks for you?*, with the last syllable descending and sounding more like *ooo ow*. Sometimes they call during the day. Their repertoire also includes clucks, grunts, and laughing sounds. When agitated, they snap their bill loudly.

OTHER BEHAVIORS. They are less aggressive, slightly smaller, and have less powerful talons than Great Horned Owls and are probably driven from favored habitats by that species. Also, they are less easily disturbed from their daytime roosting sites, typically a limb high in a tree. From this vantage point they are more likely to respond to intruders by looking down and blinking than by dashing away anxiously like Great Horned Owls. Generally, Barred Owls are curious birds that are readily lured by imitations of their call or by squeaking sounds. They respond to such sounds as far away as 50 yards.

Long-eared Owl, *Asio otus*. Princ. distrib. Texas: Locally uncommon M and WR throughout most of the state, except in the heavily forested parts of e. Texas, in woodlands and groves in prairies.

FEEDING BEHAVIOR. At night, Long-eared Owls fly back and forth low over fields or in forest clearings, alternating deep wingbeats with long glides. They locate prey (primarily small mammals but also other small vertebrates) both by sight and sound. They sometimes hover before dropping down and grabbing the animal with their talons.

VOICE. They are most vocal during the nesting season and therefore are rarely heard in Texas. Call: a series of soft, mellow coos. In flight they produce a sound by slapping their wings together below the body.

OTHER BEHAVIORS. These owls are often overlooked because they are strictly nocturnal and generally are quiet. When approached at their daytime roosts they raise their ear tufts, compress the feathers of their already slim body (they are the slimmest of all North American owls), and freeze in this posture, making them resemble an upright stump on a tree limb.

In winter, small numbers roost together in trees, some groups probably representing family units. In flight they are very agile, twisting and turning as they make their way through dense vegetation.

Short-eared Owl, *Asio flammeus*. Princ. distrib. Texas: Locally uncommon M and WR in most of Texas (except the extreme w. parts), in marshes, prairies, dunes, meadows, and other open habitats.

FEEDING BEHAVIOR. They forage during the day as well as at night, but they are especially active at dawn and dusk. They course back and forth (quarter) low over the ground, using both sight and sound to locate prey.

Their quartering behavior is more typical of a hawk than an owl; it is somewhat similar to the foraging strategy of Northern Harriers, which hunt in the same habitat. After briefly hovering over their quarry (which they can do with or without wind), they lift their wings high and drop down for the animal. They seem more adept at capturing small mammals, the principal item in their diet, than other prey, but they also consume a few birds as well as many insects.

In a second foraging method, they sit on a fence post or other perch and wait for prey.

VOICE. Silent except during the nesting season; they sometimes defend their winter territory with a loud *eee yerp*.

OTHER BEHAVIORS. Their buoyant, floppy, and mothlike flight is easily recognized. It is stronger than that of most owls. Short-eared Owls sometimes soar like buteos. They rest and sleep on the ground or on a fence post and sometimes roost communally in trees. Their courtship display—a sky dance made up of spirals and stoops—can sometimes be observed in Texas in late winter.

GOATSUCKERS AND ALLIES: ORDER CAPRIMULGIFORMES, FAMILY CAPRIMULGIDAE

Granted that our caprimulgids are boringly similar in coloration and general shape, they differ considerably in behavior. At one extreme are the two nighthawks, which constantly fly in search of insects and also perform their courtship displays in flight. At the other extreme are the much less volant Common Pauraques, which flutter up from the ground to capture low-flying insects and engage in terrestrial courtship displays. Pauraques are thought to represent the primitive caprimulgid condition.

Being obligate insectivores, goatsuckers invite comparison with the tyrant flycatchers. Both groups have bristles at the corners of the mouth and a wide gape (much wider in caprimulgids) that assist them in capturing insects in flight. Flycatchers capture, kill, and sometimes tear apart prey with their large, strong bills, whereas caprimulgids have small, weak bills and must swallow prey whole.

One consequence of having a small bill and a large gape is that parents cannot force feed their young, making it necessary for fledgling goatsuckers to reach into the parent's mouth for regurgitated food. Flycatchers use their large, sharply pointed bills to thrust entire insects into the mouths of their fledglings.

Lesser Nighthawk, *Chordeiles acutipennis.* Princ. distrib. Texas: Common M and SR in w. Texas, in desert scrub, arroyos, and arid grasslands. *See* Common Nighthawk.

FEEDING BEHAVIOR. Lesser Nighthawks are most active at dusk. They usually fly low over open areas but sometimes flutter up from a sitting position to catch insects passing by. Although active to some extent during the day, they seem less diurnal than Common Nighthawks. In urban areas they take advantage of ant swarms as well as insects that are attracted to bright lights.

COURTSHIP BEHAVIOR. The male follows the female, flying with stiff wingbeats and puffing up his white throat as he produces trills and whinnies. He does not dive like the Common Nighthawk.

NESTING. Eggs: deposited on bare ground or gravel, or, in south Texas, on the flat roofs of adobe houses. The female incubates the eggs, and both parents feed regurgitated insects to the nestlings. When the nest is threatened, parents feign injury.

VOICE. Call: a sustained and eerie toadlike trill (an earlier name was "trilling nighthawk"), that suggests a screech owl's whinny; also, a low *chuck chuck*.

OTHER BEHAVIORS. Lesser Nighthawks are minimally aggressive; they do not defend a feeding territory but instead cover large feeding areas. They sometimes migrate in small, loose flocks.

When temperatures are high, they dissipate heat by opening the mouth and fluttering the throat. When temperatures are low, or when food is scarce, they enter torpidity by lowering their body temperature.

Common Nighthawk, *Chordeiles minor.* Princ. distrib. Texas: Uncommon to common M in most of Texas; common SR throughout the state except in e. Texas forests; they occur in most open habitats, including urban areas. Where the ranges of the Common and Lesser Nighthawks overlap, the Lesser favors lower and drier habitats and the Common tends to inhabit the higher regions. The extent to which such habitat partitioning occurs in Texas birds needs to be documented.

FEEDING BEHAVIOR. They feed to a limited extent both during the day and at night, but they are mainly crepuscular (dawn and dusk) foragers. For this reason, urban areas have turned out to be ideal foraging sites because streetlights attract insects and extend the crepuscular feeding period throughout the night.

They fly in an erratic, batlike manner, capturing insects with their large, gaping mouth. Unlike Lesser Nighthawks, they typically fly relatively high instead of skimming the ground, and they rarely take insects while on the ground. They drink in flight by skimming over the water's surface.

COURTSHIP BEHAVIOR. The male ascends, with stiff and choppy wingbeats, to a high altitude, sometimes almost out of sight. Here he circles and calls repeatedly before making a spectacular plunge toward the earth. As he brings himself out of the dive he spreads his wing feathers. The rush of air through these feathers produces a loud booming sound that can be heard about a mile away. He lands near the female, spreads his tail, puffs up his throat to expose the white patch, rocks back and forth, and calls. He frequently performs this display later in the breeding season, but with shorter dives.

NESTING. Eggs: laid on bare ground, on top of a stump, or on the gravel roof of a building. The eggs are incubated by the female, who shifts her position throughout the day so that the sun is always to her back. Both parents feed regurgitated insects to the fledglings and perform a distraction display when the young are threatened.

VOICE. Call: a nasal, insectlike *peent*, uttered repeatedly in flight.

OTHER BEHAVIORS. They seem to be regularly excluded by Lesser Nighthawks from desert feeding areas. During the day they rest on fence posts and on the ground. Groups of up to 1,000 individuals migrate together, flapping and soaring at high altitudes instead of flying in their typical erratic manner.

Unlike Lesser Nighthawks, Common Nighthawks apparently do not enter torpor as a means of conserving energy (Firman et al., 1993).

Common Pauraque, *Nyctidromus albicollis.* Princ. distrib. Texas: Common PR in s. Texas, in woodlands, dense thickets, and brushy areas with openings.

FEEDING BEHAVIOR. Apparently they are most active at dawn, dusk, and on moonlit nights, flying out for insects from a branch or from the ground and occasionally flying continuously. They also pursue insects by swiftly chasing them on foot, thanks to their relatively long (but still short) legs. They also fly low along the road ahead of moving automobiles, feeding on insects visible in the beam of the headlights.

COURTSHIP BEHAVIOR. In one courtship display, two birds face each other while on the ground, rock their bodies up and down, bob their heads, and flutter up a few feet to expose the white in the wings and tail.

NESTING. Eggs: laid on bare ground, often at the base of a bush. If the nest is disturbed after hatching, the parents call and the young quickly hop to them. Both parents incubate the eggs and feed regurgitated insects to the nestlings.

VOICE. Call: a hoarse, urgent scream, *pur wee eeeeer,* only remotely sounding like "pauraque." It probably functions to advertise the territory and to attract a mate.

OTHER BEHAVIORS. At night they rest on roads; when flushed during the day they fly away on a zigzag course through trees instead of over them, rarely flying above 10 feet from the ground.

Common Poorwill, *Phalaenoptilus nuttallii.* Princ. distrib. Texas: Common M and SR in much of s. and w. Texas, in the hilly parts of arid and semiarid zones, as well as rocky outcrops on prairies.

FEEDING BEHAVIOR. They normally feed before dawn and after dusk, flying up from their ground perches to capture passing insects. Less frequently they hawk insects close to the ground, making prolonged, erratic flights with many twists and turns. They also capture prey on the ground. It has been suggested that their tendency to capture beetles reflects a need for polyunsaturated fat, which is important during torpor. Undigested parts of insects are regurgitated as pellets.

Interestingly, on moonlit nights their level of foraging activity seems to depend more on moon height than brightness.

COURTSHIP BEHAVIOR. Apparently the male advertises his territory with his call.

NESTING. Eggs: deposited on the bare ground or on leaves, often in the shade. Both parents feed regurgitated insects to their young. When threatened at the nest, the parent opens its mouth widely, tumbles, and hisses like a snake. If the nest is disturbed, they sometimes move both eggs and fledglings to a new site. Older fledglings escape by hopping and somersaulting across the ground.

VOICE. Call (usually produced by the male): a melancholy, cadenced *poor WILL low,* while the bird is sitting on the ground or a low perch; at a distance only the first two syllables (poor-will) are audible. Sometimes the female answers. At close range, a soft clucking note can be heard from flying birds. Observations in west Texas reveal that call frequency is influenced by weather, habitat, season, temperature, and moonlight (Freemyer, 1993).

OTHER BEHAVIORS. In winter they hibernate: their body temperature, heart rate, and rate of breathing decrease. At other times, they enter into a less intense state of torpidity, especially when food resources are low. Their flight is erratic and mothlike, with several flaps alternating with glides.

Chuck-will's-widow, *Caprimulgus carolinensis.* Princ. distrib. Texas: Common M and SR in e. and c. parts of the state, in river bottoms, woodlands, and farmland.

FEEDING BEHAVIOR. At night they fly out from a perch, which may be as high as a limb in a tall tree or as low as the ground. They also fly continuously along forest edges, making erratic sweeps close to the ground. They pursue large insects, but being the largest caprimulgid in North America, with an enormous, 2-inch gape, they occasionally capture (and swallow whole) birds as large as sparrows.

COURTSHIP BEHAVIOR. During the day the male droops his wings, spreads his tail, puffs up his body plumage, and struts up to the female. He also calls while making jerky movements with his body.

NESTING. Eggs: laid on the ground and probably incubated only by the female. Evidently only she feeds the young (regurgitated insects). If the nest is disturbed, she picks up the fledglings in her bill and transports them to another site.

VOICE. Call: a forceful and well-enunciated *chuck WILL'S wi DOW,* the first syllable often inaudible at a distance. (In Texas, the call is commonly but erroneously attributed to the Whip-poor-will.) Males sing from song posts, which may be fence posts, tree limbs, or other structures above the ground. When disturbed during the day or when hunting, they produce low growls or chucks. The number of calls produced per minute increases with ambient temperature.

OTHER BEHAVIORS. They spend most of the day sitting on the ground or balanced on a horizontal tree limb, their cryptic coloration blending in well with the substrate. When flushed, they fly away flapping and gliding, with mothlike wingbeats, and often return to a place near the one they just left.

Whip-poor-will, *Caprimulgus vociferus.* Princ. distrib. Texas: Common M in e. half of Texas, in deciduous, coniferous, or mixed woodlands; common SR in mountains of Trans-Pecos Texas, in pine-oak forests.

FEEDING BEHAVIOR. They forage at night, but more often at dusk and dawn, darting out from a tree limb or flying continuously near the ground at the edge of woodlands. Occasionally they flutter up from the ground to capture passing insects with their capacious mouth. They often twist and turn considerably while pursuing prey. Curiously, Whip-poor-wills also forage on the ground by rummaging among leaves and sticks — no doubt a laborious process since they have such small feet.

COURTSHIP BEHAVIOR. The female responds to the male's call by alighting near him. He takes flight, and as he circles around her, he undulates his body, bobs his head, and makes a purring sound. She responds by chuckling and strutting with her wings and tail spread and her neck held low.

In another display, he waddles up to the female, who is holding her head low. He trembles and grunts. The two birds touch bills and remain close to each other. The male also hovers in front of her and flashes his conspicuous white tail patch.

NESTING. Eggs: laid on the ground on mountain slopes or pine-oak-juniper canyons, usually in thickets or other shaded areas. Apparently only the female incubates the eggs, although both parents probably care for the young, feeding them regurgitated insects. Both perform a distraction display when the nest or young are threatened. If the female lays a second set of eggs, the male cares for the first nestlings.

In some parts of their range, nesting activity is apparently synchronized with the lunar cycle, with hatching occurring so that feeding demands by the nestlings are greatest when the moon is more than half full. At this time the parents have more time at night to forage.

VOICE. Call: a loud, rapidly reiterated *whip poor WILL*. A listener once counted more than 1,000 uninterrupted *whip poor wills* from a single bird. They sing from singing posts, which may be branches or man-made structures, or from the ground.

The call of the western subspecies (the one that nests in Texas) is faster paced than that of its eastern counterpart.

OTHER BEHAVIORS. When resting, they sit on the ground or perch lengthwise on a limb, where they are usually well concealed. They are relatively tame and can be quietly approached. In normal flight they do not show the Common Nighthawk's jerky movements; instead, they make easy sweeps, often sailing with their rounded wings held out straight. Sometimes they turn so sharply that their wings are positioned almost vertically.

Whip-poor-wills show higher levels of vocal, locomotor, and nest activity at dawn and dusk and on bright moonlit nights. They seem especially fond of taking dust baths on dirt roads.

SWIFTS AND HUMMINGBIRDS: ORDER APODIFORMES

SWIFTS: FAMILY APODIDAE

Swifts are aptly named: their streamlined bodies and long, narrow wings enable some species to fly at speeds well over 100 mph. They spend much of their day engaged in aerial activities — capturing insects in flight and dipping down to drink from the surface of ponds. They even copulate in the air. They pursue specific insects that they locate while flying, or they simply fly through swarms of insects with their mouth open.

Chimney Swift, *Chaetura pelagica.* Princ. distrib. Texas: Common M and SR throughout Texas (except the extreme s. and w. parts), in virtually all habitats where nesting sites are available, which makes them particularly common in urban areas.

FEEDING BEHAVIOR. They fly an erratic course over cities, farms, and other locations that provide a good supply of insects, which they capture on the wing, their only method of feeding. They forage in small flocks that fly relatively high but descend during wet weather to take advantage of insect concentrations. They feed at night mainly during the nesting season, perhaps in response to increased demands for food by their young.

COURTSHIP BEHAVIOR. Aerial displays include flying in synchrony, one bird closely following the other with wings held up.

NESTING. In some parts of the country they nest in colonies, but in Texas, pairs usually nest alone (see Davies, 2000). Nest: in a chimney or similar human structure, less frequently in a tree hollow. The parents construct a cup-shaped nest of twigs (collected in flight with their feet) that are held together by saliva. Their salivary glands, which are large compared with those in most birds, become even larger during the nesting season. Eggs: Both parents incubate the eggs and feed regurgitated insects to the nestlings. The tactile stimulus that elicits the begging response in the nestlings is air movement, most likely derived from the wing-flapping descent of the parents. Parents are sometimes assisted by two or three male or female adult helpers. Interestingly, helpers enhance reproductive success only if the parents are first-year nesters.

Young swifts use their strong, sharp claws to climb the vertical wall that supports the nest. They often drop from the nest before they are able to fly.

VOICE. Call: a rapid series of sharp chattering and twittering notes, uttered almost continuously. In chimneys, late at night, adult swifts and sometimes their young produce a low, roaring sound by vibrating their wings.

OTHER BEHAVIORS. Chimney Swifts are often observed flying in threes, which may represent two males chasing a female. They flap with rapid, shallow, jerky wingbeats, then sail on motionless, bowed wings (hence the comparison to a cigar with wings). The wings generally beat in unison, contrary to past claims that the birds move their wings alternately; however, when birds turn, one wing beats faster than the other.

Hundreds, sometimes thousands, of migrating swifts sometimes roost in large chimneys and comparable structures in late summer, ascending and descending in enormous clouds at dawn and dusk. Like other swifts, they feed, drink, bathe, copulate, fight, and play in flight — everything except sleep, which, although unverifiable, has also been proposed.

White-throated Swift, *Aeronautes saxatalis.* Princ. distrib. Texas: Common SR in mountains of Trans-Pecos Texas and w. Edwards Plateau, where they are also a locally uncommon WR.

FEEDING BEHAVIOR. They often forage many miles from their nesting site, usually flying high but descending to lower levels during wet weather. They capture insects while darting and swooping at extraordinary speeds.

COURTSHIP BEHAVIOR. Most courtship displays are aerial. Mating birds fly toward each other, join, and tumble down together, end over end, for 500 or more feet.

NESTING. They often nest in small colonies, apparently utilizing the same site for several years. Nest: tucked away in a vertical crevice on a high cliff. It is a shallow half-cup constructed of grass, feathers, weeds, and other materials that are glued together with sticky saliva. Supposedly the swifts are able to cling to the soft lining of their nest because their feet are structured for grasping laterally.

VOICE. Call: a shrill *je je je je je je je,* given in flight by several birds coursing together; it is harsher and louder than the Chimney Swift's call. They also produce a variety of soft twitters, especially when resting in crevices.

OTHER BEHAVIORS. They fly with flickering wingbeats that alternate with glides, sometimes sweeping their wingtips backward during a glide to suggest a flying arrowhead. This is most likely the fastest flying bird in Texas. One individual was observed escaping the stoop of a Peregrine Falcon at an estimated speed of more than 200 mph.

Although their strong claws adapt them for climbing on cliffs, they are essentially useless for supporting the birds on land or perches. After the breeding season, sometimes flocks of 200 or more roost together in crevices. They become torpid when temperatures drop or food availability decreases.

HUMMINGBIRDS: FAMILY TROCHILIDAE

Hummingbirds are exclusively New World birds, most species occurring in South America. The morphological diversity of South American hummingbirds is far greater than that of their North American relatives; however, even the ten or so regularly observed Texas species show marked differences in behavior, especially during courtship.

Hummingbirds feed on nectar and insects by hovering at flowers and inserting their long bill into the corolla. Some species will visit a feeder 15 or more times an hour if sufficient food is available. Although a functional olfactory apparatus has been demonstrated in Black-chinned Hummingbirds, the role of olfaction in the feeding ecology of hummingbirds is not known.

Evidently hummingbirds adjust their dietary intake to the sugar concentration of the food they consume. When sucrose concentrations at feeders are reduced, hummingbirds increase fluid intake by licking more frequently and by taking more fluid per lick. Hummingbirds also fly out from perches to capture passing insects, to pick insects from foliage, or to take spiders (and trapped insects) from webs.

The spectacular courtship displays of hummingbirds often utilize their unparalleled flight capabilities. Courting birds fly forward, backward, upward, downward, hover in still air, or engage in rapid, repetitive, aerial dances. Most hummingbird displays are probably ritualized intention movements.

Buff-bellied Hummingbird, *Amazilia yucatanensis.* Princ. distrib. Texas: Locally common SR in extreme s. Texas, where they are the only regularly nesting hummingbirds. They frequent dense thickets, citrus groves, clearings, and other semiopen areas, including suburban neighborhoods.

FEEDING BEHAVIOR. Typical for hummingbirds. They seem to have a preference for red tubular flowers, but they also visit mesquites and other trees.

COURTSHIP BEHAVIOR. Apparently not recorded.

NESTING. Nest: a cup of stems, bark, spiderwebs, cattle hair, and other materials, decorated on the outside with lichens and flower petals. Apparently only the female incubates the eggs and cares for the young.

VOICE. This noisy bird sings throughout the year. Call: shrill, twittering squeaks uttered in rapid succession. Extraordinarily varied vocalizations have been reported from south Texas birds (Green, 1999).

OTHER BEHAVIORS. They rest in dense thickets.

Blue-throated Hummingbird, *Lampornis clemenciae.* Princ. distrib. Texas: Rare SR in Davis and Chisos mountains, in shady canyons and wooded areas. In the Chisos, they occupy the mesic cypress-pine-oak association (Kuban and Neill, 1980).

FEEDING BEHAVIOR. Typical for hummingbirds. They often forage outside their territories, and when flowers are scarce, they can survive on insects.

COURTSHIP BEHAVIOR. Primarily vocal.

NESTING. Nest: a cup of plant fibers, spiderwebs, and moss placed on sheltered branches or under eaves and bridges. Eggs: incubated by the female, who also cares for the nestlings.

VOICE. Call: a loud, monotonous, high-pitched squeaky *seep,* uttered in flight as well as from a perch. The male perches in a tree and calls repeatedly, mainly during the breeding season. Sometimes the wings make a soft hum.

The vocalizations of this species are the most complex of the hummingbird songs so far studied. Moreover, the Blue-throated Hummingbird is the only hummingbird known in which the female has a complex song (Ficken et al., 2000).

OTHER BEHAVIORS. Being the largest hummingbird in the United States, it is not surprising that this species dominates other hummingbirds and aggressively chases them from favored feeding sites. Their flight is very swift.

Magnificent Hummingbird, *Eugenes fulgens.* Princ. distrib. Texas: Locally uncommon SR in mountains of Trans-Pecos Texas, in pines, junipers, oaks, and firs. One nest was located at 1,610 meters in the Chisos.

FEEDING BEHAVIOR. Typical for hummingbirds. They seem especially fond of wooded areas, where they fly out for insects and other prey. They sometimes follow a trap line (special route) of flowers, many of which are widely separated.

COURTSHIP BEHAVIOR. Apparently unrecorded.

NESTING. Nest: a cup built by the female from plant fibers, spiderwebs, feathers, and other materials. Eggs: incubated by the female, who also feeds the nestlings.

VOICE. Call: a scratchy squeak or twitter, and a sharp *chip* reminiscent of the Black Phoebe's call note.

OTHER BEHAVIORS. Their wingbeats are extraordinarily slow for a hummingbird and are discernible in flight. They often sail on set wings. These pugnacious and temperamental birds aggressively defend their territories, but away from their territories they are subordinate to Bluethroated Hummingbirds. It is common to see a male, perched on a limb about halfway up in a tall conifer, singing to advertise his territory.

Lucifer Hummingbird, *Calothorax lucifer.* Princ. distrib. Texas: Locally common SR in Chisos Mountains, on rocky hillsides up to 6,300 feet; one of the most common hummingbirds in Big Bend National Park in spring.

FEEDING BEHAVIOR. Typical for hummingbirds. They generally visit tubular flowers such as agaves, Chisos bluebonnets, and ocotillos. In Big Bend National Park, individuals were observed moving from one area to another to take advantage of flowers in bloom. They also visit spiderwebs for spiders and entangled insects.

COURTSHIP BEHAVIOR. The male displays while the female builds the nest and incubates the eggs. He makes short flights back and forth between two perches, with much rustling of the wings. He concludes his display by spiraling high into the air, diving steeply past the nest, then performing a series of pendulum swings in front of his mate.

NESTING. Nest: often in a cholla, agave, or ocotillo, up to 5,000 feet in elevation; it is a cup that the female constructs from plant fibers and decorates with leaves. Eggs: incubated by the female, who also feeds the nestlings. In spite of the demands of feeding the first brood, she sometimes lays a second clutch before the first fledglings have left the nest. Both parents aggressively defend their nesting and feeding territories.

VOICE. Call: a loud shriek, usually in response to a threat. Their wings make a whirring sound in flight.

OTHER BEHAVIORS. Where the two compete for food, Black-chinned Hummingbirds dominate over Lucifers.

Ruby-throated Hummingbird, *Archilochus colubris.* Princ. distrib. Texas: Common M and SR in e. half of the state, in open forested areas, gardens, and meadows.

FEEDING BEHAVIOR. Typical for hummingbirds. They have a strong preference for red and frequently visit trumpet vines and other red, tubular flowers. They also eat tree sap (from sapsucker drillings), and in the absence of flowers, they can exist entirely on insects.

COURTSHIP BEHAVIOR. The male's aerial courtship display (also referred to as an intimidation display) is noteworthy. He puffs up his red throat patch and swings slowly back and forth in a broad arc that suggests a pendulum, producing a buzz with his wings at the bottom of each dive. He then follows a shorter course back and forth in front of the perched female, who shows no obvious response at this time. Finally, both birds face each other, fly up and down, then drop to the ground to copulate. Thereafter they do not associate with each other as a pair.

NESTING. Nest: on a horizontal branch, usually hidden by leaves; it is a cup built by the female from plant fibers and spiderwebs and decorated on the outside with dead leaves. As the nestlings grow, the nest stretches to accommodate them. Eggs: incubated by the female, who also feeds the young.

VOICE. Call: rapid, squeaky chips, which are different in males and females. Their wings make a hum in flight, which produces the same low pitch in both sexes at first. After the male molts during his first year, his larger wing feathers are replaced by short, narrow, and pointed ones that produce a hum that is higher pitched than the female's.

OTHER BEHAVIORS. As the season advances, spring migrants keep pace with the blooming of flowers. Males evidently migrate before females.

The flight of most hummingbirds almost defies the imagination. This species, in particular, produces extraordinary wing movements, up to 75 complete strokes per *second*. Flight is preceded by a flick of the tail.

The courage of these hummingbirds is remarkable. They defend their territory against intruders as diverse as bumblebees, kingbirds, and eagles. Yet they themselves have been caught by frogs, dragonflies, and praying mantises.

Rubythroats like to bathe in water that collects on the surface of leaves.

Black-chinned Hummingbird, *Archilochus alexandri.* Princ. distrib. Texas: Common M and SR, mainly in w. two-thirds of the state, in towns, rivers, and open wooded areas. It is the most numerous hummingbird that nests in Texas.

FEEDING BEHAVIOR. Typical for hummingbirds, except that they probably engage more often in flycatching behavior (i.e., sitting on a perch and sallying out after passing insects). The role of olfaction in locating food sources has not been determined, but a functional olfactory apparatus has been demonstrated in this species (Goldsmith and Goldsmith, 1982).

COURTSHIP BEHAVIOR. Essentially like the display of the Ruby-throated Hummingbird, although a path described as a figure eight has been reported. During the repeated pendulum and U-shaped arcs and vertical oval flights (performed by both sexes), a sound is produced at the point nearest the ground (Pytte and Ficken, 1994).

NESTING. Nest: a deep cup of plant fibers woven by the female, who secures it to the branch of a tree and camouflages it with lichens, debris, and other materials. The nest stretches as the nestlings grow. Eggs: incubated by the female, who also feeds the nestlings. She sometimes begins building a second nest while still feeding fledglings from the first clutch of eggs.

VOICE. Call: a slurred *thew;* when chasing another bird, a series of loud chipping notes. Song: a soft, melodious, and very high-pitched warble, sung by the male in spring and unlike any sound produced by the Ruby-throated Hummingbird. Vocalizations produced during threatening encounters include one to five different note types arranged in a varying order (Rusch et al., 1996).

OTHER BEHAVIORS. Apparently males begin their spring migration before females. These hummingbirds take vigorous baths, a favorite bathing site being garden sprinklers.

Anna's Hummingbird, *Calypte anna.* Princ. distrib. Texas: Rare M and WR in w. and c. part of the state, in canyons, foothills, and river bottoms.

FEEDING BEHAVIOR. Typical for hummingbirds. They seem to depend more on insect food than any other North American hummingbird. After feeding, they fly to an exposed perch to preen while constantly turning the head from side to side.

VOICE. Call: a liquid *chick,* uttered when foraging; also, a rattling *zik zik zik* when alarmed. Song: a thin, high-pitched, and squeaky warble; this sound is heard during the breeding season, both when the bird is at rest and in flight. Although the Black-chinned Hummingbird possesses a rather rudimentary song, some ornithologists regard Anna's Hummingbird as the only North American hummingbird that truly sings. (See *Birds of North America,* No. 226, for an excellent account of vocalizations in this species.)

OTHER BEHAVIORS. They bathe in dew-covered foliage. It might be mentioned that their aerial courtship display, although not likely to be observed in Texas, is frequently regarded as the most spectacular of all of our hummingbird displays. They align their dives with the sun so as to maximize the reflection of their purple-red gorget.

Calliope Hummingbird, *Stellula calliope.* Princ. distrib. Texas: Locally uncommon M in w. Texas, along streams on mesas and broad valleys. This is the smallest bird north of Mexico, about 3 inches long.

FEEDING BEHAVIOR. Typical for hummingbirds.

VOICE. Call: a soft *tzip* when feeding, as well as other squeaks and chirps. Their vocalizations have been described as more lisping in quality than those of our other hummingbirds.

OTHER BEHAVIORS. They are often overlooked during migration because they stay low in vegetation.

Broad-tailed Hummingbird, *Selasphorus platycercus.* Princ. distrib. Texas: Uncommon to locally common SR in mountains of w. Texas, in pinyon-juniper-oak woodland (Kuban and Neill, 1980), pine-oak forests, and juniper scrub, especially on slopes and in canyons; locally uncommon M in other parts of w. Texas.

FEEDING BEHAVIOR. Typical for hummingbirds. They tend to visit red, tubular flowers, and sometimes they catch insects in the air.

COURTSHIP BEHAVIOR. The male repeatedly flies up 30–60 feet, then dives and ascends so as to describe a U-shaped pattern, all the while producing a trill with his wings. Males display like this until August, and then they may mate promiscuously with several females. In another display, both birds ascend together as high as 90 feet, then make a rapid descent.

NESTING. Nest: a cup built by the female from twigs, spiderwebs, and other materials and camouflaged with bark, moss, and lichens. Eggs: incubated by the female, who also feeds the nestlings and sleeps on the nest until the young are about a week or two old. If nesting is successful, females are likely to return to the same nesting site the following year.

VOICE. Call: a thin *tzip.* Broadtails are easily distinguished from other hummingbirds by their shrill, metallic, rattling trill, which vaguely resembles the rattling sound of a cicada. Males produce the sound by spreading the tapered tips of the first two outer primaries to form slots that air rushes through.

OTHER BEHAVIORS. The male sits on a high perch and defends his territory by chasing off intruders. The buzzy trill produced by his extremely fast wing movements (50 wingbeats per second) appears to be an important component of his territorial defense.

Broadtails bathe on flat rocks in mountain streams. Their flight is exceptionally swift. They enter a nocturnal torpidity in response to an inadequate energy intake and to low environmental temperatures (*Birds of North America*, No. 16).

Rufous Hummingbird, *Selasphorus rufus.* Princ. distrib. Texas: Common M in w. Texas, in mountain meadows as well as pinyon-juniper-oak woodland (Kuban and Neill, 1980); uncommon WR on the coast and in extreme s. Texas (Phillips, 1998). They are probably the most numerous migrant hummingbird in the highlands.

FEEDING BEHAVIOR. Typical for hummingbirds. They are strongly attracted to the color red, not only favoring red tubular flowers but also investigating inanimate objects such as red boxes, handkerchiefs, and pieces of paper. They prefer elevated feeders over those near the ground (Blem et al., 1997).

Rufous Hummingbirds are exceptionally aggressive when defending their feeding sites and readily chase off birds considerably larger than themselves. Early in the morning they begin visiting flowers at the periphery of their territory. This is an effective strategy because those flowers, being the farthest from the center of the territory and therefore the least easily defended, are best harvested early, before neighboring hummingbirds have a chance to visit them later that day (Hurley et al., 2001).

VOICE. Call: a sharp, double *bzeeep*; also, squeaky notes heard from migrants as they feed. In flight, their wings make a vibrating or rattling sound.

OTHER BEHAVIORS. They fly swiftly and with remarkable agility. In spring, males migrate slightly earlier than females.

KINGFISHERS AND ALLIES: ORDER
CORACIIFORMES, FAMILY ALCEDINIDAE

Kingfishers have large heads, large bills, and short tails. Our species capture fish and other aquatic animals by plunging headfirst into water. To signal aggressiveness they raise their head feathers or assume postures that emphasize their large bill.

Ringed Kingfisher, *Ceryle torquata.* Princ. distrib. Texas: Locally common PR in s. Texas, frequenting rivers, lakes, and lagoons, in particular the Rio Grande.

FEEDING BEHAVIOR. Essentially like the Belted Kingfisher, although they generally dive from higher perches, hover less frequently, and take larger fish. They follow a specific route up and down the river, landing on lookout perches every few hundred yards.

COURTSHIP BEHAVIOR. Apparently unrecorded.

NESTING. Nest: a horizontal burrow (up to 8 feet in length) in a dirt bank, dug by both parents and enlarged at the end to form a nest chamber. Occasionally burrows are located far from water. Both parents incubate the eggs, and probably both feed the nestlings.

VOICE. Call: a single, measured *tzaack,* given loudly in flight; when threatened, this note escalates into a rattle.

OTHER BEHAVIORS. Ringed Kingfishers occur in the same general habitats as Belted Kingfishers (which generally perch lower) and Green Kingfishers (which perch even lower). When coursing rivers they usually fly higher than Belted Kingfishers.

Belted Kingfisher, *Ceryle alcyon.* Princ. distrib. Texas: Uncommon to locally common SR in most parts of Texas, less common on the coast and in s. Texas, and wintering in e., s., and n.e. parts of the state. They generally require clear water that is deep enough so it is not choked with vegetation; however, during the breeding season their range is also limited by the availability of dirt banks in which to excavate nesting burrows.

FEEDING BEHAVIOR. Belted Kingfishers hunt both during the day and night. They survey the water from a low perch at the bank of a river, and after spotting prey (usually a fish), they plunge into the water and capture it with their bill. They often hover momentarily before diving. They fly back to their perch, beat the fish on a limb, juggle it in the air, then swallow it headfirst. Other food items include frogs, tadpoles, crayfish, and terrestrial animals such as lizards, young birds, and small mammals — not really surprising since several Old World members of this family forage exclusively on land.

Digestion is amazingly fast: the head of the fish begins digesting in the stomach when the tail is still in the throat. They regurgitate undigested bones, scales, and other body parts as pellets.

COURTSHIP BEHAVIOR. Primarily courtship feeding by the male, who also makes a series of rapid, prolonged mewing calls. Postcopulatory behavior is unique: the female follows the male as they both soar and dip close to the water.

NESTING. Nest: a grass or leaf saucer at the end of a horizontal burrow 3–6 feet long. Both parents excavate the burrow in a vertical dirt bank, usually one in which the soil has a high sand and low clay composition. Apparently their fused toes aid them in digging.

Both parents incubate the eggs and feed the nestlings, giving them regurgitated fish at first and later whole fish.

VOICE. Call (heard all year): a loud, vigorous, raucous rattle, easily detected half a mile away. Mates recognize each other's specific call.

OTHER BEHAVIORS. Young kingfishers must be taught how to fish. For about 10 days, the parents drop food items into the water for the young birds to retrieve. Once the nestlings are able to forage independently, the parents drive them from the feeding territory.

Outside the nesting season, males and females occupy and vigorously defend separate territories. Kingfisher territories are linear strips of a riverbank or lakeshore having a length inversely proportional to the availability of fish in the adjacent water.

Their flight is distinctive: five or six quick, powerful wingbeats that alternate with a glide. Kingfishers fly low over water and high over land. They have been seen escaping Peregrine Falcons and other raptors by diving into water. When threatened, they spread their wings, raise their white eyespots, and scream loudly.

The significance of the female's brighter plumage (her chestnut-colored underparts make her more colorful than the male) is not known. Unlike phalaropes, her brighter plumage is not correlated with typical male behavior.

Green Kingfisher, *Chloroceryle americana.* Princ. distrib. Texas: Locally uncommon PR in c. and s. Texas and on the Rio Grande in w. Texas, along shaded, clear rivers and quiet backwaters. During the nonbreeding season, when they are not dependent on dirt banks for nesting burrows, many individuals move to new fishing areas (Wauer, 2000).

FEEDING BEHAVIOR. These small kingfishers perch alone on a limb, snag, or rock that is 3–6 feet above the water and often concealed by shoreline vegetation. From this vantage point they search the water for small fish or aquatic insects while repeatedly jerking their tail up and down.

They pursue prey by flying out and diving headfirst into the water. Being the smallest of our three kingfishers, they are capable of fishing in water too shallow for the other two kingfishers. Unlike Belted Kingfishers, they rarely hover before diving.

It is said that Green Kingfishers sometimes stray far from water, feeding for insects by sallying out from a perch like flycatchers. Evidently this behavior has not been reported from Texas.

COURTSHIP BEHAVIOR. Apparently unrecorded.

NESTING. Nest: in a nest chamber at the end of a burrow 2–3 feet long, which, unlike that of the other two kingfishers, has its entrance concealed by vegetation. It is located above water and toward the top of a dirt bank, unfortunately easily accessible by fire ants, by far the worst threat to their reproductive success.

Both parents incubate the eggs, and both feed regurgitated fish to the nestlings.

VOICE. Call: a *cheep* in flight; also, a shrill, rattling twitter. When alarmed or excited, they utter a *tick tick tick tick,* like the sound of striking two small pebbles together.

OTHER BEHAVIORS. They fly low over the water, swiftly, resolutely, and with quick wingbeats. They are sometimes driven from their territory by Belted Kingfishers.

WOODPECKERS AND ALLIES: ORDER
PICIFORMES, FAMILY PICIDAE

Morphologically, woodpeckers are beautifully engineered for scaling tree trunks and pecking into the bark: their strong feet and stiff tail feathers brace them securely against the tree, their toes are structured for efficiently gripping, and their robust skulls are designed to absorb the impact of hammering on hard surfaces. Most species peck through dead, dying, or diseased bark to expose burrows of wood-boring insects, then use their long tongue to extricate the insects from the deep recesses.

Not surprisingly, insects dislodged from the bark sometimes fall before the woodpecker can seize them with its bill. To capture falling insects, some woodpeckers employ wing catching: they actually reach out with a wing to catch the insect before it falls to the ground. Wing catching has been observed in Downy, Hairy, Red-bellied, and Pileated Woodpeckers. Woodpeckers also eat other insects, including ants.

Several species cache seeds and nuts, and they do this in different ways. Red-bellied Woodpeckers are scatter hoarders, caching acorns in dispersed sites throughout a large area, whereas Red-headed Woodpeckers store nuts in central larders. Accordingly, the volume of the hippocampal complex (one part of the brain important for memory) is relatively larger in the Red-bellied Woodpecker, suggesting that it is better equipped biologically for memorizing geographic locations where acorns could be stored (Volman et al., 1997).

Some male woodpeckers announce their territory by drumming; in some species, males and females drum in duet.

Woodpeckers nest in tree holes and lay white eggs that both parents incubate.

Red-headed Woodpecker, *Melanerpes erythrocephalus.* Princ. distrib. Texas: Locally common SR in e. half of Texas and e. half of the Panhandle, in groves, orchards, urban areas, and other places where large, scattered trees predominate. They seem to prefer burned and cut-over habitats instead of undisturbed wooded areas. Families often select an area and remain there for several years without colonizing nearby sites that appear equally as suitable.

FEEDING BEHAVIOR. They are generally regarded as the most omnivorous of our woodpeckers, consuming insects, nuts, berries, and occasionally even bark. Less frequently they capture small birds and mammals.

Like other woodpeckers, they drill for grubs in dead wood, but they employ this method of foraging less than most woodpeckers. More often, they dart out from a perch to capture terrestrial or flying insects. In fall they gather acorns and other nuts that they store for winter in holes and crevices; yet unlike Acorn Woodpeckers (which do the same), they do not dig their own storage cavities. Their caches are sometimes robbed by jays and European Starlings.

COURTSHIP BEHAVIOR. During his courtship display the male sits horizontally, stretches his neck forward, sleeks his plumage, and humps his shoulders.

NESTING. Nest: a hole in a dead trunk or limb, from just above the ground to as high as 65 feet. It is usually excavated by the male, who sometimes selects his previous winter roosting cavity as the nesting site. After the male shows the female a potential site, she taps on the tree to signal that she accepts it.

Natural cavities seem critical for nesting success, as Red-headed Woodpeckers do not accept birdhouses, and young birds in holes in creosote-treated utility poles often die. Eggs: incubated by both parents, the male usually at night. Both parents feed the nestlings, even after the female has laid a second clutch in the same or in another cavity.

VOICE. Call: a harsh, scolding, henlike *kwurrk*, as well as *kweeer* and *churr churr*.

OTHER BEHAVIORS. The male advertises his territory by drumming and calling and confronts intruders by spreading his wings and tail and bowing. These woodpeckers are relatively unwary of people.

Acorn Woodpecker, *Melanerpes formicivorus.* Princ. distrib. Texas: Common PR in mountains of w. Texas and the s.w. edge of the Edwards Plateau, in pine-oak woodlands. Evidently their range is limited by the availability of acorns and storage trees.

FEEDING BEHAVIOR. They rarely drill for insects, depending, instead, on a supply of acorns (about half their annual diet), insects (especially ants), fruits, seeds, and bird eggs. They often forage in small groups, gleaning insects from the surface of trees or sallying out to capture them in flight. They also dig pits in the bark (for sap) and visit pits made by sapsuckers.

In fall they store acorns in holes they excavate in trees. These so-called granary trees (many of which are used for several generations) may be sculptured with thousands of holes that apparently are not deep enough to kill the tree. Birds feed on these acorns throughout the winter, but they must regularly defend their caches against jays and squirrels.

COURTSHIP BEHAVIORS. Males noisily fly through trees, performing showy antics that include bowing and wing spreading.

NESTING. They nest in loose colonies of one to seven males, one to three females, and possibly several nonbreeding helpers. The helpers, usually offspring from previous clutches, are usually more numerous in years when the chances of finding a vacant territory are low. Nest: excavated in a dead tree by both parents, with assistance from helpers. Several females sometimes lay eggs in the same nest (with as many as 17 eggs in a nest), and a dozen or more birds may incubate the eggs and care for the nestlings. European Starlings are notorious for invading the cavities and evicting the woodpeckers.

VOICE. This is a noisy bird. Call: a raucous, parrotlike *whack up whack up*, often used to call a mate.

OTHER BEHAVIORS. Both males and females drum during territorial encounters. Their flight is less consistently undulating than that of most woodpeckers.

This, our most social woodpecker, has a complex social structure that is only partly understood and not easily summarized here. See *Birds of North America*, No. 194, for an excellent discussion of its social and courtship behavior.

Golden-fronted Woodpecker, *Melanerpes aurifrons.* Princ. distrib. Texas: Common PR in s., s.w., and w. Texas. They occur in mesquite scrub, river bottoms, and most open woodlands in arid and semiarid regions, including urban parks and gardens.

FEEDING BEHAVIOR. Omnivorous, feeding on insects, nuts, berries, acorns, and a wide variety of other food items. They probe for insects beneath the bark of tree trunks, less often foraging on the ground or capturing insects in flight. They cache food in bark crevices.

COURTSHIP BEHAVIOR. The male advertises his territory with loud calls. He also drums, but not as frequently as many woodpeckers, perhaps because drumming sounds do not carry far in the open areas he inhabits. Birds have been seen exhibiting reverse mounting, as well as synchronized tapping just below the entrance to the cavity they are excavating (Husak, 1996).

Other displays resemble those of Red-bellied Woodpeckers, which sometimes hybridize with Golden-fronted Woodpeckers where the ranges of the two species overlap.

NESTING. Nest: in a live or dead tree trunk (or utility pole), 6–20 feet above the ground, which is excavated by both parents. Eggs: incubated by both parents, who also feed the nestlings.

VOICE. This is a noisy bird that often calls in flight. Call: a sharp *chuck,* a loud rolling *kar r r r r r,* and a flickerlike *keck keck keck,* as well as several variations of these calls. Generally, these sounds are louder and harsher than the Red-bellied Woodpecker's.

OTHER BEHAVIORS. Territorial behavior has been described by Husak (2000).

Red-bellied Woodpecker, *Melanerpes carolinus.* Princ. distrib. Texas: Common to locally abundant PR in e. half of Texas, in woodlands, orchards, and urban areas.

FEEDING BEHAVIOR. Their diet includes a large variety of food items, including acorns, fruits, and seeds, as well as many insects. They glean tree trunks and larger limbs, perching among branches for fruits and berries and sallying out from perches to capture flying insects. Occasionally they drop to the ground to feed.

Red-bellied Woodpeckers sometimes prey on vertebrates, including House Wren nestlings, which they remove from the nest and carry away in their bill (Neill and Harper, 1990). They cache a limited number of nuts and seeds in bark crevices, but the extent to which these items are utilized in winter is not known.

COURTSHIP BEHAVIOR. Courtship displays include bowing, spreading wing and tail feathers, and raising the crest. Mutual tapping and reverse mounting are also common. They sometimes hybridize with Golden-fronted Woodpeckers.

NESTING. Nest: a cavity excavated in a dead tree, fence post, or utility pole, typically below 50 feet. The female selects one of several cavities that have been excavated by her mate. Eggs: incubated by both parents. Both parents feed the nestlings, sometimes for as long as 6 weeks after they leave the nest. European Starlings sometimes usurp their nesting cavities.

VOICE. Call: a soft, rolling, scolding *ker r r r r r,* repeated several times, as well as *chuf chuf chuf* and other sounds.

OTHER BEHAVIORS. In defending their territory, they spread their wings, fly slowly in a floating manner, and raise their head feathers. In winter, especially when food resources are scarce, small flocks wander in the company of sapsuckers.

Yellow-bellied Sapsucker, *Sphyrapicus varius.* Princ. distrib. Texas: Uncommon to locally common M and WR throughout Texas, except the extreme w. part. They occupy virtually any habitat that has sufficient trees for sap, including urban areas.

FEEDING BEHAVIOR. Sapsuckers drill tiny, evenly spaced holes (sap wells) in tree trunks and limbs. The holes are square or round, and in a vertical or horizontal series. From these wells they take sap, insects drawn to the sap, and bits of wood. During the nesting season they guard these wells against other birds, including hummingbirds, but it is not known if wintering birds in Texas do this. Sapsuckers also eat berries and fruits, dash out for flying insects, and glean insects from tree trunks. They rarely feed on the ground.

VOICE. Call: a catlike, downward-slurred *kee yew* and various other notes.

Red-naped Sapsucker, *Sphyrapicus nuchalis.* Princ. distrib. Texas: Rare PR in Guadalupe Mountains, migrating and wintering locally in other parts of Trans-Pecos Texas (see Lockwood and Shackelford, 1998).

This species so closely resembles the Yellow-bellied Sapsucker that until recently the two were considered races of the same species. Apparently their behavior is identical.

Ladder-backed Woodpecker, *Picoides scalaris.* Princ. distrib. Texas: Common PR in w. two-thirds of Texas, in mesquite scrub, by brush-lined streams, and in most dry, semiopen habitats, including those that are sparsely vegetated; they rarely occur above 5,000 feet.

FEEDING BEHAVIOR. The pair forage together, visiting sites as diverse as tall trees, cacti, yuccas, and tall weeds. They glean for insects and, less frequently, fruits and berries. The larger male, with his stouter bill, probes and pecks on trunks and larger limbs; the female more often concentrates on outer twigs, bushes, and higher branches in trees, where she gleans bark surfaces. Both males and females sometimes drop to the ground to search for insects.

COURTSHIP BEHAVIOR. Courtship behavior seems to be minimal. Some of the male's territorial displays—bobbing and turning the head, spreading the wings and tail, and raising the head feathers—are also part of his courtship behavior.

NESTING. Nest: a cavity located in a tree, yucca stalk, fence post, or similar structure, usually less than 20 feet above the ground. Both parents incubate the eggs and feed insects to the nestlings.

VOICE. Call (often in flight): a shrill, ringing *tcheek,* louder and harsher than the Downy Woodpecker's, and a rapid series of hoarse, descending *kehs*, also harsher and more run together than the Downy's comparable call. Drumming occurs in spring.

OTHER BEHAVIORS. The pair stay together most of the year.

Downy Woodpecker, *Picoides pubescens.* Princ. distrib. Texas: Common PR in e. half of Texas, and locally uncommon PR to the w., in forests, farms, orchards, and other wooded areas. They readily make themselves at home in suburban yards, more so than Hairy Woodpeckers, and in this habitat they often become very tame.

This is the smallest woodpecker in North America. It is probably more closely related to the Ladder-backed Woodpecker than to the Hairy Woodpecker, in spite of its superficial resemblance to the latter species.

FEEDING BEHAVIOR. Downy Woodpeckers forage on substrates as diverse as tall trees along river bottoms and weed stalks in meadows. They maneuver with agility, sometimes hanging upside down on small limbs to take food items inaccessible to other woodpeckers. Males and females forage somewhat differently, males apparently preferring smaller branches in the upper canopy. Both males and females spend more time gleaning in summer and more time tapping and excavating in winter. In fall and early winter the members of a pair occupy separate feeding territories.

Downy Woodpeckers join mixed flocks of chickadees, nuthatches, titmice, and other small species. In addition to insects, they consume seeds, berries, and suet.

COURTSHIP BEHAVIOR. Pairs sometimes remain mated for 4 or more years. In late winter, when they still occupy separate feeding territories, the male and the female take turns drumming loudly on dead limbs. This mutual display evidently prompts the male to enter his mate's territory.

Birds also raise their crest, wave their bill, fly in a stilted, floating manner, dance, and call in duet. Some of these displays are also used to defend the territory.

NESTING. The parents excavate a new cavity each year, typically 15–30 feet above ground in a dead tree. The entrance is often camouflaged with lichens. Both parents incubate the eggs and feed insects to their young, but males apparently do most of the brooding.

VOICE. Call: a sharp, abrupt *peek,* often uttered after feeding; also, a staccato roll of about a dozen musical *tschicks* that drop in pitch at the end. Compare the Ladder-backed Woodpecker's call.

OTHER BEHAVIORS. Downy Woodpeckers drum on trees to advertise their territory. They roost but do not nest in birdhouses, and in winter they may excavate roosting cavities separate from their past nesting cavities.

Hairy Woodpecker, *Picoides villosus.* Princ. distrib. Texas: Uncommon to locally common PR in e. half of Texas; locally uncommon M and WR to the w. They also nest in the Guadalupe Mountains and in the Panhandle. They inhabit forests, river bottoms, and swamps. Compared to Downy Woodpeckers, they favor more secluded timberlands, require larger trees, and venture less frequently into urban areas, although they sometimes move to more open areas in fall and winter.

FEEDING BEHAVIOR. Hairy Woodpeckers pound through bark to excavate wood-boring insects; they also expose insects by probing into cracks and holes and by scaling off bark. Sometimes they strike into the bark and pause momentarily, possibly to feel or hear insects beneath the bark.

These woodpeckers are specialized for pounding on bark. They spend relatively little time gleaning bark and foliage and only infrequently engage in alternative foraging styles, such as flycatching for insects, storing nuts, hopping on the ground in search of insects, or drilling through the bark for sap. On the other hand, they readily take advantage of sap already exposed by sapsuckers, as well as seeds, meat scraps, suet, bananas, and peanut butter at bird feeders.

Males forage more deliberately and tenaciously than females and concentrate longer in one location. Females forage higher in the trees, but in winter both members of the pair forage at lower levels. Males and females usually maintain separate feeding territories in winter, then pair up again before spring, often with their previous mate.

COURTSHIP BEHAVIOR. Some courtship displays are mutual, as when both birds drum in duet. In her winter feeding territory the female taps at both suitable and unsuitable nest sites, then performs a fluttering display flight to attract the male. The pair perform other displays that seem to be ritualized aggressive and appeasement behaviors.

NESTING. Nest: in a tree cavity excavated by both parents, up to 60 feet above the ground. Both parents incubate the eggs and bring insects to the nestlings. The female rarely searches beyond hearing distance of the nestlings, whereas the male ventures farther from the nest. He brings back insects less frequently, but he usually carries more insects in a trip.

VOICE. Call: a loud *peek,* louder and shriller than the Downy's; it is a rattling call that is more run together than the Downy's and suggests to some the call of a Belted Kingfisher.

Both sexes drum; during the ritual flight that follows they beat their wings against their thighs to produce a loud sound.

OTHER BEHAVIORS. Their flight is graceful and bounding. Sometimes the female engages in a floating, fluttering flight. Hairy Woodpeckers are noisier and more active than Downys, tapping louder and more often. They are also less bold and will more readily disappear behind a tree trunk when approached.

Red-cockaded Woodpecker, *Picoides borealis.* Princ. distrib. Texas: Locally uncommon PR in e. Texas. They require a habitat now scarce in Texas, a mature pine forest (with trees at least 80 years old) that supports a very open understory regularly exposed to fire. Although the e. Texas population is one of the largest in existence, it currently consists of fewer than 140 active clusters (Craig Rudolph, pers. comm.).

Birds do not persist in an area once logging removes the older pines. Observations in e. Texas demonstrate that the reduction of hardwood midstory around cavity trees, preferably by fire, is necessary for persistence of the birds (Conner et al., 1996).

FEEDING BEHAVIOR. Family groups forage together, flying restlessly from tree to tree, noisily calling to each other, and briefly chasing each other as they busily flake off bark in search of insects, especially ants and beetles. They often hop on the ground or along horizontal tree branches. In addition to insects, they eat other invertebrates, berries, fruits, and nuts. They have been known to visit corn fields next to forests in order to extract worms from the ears.

Males concentrate on upper trunks and branches, sometimes hanging upside down like nuthatches; females seem to prefer lower trunks.

COURTSHIP BEHAVIOR. They drum more frequently than most of our woodpeckers. The male's courtship display includes bobbing, swinging his head, and fluttering. Wing spreading seems to be an important display that has several functions, including courtship.

NESTING. Nest: in large, live pines that have a soft center due to infection by the red heart fungus, another requirement not met in many otherwise potential nesting sites. They excavate, and keep open, small holes next to the entrance of the nesting cavity. These wells exude resin that drips down the trunk, hardens, and reduces the ability of rat snakes to climb to the nest and roost cavities. The resin turns white and identifies trees where the woodpeckers nest. Nests are sometimes utilized by many generations, perhaps because ideal nesting sites are rare.

Helpers assist parents in incubating the eggs. Generally, helpers are males from the same family and usually birds from the previous breeding season. Helpers also assist in feeding the nestlings. Interestingly, helpers do not enjoy greater reproductive success than nonhelpers, suggesting that (at least in this species) the experience of helping does not increase reproductive potential (Khan and Walters, 1997).

VOICE. Call: a short *yenk yenk*, like the nuthatch's; also, a longer call that has been compared to the sound of a young bird begging for food. Tongue drumming, a soft sound suggesting a rattlesnake, is produced by rapidly vibrating the tongue on tree surfaces.

OTHER BEHAVIORS. They live in isolated groups (clans or clusters), which include a pair of adults and up to four other, usually related individuals. Clusters feed together, defend a feeding territory (100–200 acres) all year, and participate in nesting activities.

Northern Flicker, *Colaptes auratus.* Princ. distrib. Texas: Common M and WR, and local PR, throughout most of the state, in almost every habitat except forests devoid of open areas and prairies and deserts devoid of trees.

Evidently the eastern (Yellow-shafted) race and the western (Red-shafted) race differ in behavior very little, if at all, although the Red-shafted's call is slightly coarser, rougher, and heavier.

FEEDING BEHAVIOR. Flickers consume enormous numbers of ants, more than any other North American bird, about 5,000 being the highest count in the stomach of a single bird. They gather ants while hopping on the ground, or they simply station themselves by an ant hill. Their enlarged salivary glands produce sufficient saliva to neutralize the formic acid they ingest.

They also forage for insects and spiders on tree trunks and, less frequently, sally out from a perch to capture passing insects. Occasionally they eat fruits and berries (poison ivy being a favorite), and they often perform odd maneuvers at the tips of limbs to reach these food items.

Their slightly decurved and less robust bill gives them an advantage in picking up insects and other small animals. The price they pay for having this bill shape is that they cannot hammer through bark as effectively as woodpeckers that have the more typical chisel bill (for example, the Hairy Woodpecker).

COURTSHIP BEHAVIOR. Many of their noisy and active courtship displays are also used to advertise territory. The male swings his head back and forth, spreads his tail, and flicks his wings to display his brightly colored plumage. He also engages in considerable calling and drumming. While facing each other, the pair bob their heads and touch bills.

NESTING. Nest: a cavity in a dead tree or post, the side of a house or barn, or rarely in a burrow in the ground. When selecting a nest site, the pair engage in ritualized tapping. Both birds excavate the cavity.

Both parents incubate the eggs, the male part of the day and throughout the night; both feed regurgitated insects to the nestlings. They rotate guard duty at the nest, and later they lead the young to good foraging areas.

VOICE. Call: a loud *wick wick wick,* or *wicker wicker wicker,* also rendered as *flicker flicker flicker* (although the name "flicker" is said to be derived from their habit of flicking their wings); also, a far-reaching *kee yer,* which apparently serves to maintain contact between members of the same family. In spring they drum rapidly on logs, tin roofs, and even garbage cans.

OTHER BEHAVIORS. The visual releaser for aggression among males is the moustache, and a male will attack his mate if a moustache is painted on her.

Unlike most woodpeckers, flickers often perch crossways on limbs. They go to roost before sundown, spending the night in tree cavities, under eaves, or in other secure places. Sometimes they drill a hole in the side of a wooden cabin and sleep in the wall.

Their nesting cavities are often taken over by owls, European Starlings, and other birds, as well as by squirrels; however, they themselves sometimes usurp kingfishers and swallows from burrows. In the aggressive, ritualized flicker dance, which the pair employ to defend their breeding territory, one member of the pair engages in a fencing duel with the intruder, using the bill as a foil.

Pileated Woodpecker, *Dryocopus pileatus.* Princ. distrib. Texas: Locally common PR in e. and c. Texas, mainly in deciduous or mixed forests. They sometimes inhabit urban parks, but always where there are large feeding areas.

FEEDING BEHAVIOR. The stout bill enables them to dig deep into rotten wood and probe for ant nests (especially nests of carpenter ants). They leave rectangular holes that are 4–5 inches deep, some quite close to the ground.

Pileated Woodpeckers eat enormous quantities of insects (75 percent of their diet), the remaining food items being primarily fruits and berries that they gather while clambering about in small branches. The horned passalus (*Popilius*), a large beetle, seems to be an important food item in east Texas (Conner, 1982).

COURTSHIP BEHAVIOR. The male raises his crest, spreads his wings (which exposes the white wing patches), swings his head, and bobs. His aerial displays include circling.

NESTING. Nest: a cavity in a dead tree or utility pole, usually less than 80 feet above ground and often facing east or south on a bark-free surface. Both parents excavate the cavity and prepare a new cavity each year. Both parents incubate the eggs, and both feed regurgitant to the nestlings. The nestlings may remain with their parents for several months after hatching.

VOICE. Generally silent except in spring. Call: loud and repetitive, suggesting a flicker. One is a deep, full-throated *cuk cuk cuk cuk cuk* that rises in pitch, then descends. Both males and females produce a *yucka yucka yucka* sound.

The resonant drumming, used to advertise the territory, is distinctive: it speeds up, then fades away after 4–5 seconds. Both sexes drum (the female less), often before they go to roost at night.

OTHER BEHAVIORS. These imposing, crow-sized woodpeckers fly with swift, powerful wingbeats. They lack the strongly undulating pattern typical of most woodpeckers. They are shy and quickly flee from human intruders.

They maintain year-round territories and pair bonds. Each member of the pair excavates one or more roosting cavities not too far from the nesting cavity. Territorial displays include crest raising, head bobbing, and wing spreading.

PASSERINE BIRDS: ORDER PASSERIFORMES

TYRANT FLYCATCHERS: FAMILY TYRANNIDAE

Our North American flycatchers forage chiefly by flycatching: darting out to capture insects on the wing, then returning to the same perch to swallow their prey. Less frequently they hover for insects at the tips of branches, or glean insects from leaves and twigs as they slowly work their way through foliage. Some flycatchers utilize all three foraging techniques. After consuming prey, many species regurgitate the indigestible parts as pellets.

Most flycatchers project a pugnacious nature and a readiness to attack almost any animal, including hawks, large mammals, and humans. Given this temperament, it is not surprising that occasionally small species such as the Vermilion Flycatcher pursue and capture prey items as unlikely as fish.

Many flycatchers have a special dawn song known only to birders who begin their day very early. In some species it is better termed a predawn song, as often it concludes before the first hint of sunrise.

See the introductory notes for the family Caprimulgidae for a comparison of these two insectivorous families.

Northern Beardless-Tyrannulet, *Camptostoma imberbe*. Princ. distrib. Texas: Rare PR in extreme s. Texas, in mesquite thickets and woodlands, and almost always near water (see Brush, 1999).

FEEDING BEHAVIOR. These secretive birds flick their tail as they glean insects from shrubs and lower tree branches. They also flycatch, especially in summer, and sometimes eat seeds and berries. Their general behavior is more like a vireo or kinglet than a typical flycatcher.

COURTSHIP BEHAVIOR. Largely unknown; the male advertises his territory by singing.

NESTING. Well hidden in mistletoe or a clump of leaves. Nest: a round structure of grass and weeds, suggesting a baseball in size and shape, that is entered through an opening near the top.

VOICE. Call: a loud and repetitive piping, rendered as *pee yeerp*, possibly the best field mark since in appearance this species closely resembles an immature Verdin (found in the same habitat). The thin, high song consists of three to five descending *eees*.

Olive-sided Flycatcher, *Contopus cooperi.* Princ. distrib. Texas: Uncommon SR in Guadalupe Mountains, less common elsewhere in w. Texas; rare to common M throughout Texas. They favor open woodlands, especially those harboring dead trees.

FEEDING BEHAVIOR. These big-headed flycatchers sit upright and alert on conspicuous perches, especially high dead limbs, from which they sally out to capture honeybees, winged ants, wasps, and other insects.

COURTSHIP BEHAVIOR. Largely unknown; the male's display includes chasing the female through the treetops.

NESTING. Nest: an open cup of twigs and grass, well concealed and high up. Eggs: incubated by the female. Both parents feed the nestlings. The male persistently chases (or attempts to chase) all intruders from his territory, including humans.

VOICE. Call: *pip pip pip,* uttered at regular intervals, used to express alarm. Song: rendered as *Quick! three beers,* sung repeatedly in spring to advertise the territory.

Western Wood-Pewee, *Contopus sordidulus.* Princ. distrib. Texas: Locally common SR in the Trans-Pecos, mainly in pine-oak mountain forests and canyons; common M e. to the Edwards Plateau. *See* Eastern Wood-Pewee.

FEEDING BEHAVIOR. Probably indistinguishable from that of the Eastern Wood-Pewee.

COURTSHIP AND NESTING BEHAVIORS. *See* Eastern Wood-Pewee.

VOICE. Call: a rapidly chattered *pit pit pit.* Song: a descending, nasal *peeer,* recalling the Common Nighthawk's *peent* and generally heard only at dusk and dawn.

Eastern Wood-Pewee, *Contopus virens.* Princ. distrib. Texas: Common SR in e. part of the state and common M w. to the Trans-Pecos. They favor woodlands, river bottoms, and other habitats that support tall trees. The Western Wood-Pewee seems to tolerate more open habitats than its eastern counterpart, and it generally avoids dense woodlands.

FEEDING BEHAVIOR. These likable birds sit patiently on dead branches or other exposed perches in the midlevel section of the forest; from here they fly out to snap up passing insects. They also hover for insects on branches and leaf clusters. When flying insects are scarce, they forage for berries and ground-dwelling invertebrates.

COURTSHIP BEHAVIOR. The only important courtship behavior seems to be a courtship chase, which evidently has been described only for the Eastern species.

NESTING. Both Eastern and Western Wood-Pewees use grass, stems, and other plants as nesting materials. The Eastern usually attaches lichens to the outside, making the nest appear from below like a knot on a limb.

The Western's nest is deeper and more compact than the Eastern's, but both nests strike one as being too small for the size of the bird. Both nests are probably constructed by the female. Eggs (both species): incubated by the female. Both parents bring food to the nestlings.

VOICE. Call (Eastern): a sharp *chip*. The song, a thin, contemplative *pee uh wee pee wu,* discreetly interrupts the silence of the forest. It is generally sung at dawn and dusk and is one of the last songs heard in the evening.

Yellow-bellied Flycatcher, *Empidonax flaviventris.* Princ. distrib. Texas: Uncommon to common M in e. half of Texas, sometimes migrating in large numbers. They inhabit dense vegetation, shrubby streams, and brushy thickets by fields and roads.

FEEDING BEHAVIOR. From a perch in the low or middle level of a tree, these shy and retiring birds sally out for flying insects or hover briefly for invertebrates that crawl on leaves and stems. They often forage less than a couple of feet from the ground, so normally they are not easy to locate. They take large numbers of spiders and red ants, and in fall they consume many seeds and berries.

VOICE. Call: a soft, plaintive *peah peah.* Song: a simple *per wee.* Migrants in Texas are generally silent.

Acadian Flycatcher, *Empidonax virescens.* Princ. distrib. Texas: Common M and SR in e. Texas, w. to the Edwards Plateau, in bottomlands, mature deciduous forests, and swamps.

FEEDING BEHAVIOR. Their foraging behavior is typical of most *Empidonax* flycatchers: they generally dart out from perches in the low to middle level of the forest, but they also hover next to branches and leaves. They occasionally eat seeds and berries.

COURTSHIP BEHAVIOR. The male hovers by the perched female or pursues her in swift, erratic courtship flights. The pair bond is generally long-term, and the two birds often maintain a strong fidelity to the previous year's breeding territory.

NESTING. Nest: a cuplike structure probably built by the female. Eggs:

incubated by the female. In some years she leaves the territory to incubate a second clutch before the first fledglings have developed, leaving the male to care for the first brood.

VOICE. Call: a spiritless *peek*, sometimes accompanied by a twitch of the tail, and a series of twittering notes produced in flight. Song: a forceful *spit see.*

Alder Flycatcher, *Empidonax alnorum.* Princ. distrib. Texas: Uncommon M in e. part of the state and on the coast, where they occupy open country and mixed woodlands, often near water.

FEEDING BEHAVIOR. Alder Flycatchers perch in a tall shrub or low tree, from which they fly out to catch insects on the wing. They also hover for insects and spiders that crawl on leaves and branches, and occasionally eat seeds and berries.

VOICE. Call: a flat *kep.* Song: rendered as a husky *fee BEE oh.* Unlike the songs of most passerines, it is entirely innate rather than learned or partly learned.

Willow Flycatcher, *Empidonax traillii.* Princ. distrib. Texas: Rare local SR in the Trans-Pecos, in deciduous woods and thickets; locally uncommon M throughout Texas.

FEEDING BEHAVIOR. Essentially like the Alder Flycatcher's.

VOICE. Call: a thick *whit.* Song: sometimes rendered as *witch brew.* Unlike most passerines, Willow Flycatchers have an innate rather than learned or partly learned song. Both members of the pair sing to advertise their territory.

Least Flycatcher, *Empidonax minimus.* Princ. distrib. Texas: Common M throughout most of Texas, especially in c. and e. Texas, along streamsides and in mixed woods and parks. They usually occupy fairly open areas.

FEEDING BEHAVIOR. They survey their surroundings from dead twigs or other perches at low to middle levels of a tree, then fly out to capture passing insects or hover next to foliage for tree-dwelling insects and spiders. They also creep along branches in search of insects, not a common foraging technique for North American flycatchers. They occasionally eat seeds and berries.

VOICE. Call: a thin *pit.* Song: an emphatic *che BECK,* uttered as the bird flicks its tail and jerks its head upward; the song is occasionally heard in Texas in spring.

Hammond's Flycatcher, *Empidonax hammondii.* Princ. distrib. Texas: Locally uncommon M in Trans-Pecos Texas, in lower woodlands and thickets, canyons, and on mountain slopes, especially along streams.

FEEDING BEHAVIOR. Hammond's Flycatchers are quieter and less active than the other *Empidonax* flycatchers. They sally out from perches that range from the tops of trees to the lowest branches. They also pursue insects by hovering and sometimes by dropping to the ground.

VOICE. Interestingly, males and females have different calls. The male utters a sharp *pip,* the female a *tweep.* Song: almost never heard in Texas; it is commonly rendered as *chi pit brrk krrip.*

Gray Flycatcher, *Empidonax wrightii.* Princ. distrib. Texas: Locally common SR in Davis Mountains and uncommon M in Trans-Pecos Texas, in mixed woodlands and canyon streams.

FEEDING BEHAVIOR. They sit quietly in bushes and in the lower levels of trees; from these perches they capture insects and spiders by flycatching, hovering, or dropping to the ground.

VOICE. Call: a colorless *whit.* Song (rarely heard in Texas): *chuiwip cheeah.*

OTHER BEHAVIORS. They dip the tail slowly, then quickly raise it.

Dusky Flycatcher, *Empidonax oberholseri.* Princ. distrib. Texas: Common M in Trans-Pecos Texas, in mesquite woodlands, arroyos, and along mountain slopes.

FEEDING BEHAVIOR. In typical *Empidonax* fashion, they fly out from a dead branch or other exposed perch to capture insects on the wing. They also hover next to trees and shrubs and occasionally drop to the ground. Apparently they feed entirely on insects.

VOICE. Call: a dry *whit.* Song: a soft *chipit chuwee psee,* which is probably sung only rarely by migrants.

OTHER BEHAVIORS. They sometimes flick the tail when perched.

Cordilleran Flycatcher, *Empidonax occidentalis.* Princ. distrib. Texas: Uncommon SR in Guadalupe, Davis, and Chisos mountains, in mountain forests and shady canyons. This and the Pacific-Slope Flycatcher were formerly lumped as a single species, the Western Flycatcher (*E. difficilis*). In w. United States, in localities where either this species or Hammond's Flycatcher is an abundant SR, one species is abundant and the other is usually rare or absent.

FEEDING BEHAVIOR. From a shaded perch often high in the tree (or lower in trees near streamsides), these small birds dart out for flying insects; they also hover for insects and spiders that crawl about on leaves and twigs, and they occasionally eat seeds and berries.

COURTSHIP BEHAVIOR. Apparently not recorded.

NESTING. Nest: a cup of rootlets, moss, weed stems, and other materials that the female constructs on a tree trunk, ledge, or stump. Eggs: incubated by the female. Both parents bring food to the nestlings.

VOICE. Call (sex-specific): the male's is a *pit peet,* the female's a thin *tseet.* Song: *tezeep sip tseet,* the last note highest; birds generally sing at daybreak.

Black Phoebe, *Sayornis nigricans.* Princ. distrib. Texas: Rare to locally uncommon PR in the Trans-Pecos and parts of the Edwards Plateau. They favor streams, rivers, and lakes in arid and semiarid regions, especially where willows, sycamores, and cottonwoods border waterways.

FEEDING BEHAVIOR. Black Phoebes dart out from a low perch to hawk insects near the ground, sometimes hovering at trees or shrubs and occasionally flying out over water to capture floating insects. Occasionally they capture small fish, and in winter they glean insects from the ground.

COURTSHIP BEHAVIOR. After ascending into the air, fluttering and singing, the male slowly drops down to a perch.

NESTING. Only where mud is available nearby. Nest: an open cup of mud pellets, hair, and plant material plastered to a cliff, bridge, eave of building, or similar structure, probably built by the female. Eggs: incubated by the female. Both parents feed the nestlings.

VOICE. Call: a sharp *tsip.* Song: a thin, repetitious *ftee wee, ftee wee,* the first pair of notes rising, the second falling.

OTHER BEHAVIORS. They wag their tail by quickly moving it downward and then raising it slowly. Black Phoebes fly with slow wingbeats that give the birds a mothlike appearance. Males and females maintain separate feeding territories in winter.

Eastern Phoebe, *Sayornis phoebe.* Princ. distrib. Texas: Common to uncommon SR in n. two-thirds of Texas (except the Panhandle); locally common WR in e. two-thirds; locally common M throughout most of the state (except the extreme w. parts). They frequent farms, streamsides, forest edges, urban areas, and other semiopen habitats, usually with water nearby.

FEEDING BEHAVIOR. They normally fly out from a perch to capture insects on the wing, especially wasps, beetles, and ants, but they also drop to the ground or hover briefly for insects crawling about on leaves and twigs. Fruits and berries seem to be an important component of the winter diet.

COURTSHIP BEHAVIOR. The male seems to advertise his territory only by singing.

NESTING. Nest: built by the female, an open cup of moss, grass, and leaves that is glued to a sheltered surface with mud. She sometimes uses the previous year's nest or builds a new nest on top of the old one. Females roost in the nest several weeks before laying their eggs, possibly to conserve energy by sleeping in a warm structure. Eggs: incubated by the female. Both parents feed the nestlings.

VOICE. Call: an abrupt *chip.* Song: repeated in bouts of varying lengths, an emphatic, husky *fee bee,* the *bee* alternately higher and lower than the first note; in flight, *bee bee bee.* Males sing throughout the day, but especially at dawn.

OTHER BEHAVIORS. Their flight is agile and airy. Upon alighting, they wag their tail slowly, making sweeps that are usually vertical but sometimes lateral. Tail-wagging under these circumstances might be a displacement behavior rather than a balancing mechanism.

These solitary birds are fairly tame. They adapt well to human presence and readily habituate to most noises and movements. Audubon banded this species in what was the first recorded bird-banding experiment in America.

Say's Phoebe, *Sayornis saya.* Princ. distrib. Texas: Locally common PR in much of w. Texas, including the Southern High Plains. They prefer brushy pastures, canyons, and prairies and usually avoid dense woodlands. They are attracted to habitats more arid than those preferred by the Eastern Phoebe and Black Phoebe and are quite at home on vast, inhospitable grasslands that are devoid of trees and shrubs.

FEEDING BEHAVIOR. In their sparsely vegetated feeding areas, these hardy birds perch on virtually any prominence that provides a view of the surrounding area. From such vantage points, which include rocks, tall weeds, and dry cow manure, they sally out after insects, often hovering low over the grass before dropping down to capture prey. They also eat other terrestrial invertebrates, including spiders and millipedes.

COURTSHIP BEHAVIOR. Apparently the male advertises his territory only by singing.

NESTING. By necessity (because of the often harsh and variable environment in which they nest) Say's Phoebes are opportunistic in selecting a nest site (cf. Black Phoebe). Although they apparently prefer eaves, bridges, and other structures, they readily utilize cliffs, tree cavities, openings in dirt banks, or old nests of other species. Nest: probably built by the female; unlike other phoebes, they rarely incorporate mud into their nests. Eggs: incubated by the female. Both parents care for the nestlings.

VOICE. Call: a descending whistled note. Song: a melancholy, descending *phee ur*, uttered as they flick the tail and raise the crest. Another song, a *pit see ur* followed by a trill, is normally sung in flight.

OTHER BEHAVIORS. They are seen singly or in pairs. When unmolested, they become quite tame, but in general they seem more restless and wary than Eastern Phoebes. Their flight is strong and erratic, and when perched they rapidly dip and fan their tail. They do not have the same need for nearby water as the other two phoebes.

Vermilion Flycatcher, *Pyrocephalus rubinus*. Princ. distrib. Texas: Locally uncommon PR in s. and w. parts of the state, in pastures, deserts, clearings, and other open areas, generally with water nearby (Brush, 1999).

FEEDING BEHAVIOR. These stunning birds normally select a low to moderately high perch from which to survey their surroundings. They dash out for passing insects, then immediately return to the same or a nearby perch. They also hover or drop to the ground. Near streams and ponds they fly out for insects that are floating on the water's surface (and rarely small fish swimming near the surface). Bees are an important dietary component when hives are nearby.

COURTSHIP BEHAVIOR. With breast feathers fluffed, crest erect, and tail lifted, the displaying male makes a spectacular ascent high into the air. While doing this he rapidly vibrates his wings and sings. At the peak of the ascent he hovers briefly (like a butterfly), then flutters down to a perch. This behavior is triggered by the presence of his mate or an intruder, as well as by light aircraft flying overhead.

NESTING. Nest: built by the female at a moderate height in the horizontal fork of a tree, often in the vicinity of water. Eggs: incubated by the female. The male feeds his mate while she is incubating the eggs. He is fearless when defending the nesting area and does not hesitate to chase or dive at most intruders.

Both parents feed the nestlings. When, as sometimes occurs, the female begins a second nest before the first brood leaves the nest (clutch overlap), the male tends to the first clutch by himself.

VOICE. Call: a piercing *peet*. Song: a loud, energetic *pit a see*, uttered while jerking the head; it is frequently sung at night. In flight, birds produce a series of soft, sweet, twittering notes.

OTHER BEHAVIORS. When perched, they pump the tail gently, like a phoebe. These relatively tame birds are usually seen in pairs, even though they may be foraging several hundred yards apart.

Ash-throated Flycatcher, *Myiarchus cinerascens.* Princ. distrib. Texas: Locally common SR in w. half of Texas, in mesquite woodlands, juniper hillsides, desert grasslands, thorn scrub, and other semiarid habitats.

FEEDING BEHAVIOR. They forage singly or in pairs. In contrast to most of our flycatchers, they do not normally fly out to capture insects in midair; instead they dart out from a low perch, hover by a tree or bush, and pick insects from the leaves or stems. Also, they drop to the ground for prey, then return to a different perch. They sometimes eat fruits and berries.

COURTSHIP BEHAVIOR. Apparently the only recorded courtship behavior is the male's song, which he also uses to advertise his territory.

NESTING. Possibly because they tolerate spartan habitats, Ash-throated Flycatchers are opportunists when selecting nest sites and nest materials. They readily occupy drain pipes, birdhouses, Cactus Wren nests, and other sites. Nest: constructed by both parents from grass, twigs, weeds, snake skins, horse hair, cow dung, and other materials. Eggs: incubated by the female. Both parents feed the nestlings.

VOICE. Call: an abrupt *pwitt.* Song: a rasping, reproachful *prrit* and *kuh wheer.* The dawn song is a rapid series of triplets rendered as *tee fa yu.*

Great Crested Flycatcher, *Myiarchus crinitus.* Princ. distrib. Texas: Locally common SR and M in e. and s. Texas, in orchards, parks, and edges of deciduous forests.

FEEDING BEHAVIOR. These handsome birds capture insects on the wing with quick, erratic flight movements. Their extraordinary ability to maneuver in flight allows them to chase down and capture small lizards. Less frequently they search tree trunks for insects and spiders, and occasionally they eat berries and small fruits.

COURTSHIP BEHAVIOR. The male repeatedly flies at the female, and after forcing her to retreat to a hole, he hovers nearby.

NESTING. Nest: a bulky structure that both parents construct. In the absence of natural cavities, they appropriate man-made structures such as drain pipes, birdhouses, and hollow fence posts. Eggs: incubated by the female. Both parents bring food to the nestlings.

Great Crested Flycatchers attack squirrels, woodpeckers, and other small animals that enter holes and cavities; however, unlike Scissor-tailed Flycatchers, which must defend an exposed nest, they do not attack hawks, possibly because hawks present no particular threat to cavity nesters. They ignore many intruders on the basis of song. When models of orioles, vireos, and other birds were placed in the territory, the flycatchers attacked them only if the Great Crested's own song came from speakers inside the models.

VOICE. Call: a loud *prrrrrreet* that quickly reveals the presence of this noisy bird. It has been compared to the Red-headed Woodpecker's call. Song: an imposing *wheeeep*, ascending steeply in pitch and often followed by scolding notes.

OTHER BEHAVIORS. When engaging in territorial fights, males peck and claw at each other's breast and tail feathers, sometimes with such intensity that they tumble together to the ground.

Brown-crested Flycatcher, *Myiarchus tyrannulus.* Princ. distrib. Texas: Common SR in s. Texas, in large mesquites, hackberries, and other trees typical of river-bottom woodlands. Where their ranges overlap in s. Texas, the smaller Ash-throated Flycatchers seem to prefer the lower and more open stands of mesquite and cactus scrub.

FEEDING BEHAVIOR. These conspicuous, aggressive, kingbird-sized flycatchers feed on large insects, occasionally taking animals as large as hummingbirds and small lizards. Their behavior is basically like that of the similar Great Crested Flycatcher: they fly out for passing insects, hover for small animals that crawl on leaves and twigs, or descend to the ground for prey. They also eat fruits and berries.

COURTSHIP BEHAVIOR. Apparently not recorded, but probably resembles the Great Crested Flycatcher's.

NESTING. In towns, river bottoms, and other areas that contain large trees. Nest: in a natural cavity and built by both parents from hair, feathers, plant fibers, and other soft materials.

Like the Great Crested Flycatcher, they more often defend their territory against woodpeckers, wrens, and other cavity-nesting birds than against hawks, which are largely ignored. They are among the few native birds that successfully defend their nests against European Starlings.

VOICE. Call: an abrupt *whitt* or *wheeriiirrp*. Song: a husky *whit will do*, sung repeatedly before dawn.

OTHER BEHAVIORS. The demeanor of this bird is distinctive: active, aggressive, quarrelsome, and very loud, especially during the nesting season.

Great Kiskadee, *Pitangus sulphuratus.* Princ. distrib. Texas: Locally common PR in s. Texas, in orchards, towns, and open woodlands, usually near water.

FEEDING BEHAVIOR. Ordinarily, these high-profile, tropical flycatchers launch themselves from a conspicuous perch and dash after large insects; however, their large size (they are the most massive of our flycatchers) equips them to capture lizards, frogs, baby birds, and mice. Moreover, they sometimes plunge like kingfishers into the water for tadpoles and small fish that swim near the surface. They do not immerse themselves as deeply as kingfishers, and after about three dives they must pause to dry off. They also pick many berries while hopping in trees and shrubs.

COURTSHIP BEHAVIOR. Apparently not recorded.

NESTING. Nest: usually buried in dense foliage, a large, bulky structure that is oval or football-shaped, with an entrance on the side. Incubation and parental behavior are poorly known, but both parents vigorously defend the nest against intruders.

VOICE. Call: a nasal *whenk* that suggests a caricature of a crying baby; also, a variety of chattering notes. Song: a loud, deliberate *kis ka dee* (to some ears, *kiss me dear*), which indelibly registers one's first trip to the tropics or subtropics.

Tropical Kingbird, *Tyrannus melancholicus.* Princ. distrib. Texas: Rare PR in the lower Rio Grande Valley, in river groves, open wooded areas, mesquite woodlands, and towns.

FEEDING BEHAVIOR. With typical kingbird dexterity, they pursue passing insects by flying out with rapid wingbeats, hovering next to twigs or leaves, or dropping to the ground. Atypical for flycatchers, they circle like swallows, especially at dusk. They also eat fruits and berries.

VOICE. Call: rapid, staccato notes that rise in pitch. Song: a series of high notes (rendered as *pit it it it it*) followed by higher trills, usually sung at dawn.

OTHER BEHAVIORS. Unlike most of their relatives, which are generally solitary, these kingbirds form loose flocks during the nonbreeding season. They are not particularly shy around people.

Couch's Kingbird, *Tyrannus couchii.* Princ. distrib. Texas: Common SR and uncommon WR in s. Texas, in open wooded areas and often near water. Couch's Kingbird is so closely related to the Tropical Kingbird that differences in behavior between the two species (other than vocalizations) are probably too slight to be detected in the field.

VOICE. Call: a buzzy *chee kweear,* slightly higher than the corresponding call of Cassin's Kingbird, and a *preeeeeeeeee,* suggesting (to some listeners) a policeman's whistle.

Cassin's Kingbird, *Tyrannus vociferans.* Princ. distrib. Texas: Uncommon SR in Trans-Pecos Texas (locally common in Davis Mountains); they generally occur at higher elevations and in denser woodlands than the Western Kingbird.

FEEDING BEHAVIOR. They capture insects in typical kingbird manner, flying out from a perch, hovering, and dropping to the ground. Apparently this species consumes more berries and fruits than any of our other flycatchers, but this tendency may depend on insect availability instead of a preference for plant food items.

COURTSHIP BEHAVIOR. The male displays with a rapid, frenetic, zig-zag courtship flight. At other times the courting pair perch together on a limb and both birds call energetically while quivering their wings.

NESTING. Nest: a bulky cup of leaves, weeds, twigs, and debris. The sides of the nest are sometimes decorated with feathers. Eggs: incubated by the female. Both parents feed the nestlings.

VOICE. Call: the coarse, nasal *chi kweer* is similar to the Western Kingbird's but lower in pitch; also a loud *ka deer.* These are the most vociferous of our common kingbirds, possibly because they live in denser habitats where visual signals are less effective. Song: a strident *berg berg berg berg,* sung just before dawn and sometimes at night.

OTHER BEHAVIORS. They are less excitable and less active than Western Kingbirds, but like Westerns, they readily drive hawks and ravens from their territory. In Mexico, large groups of wintering birds roost together at night.

Western Kingbird, *Tyrannus verticalis.* Princ. distrib. Texas: Common SR throughout Texas, except the forests of e. Texas, and common M in most parts of the state; they are found in parks, roadsides, ranches, and other open areas as long as trees or utility poles are available. Of our three common kingbirds (the Western, Eastern, and Cassin's), the Western is most inclined toward open habitats.

Where Cassin's and Western Kingbirds overlap in distribution, Cassin's nests predominantly in riparian creek habitat and the Western in more open desert habitat (Blancher and Robertson, 1984). On the other hand, competition for food resources appears unimportant for maintaining habitat separation between these two species, as both forage in similar manners and feed their young approximately the same kinds and sizes of insects.

FEEDING BEHAVIOR. Western Kingbirds forage energetically and tirelessly. They search their surroundings from a conspicuous perch, then fly out to capture insects on the wing. They sometimes hover momentarily before dropping to the ground for prey. They also regularly consume fruits and berries.

COURTSHIP BEHAVIOR. In his courtship flight the male flies along a vertical zigzag course while chattering repeatedly and emphatically.

NESTING. Nest: a cup of weeds, grass, feathers, animal fur, and other plant and animal materials. Nests are usually placed in trees, but on treeless plains birds place them on the cross arms of power poles (Seyffert, 2001). The female incubates the eggs, and both parents feed the nestlings. Reproductive success fluctuates with environmental conditions. When insect biomass is relatively high, Western Kingbirds have earlier clutch initiation dates, larger clutch sizes, and higher nesting growth rates.

VOICE. Call: a sharp *whip*, as well as numerous sharp, high-pitched chatters. Song: usually sung at dawn. It is a more structured and more pleasing version of the chatters and squeaks heard throughout the day.

OTHER BEHAVIORS. Western Kingbirds so readily habituate to human activity that traffic and noise rarely deter them from nesting on a particular tree limb or utility pole. Although quick to chase crows and hawks from their nesting areas, and to dive-bomb human intruders, they are not as pugnacious as Eastern Kingbirds.

They migrate in small flocks. Even during the nesting season they more readily tolerate the presence of their own species than do Eastern Kingbirds, and sometimes more than one pair will nest in the same tree.

Eastern Kingbird, *Tyrannus tyrannus.* Princ. distrib. Texas: Locally common SR in e. part of Texas and e. Panhandle; locally common M throughout Texas, in open country, woodland edges, urban areas, and along roads.

FEEDING BEHAVIOR. Eastern Kingbirds dash out from a conspicuous perch to capture flying insects. They also flutter next to a tree or bush and snatch insects from the foliage, drop down to pick up terrestrial insects, or gracefully swoop to the surface of a pond or stream to take floating insects. They supplement their diet by picking berries while hovering next to a tree or bush.

Compared with other kingbirds, their feeding rates are more costly in terms of time and energy, but important benefits such as greater reproductive success accrue from this extraordinary expenditure of energy (Maigret and Murphy, 1997).

COURTSHIP BEHAVIOR. The male's spectacular aerial acrobatics include zigzags, backward somersaults, and quick ascents and descents, all seemingly orchestrated to display the bird's orange crown patch and white tail band.

NESTING. In an isolated tree, sometimes in bushes by water. Given a choice, they select habitats with denser vegetation, evidently to avoid nest predators. Nest: a bulky cup of twigs, grass, weed stalks, and other plant materials constructed by the female with help from her mate. Eggs: incubated mostly, if not entirely, by the female. Egg laying begins earlier in years when insects are abundant. Nestlings are fed by both parents, who attend to them longer than usual for a flycatcher, often more than a month.

These spirited and highly territorial birds—certainly among the most fearless and aggressive of our flycatchers—readily attack all intruders, including hawks and other birds (even other kingbirds), as well as humans. They sometimes alight on the back of a hawk, and at least one bird was seen chasing a low-flying airplane that happened over its territory.

VOICE. Call: a strident *tzee*, as well as a sputtering series of harsh *tzeeps*. Song: a series of repeated sharp notes with a phoebelike conclusion, sung just before dawn.

OTHER BEHAVIORS. They are remarkably agile in flight, maneuvering with short, quick wingbeats that sometimes appear as a flutter or quiver. They migrate in loose flocks that contain up to 70 birds.

Scissor-tailed Flycatcher, *Tyrannus forficatus.* Princ. distrib. Texas: Common SR throughout Texas (less common in the Trans-Pecos), in open areas, roadsides, and farmlands, but generally avoiding cities.

FEEDING BEHAVIOR. These flashy birds are unrivaled among passerines in their ability to maneuver in flight. They dart out from a conspicuous perch and appear to snap up flying insects with little effort.

They probably forage more for terrestrial invertebrates than do the other members of the family. When doing this, they hover low over the ground before dropping down for their prey; when on or near the ground they usually keep their tail lifted and away from the dirt. Unlike many flycatchers, they rarely consume fruits and berries.

In some parts of Texas they have learned to follow Wild Turkeys and take advantage of insects that are flushed.

COURTSHIP BEHAVIOR. The male arrives earlier in spring than the female. In one of his courtship displays, he performs a stunning sky dance. As he gracefully opens and closes his forked tail, he lifts himself to more than 100 feet, where he deftly engages in zigzag twists, turns, sharp plunges, and backward somersaults, chattering excitedly all the while. Apparently this display is also used to advertise his territory (Regosin and Pruett-Jones, 2001).

NESTING. Nest: bulky and roughly built, in a conspicuous position in a tree or shrub, or where wires connect to utility poles. As many as 90 percent of the nests in one south Texas locality were placed in mesquite trees (Nolte and Fulbright, 1996); near Post, Texas, a pair of Scissortails built a nest on the arm of a working oil pump and successfully raised their young in spite of the incessant up and down movements. Nesting materials include weeds, rootlets, and grass. Males do not engage in nest building, incubation, or brooding the young, but both parents feed the nestlings and aggressively defend their territory.

VOICE. Call: rapid *ke quees* and a sharp *kek,* as well as frequent fussing and scolding throughout the day. Song: a dawn (more aptly, predawn) song that is delicate, eerie, and mesmerizing. It is usually delivered only in the very early morning before there is enough light to see the bird. It begins slowly and deliberately with a dry *chit chit chit chit chit,* then gains speed and rises in pitch to end in an agitated but musical *chicka chicka chicka CHICK.*

OTHER BEHAVIORS. Although Scissortails are usually observed singly, in pairs, or in family units, migrants sometimes roost communally in congregations containing more than 250 birds. This tendency to flock in such extreme numbers (roosting trees) is apparently unique among our flycatchers.

These are touchy birds: they mercilessly harass hawks and ravens that happen to fly over their territory. Their flight is strong, direct, and rapid.

Anting has been reported by Husak and Husak (1997).

SHRIKES: FAMILY LANIIDAE

Loggerhead Shrike, *Lanius ludovicianus.* Princ. distrib. Texas: Locally common PR in many parts of Texas, more common in winter. They favor semiopen terrain with sufficient perches and readily utilize utility wires, trees, and fences. The principal habitat of wintering shrikes on the coast is grasslands. Habitat suitability for shrikes is improved by providing hunting perches that allow the birds to forage in otherwise unsuitable areas.

FEEDING BEHAVIOR. They drop from a perch and with their large, hooked bill capture and kill insects, rodents, and small birds, repeatedly biting vertebrates to sever the neck vertebrae. The Northern Cardinal is the largest bird known to be carried into the air by a Loggerhead Shrike.

While flying back to a perch they often transfer larger quarry from their bill to their feet. They consume small insects in one gulp but hold larger prey items with their feet before tearing off pieces to eat. They store grasshoppers and other food items by impaling them (head up) on thorns, barbed wire, or similar structures. Shrikes possess a remarkable memory for locating their larders. In Texas a frog impaled for 8 months was revisited and eaten by a Loggerhead Shrike.

They also search for prey by hopping on the ground and entering bushes. In such cases they capture only birds that they flush from the protection of the dense vegetation. They take these in flight with their strong bill.

Their feet do not compare in strength or killing effectiveness to the powerful, raptorial feet of hawks and owls, although they are stronger than those of most passerines. They sometimes use their feet to knock birds out of the air. The shrike's eyesight is extraordinary, as good as that of many hawks. They can spot, pursue, and capture a bumblebee at more than 100 yards and can identify another shrike as far as 3,000 feet away.

The closely related Northern Shrike lures small passerines with its call, then captures them, a behavior apparently unreported in the Loggerhead Shrike.

Shrikes were viewed with distaste in the past because they recalled moments in human history when specially trained torturers impaled people who were conquered in war.

COURTSHIP BEHAVIOR. The male feeds the female and performs short courtship flights about 20 feet in front of her.

NESTING. In a dense shrub or tree, often concealed among thorny branches. In southwestern Oklahoma, almost one-third of all nests were in osage orange (*Maclura pomifera*). Nest: a solid, bulky cup of grass, weeds, twigs, and other materials, built by both parents. The female incubates the eggs. At this time her mate brings her food, including items such as insects that have been stored on thorns. Both parents feed the nestlings.

VOICE. Call: a harsh scream, *sheeer chweee chweee chweee,* and a *shank shank.* Song: sweet, reflective, like a thrasher's, with trills, whistles, and liquid warbles. It is usually sung in spring, but softer versions are often heard on cool, sunny autumn days.

OTHER BEHAVIORS. Their flight pattern is diagnostic: they drop from the perch, sweep low over the ground in an undulating flight (like a wood-pecker), then suddenly swoop up to the next perch.

When their territory is threatened, they perform an aggressive wing-fluttering display. Males and females hold different feeding territories in winter.

Northern Shrike, *Lanius excubitor.* Princ. distrib. Texas: Rare WR in the Panhandle. Its behavior is essentially like the Loggerhead Shrike's, except that its slightly larger size probably enables it to take larger prey. The vocalizations of the two species apparently differ only slightly. (*See* "Feeding behavior" under Loggerhead Shrike.)

VIREOS: FAMILY VIREONIDAE

Their generally plain coloration, as well as a tendency to inhabit treetops and dense thickets, makes these small birds difficult to observe. Most are incessant singers, so eventually they can be located by tracking their song.

They move slowly and deliberately through foliage searching for insects and berries, sometimes sallying out for passing insects. Many courtship displays involve swaying the body and spreading the plumage.

White-eyed Vireo, *Vireo griseus.* Princ. distrib. Texas: Uncommon to common M and SR in e. half of Texas, becoming less common to the w.; locally common WR in extreme s. part of the state. They frequent thickets and other densely vegetated habitats, usually near water.

FEEDING BEHAVIOR. They move actively in trees and shrubs, typically low in the denser parts, where they glean for insects; they sometimes hover to capture insects on leaves. They also eat berries and small fruits.

COURTSHIP BEHAVIOR. The male utters a whining *yip yip yah* while spreading his tail and fluffing his plumage.

NESTING. Typically low in a tree or shrub. Nest: constructed by both parents from roots, twigs, shreds of bark, lichens, and other materials. Both parents incubate the eggs and bring food to the nestlings.

VOICE. Call: a single, loud whistle, a harsh mewing note, and a short *tik.* Song: husky, abrupt, and unlike that of most vireos. One common rendition for Texas birds is *chick! ticha wee o chick,* but the song varies geographically. Moreover, individual birds have several songs that they alternate while singing. They also appropriate other birds' vocalizations into their songs.

OTHER BEHAVIORS. These inquisitive birds respond to squeaking by approaching and scolding. Unlike most vireos, they are lively and not particularly shy, but they still can be difficult to locate even when they are singing. Their flight is quick and flittering.

Bell's Vireo, *Vireo bellii.* Princ. distrib. Texas: Common SR in most of w. half of Texas, becoming increasingly rare away from the Trans-Pecos and Edwards Plateau; locally common M in most parts of the state. They favor fencerows, mesquite scrub, river bottoms, and brushy habitats.

FEEDING BEHAVIOR. They feed almost entirely on insects, remaining low (usually below 12 feet) and gleaning leaves and twigs; occasionally they hover next to foliage or fly out to catch insects in midair.

COURTSHIP BEHAVIOR. The male chases the female in an active courtship flight. He leaps and flutters in front of her and then follows her, spreading his tail and singing as he does so. During the early stages of nest building, both males and females posture and display.

NESTING. Typically very low in a tree or shrub. Nest: a small, hanging cup constructed by both parents, who periodically display to each other during the early stages of construction. They utilize grass, plant fibers, strips of bark, and other materials that are held together partly by spiderwebs. Both parents incubate the eggs and feed the nestlings.

Cowbirds commonly deposit eggs in their nests; sometimes the vireos destroy the cowbird eggs or build a second floor over them.

VOICE. Call: a harsh, scolding *toh wheeo skii,* given especially when disturbed. Song: an unmusical chatter, suggesting a White-eyed Vireo in quality, but more jumbled; it has been rendered as *keedle keedle kee? keedle keedle koo,* as though asking a question, then answering it.

The male sings incessantly to announce his territory, regardless of time of day or climatic conditions. When the July temperature in central Texas exceeds 100 degrees, and most birds are seeking shade rather than someone else's territory, the loquacious little Bell's Vireo deserves at least some credit for keeping his defenses up.

OTHER BEHAVIORS. They can be quite tame if approached slowly. Flight is quick and jerky.

Black-capped Vireo, *Vireo atricapillus.* Princ. distrib. Texas: Locally uncommon SR in c. and n. Texas, particularly on the Edwards Plateau, in oaks, sumacs, persimmons, and other trees and shrubs.

FEEDING BEHAVIOR. They search for insects in dense, low vegetation, hopping restlessly from branch to branch and sometimes hanging upside down to glean from the underside of a leaf. Their diet includes berries.

COURTSHIP BEHAVIOR. The male sings as he pursues the female. He also performs a brief song-flight display that includes energetic wing fluttering.

NESTING. In dense vegetation, generally a bushy tree or shrub. Nest: a small, hanging cup constructed by both parents of grass, weeds, leaves, and bark that is held together with spiderwebs. Both parents incubate the eggs.

The 2-week incubation period (a relatively long time for small birds) allows more time for them to be parasitized by cowbirds, and in some parts of Texas cowbirds are a serious threat to Black-capped Vireos. Both parents participate in incubating the eggs and feeding the nestlings. Sometimes the female attempts to nest a second time, leaving the male to care for the first brood.

VOICE. Call: a raspy squeak. Song: emphatic and hurried, sometimes rendered as *Come here! Right now, quick!* It has a husky quality that suggests the songs of the White-eyed and Bell's Vireos.

OTHER BEHAVIORS. Elusive and not easily located, they usually appear only briefly in response to squeaking, then flit back into dense foliage. This is probably our liveliest vireo.

Gray Vireo, *Vireo vicinior.* Princ. distrib. Texas: Locally uncommon SR in Chisos, Davis, and Guadalupe mountains, and in dry mountain scrub e. to the Edwards Plateau. This habitat, mainly juniper-pinyon and oak scrub associations, plains scrubland, and chapparal, contrasts with the brushier areas and river bottoms favored by Bell's Vireo (Bryan and Lockwood, 2000).

FEEDING BEHAVIOR. When Gray Vireos search for insects, their principal food item, they tend to stay in the upper part of trees that are less than 10 feet tall. They only occasionally come to the ground. They move about jerkily, like a wren, and flick their tail more like a gnatcatcher than a vireo.

COURTSHIP BEHAVIOR. Apparently unrecorded.

NESTING. Nest: a rounded cup constructed of grass, leaves, weeds, and other plant parts, built by both parents, who weave it onto a twig in a shrub or tree, sometimes as low as a foot from the ground. Both parents incubate the eggs and bring food to the nestlings.

Gray Vireos are regularly victimized by cowbirds. They often respond to intrusions by laying a second floor of fine grass over the cowbird eggs.

VOICE. Call: a harsh, scolding *schray.* Song: liquid whistles, recalling the Blue-headed Vireo's song. It has been rendered as *chee bee, chee bee chick veer chee bur* and is considered one of the finest of the western vireo songs.

OTHER BEHAVIORS. Like Black-capped Vireos, they are shy but very active. With a little patience one can approach them rather closely. Their flight is quick and usually from bush to bush. They defend feeding territories in winter.

Yellow-throated Vireo, *Vireo flavifrons.* Princ. distrib. Texas: Uncommon to common M and SR in e. half of Texas, locally common to the w., in deciduous woodlands with tall trees, including river bottoms and residential areas. They generally avoid habitats with dense undergrowth, as well as coniferous forests.

FEEDING BEHAVIOR. These vireos move slowly, methodically, and deliberately. They make their way along limbs high in the treetops, picking insects from twigs and leaves. They also eat berries, especially in the fall.

COURTSHIP BEHAVIOR. The male leads the female to potential nest sites, where he performs ritualized nest-building behavior by crouching horizontally, yet without holding nest-building materials in his bill. Frequently a behavior loses one or more components when it evolves into a display. In this case, holding nesting material in the mouth was lost, but the posture remained.

NESTING. Nest: built by both parents from weeds, leaves, grass, and plant fibers. It is bound with spiderwebs and decorated on the outside with mosses and lichens. Both parents incubate the eggs and feed the nestlings, each parent caring for half of the fledglings.

VOICE. Call: a harsh scolding note. Song: suggests the Red-eyed Vireo's, but is slower, huskier, and more deliberate. The rich, reedy notes are slurred together and at frequent intervals contain a distinctive, two-note descending slur. To some listeners, the total effect suggests questions followed by answers. Others hear the phrase *Dee ar ee Come here,* repeated several times, and concluding with *three-eight.* Birds often continue singing into September.

OTHER BEHAVIORS. Typical of most vireos, they move slowly among dense foliage high in the treetops and are not easily located. Around the nest they can become fearless.

Plumbeous Vireo, *Vireo plumbeus;* **Cassin's Vireo,** *Vireo cassinii;* **Blue-headed Vireo,** *Vireo solitarius.* Princ. distrib. Texas: Plumbeous, uncommon SR in mountains of Trans-Pecos Texas in juniper-pine oak woodlands above 5,000 feet; Blue-headed, uncommon to common M throughout most of Texas, primarily in e. Texas, occurring in wooded or brushy habitats; Cassin's, rare M in w. Texas. These three species were formerly regarded as a single species, the Solitary Vireo, *Vireo solitarius*. Apparently no differences in their behaviors have been reported.

FEEDING BEHAVIOR. They move about deliberately while searching for insects, typically high in trees but occasionally in lower bushes, flying out occasionally for passing insects. They eat very few berries except on their wintering grounds.

COURTSHIP BEHAVIOR. The male fluffs his yellow flank feathers as he bows, bobs, and sings to the female. During his nest-building display he crouches and holds his body horizontally. He sometimes builds one or more nests before he is paired, a behavior that seems more closely related to pair formation than to nesting.

NESTING. In a horizontal fork of a tree branch and typically lower than 12 feet above ground. Nest: a bulky open cup built by both parents from grasses, rootlets, weeds, and other materials and decorated on the outside with pine needles, moss, and pieces of paper. Both parents incubate the eggs. The incubating female is a close sitter and will allow a person to feed and stroke her. Both parents feed the nestlings.

VOICE. Call: a nasal, trumpeting *seea weep*. Song: somewhat vireolike. It is a series of short phrases resembling the Red-eyed Vireo's, but higher pitched, clearer, and slower.

OTHER BEHAVIORS. Not particularly shy and even fearless around the nest; however, often they are not easy to locate because of their preference for treetops and dense vegetation. Their flight is strong and swift.

Hutton's Vireo, *Vireo huttoni.* Princ. distrib. Texas: Locally common SR in mountains of Trans-Pecos Texas, in canyons and along rocky slopes. They favor low trees and shrubs (oaks, junipers, and pinyon pines) over thickets and dense understory.

FEEDING BEHAVIOR. Hutton's Vireos glean twigs and foliage for insects, occasionally pausing as if to listen for insects. They sometimes hover briefly as they pick prey from the foliage. They also eat a few berries.

COURTSHIP BEHAVIOR. The male fluffs his feathers, fans his tail, and gives a whining or snarling call.

NESTING. Nest: a round, cup-shaped structure built by both sexes from lichens, moss, grass, and bark fibers that are bound with spiderwebs. Birds cover the nest with plant down and spider egg cases. Both parents incubate the eggs and feed the nestlings, sometimes bringing them food up to 3 weeks after they have fledged.

VOICE. Call: a low *whit whit*. Song: a single note repeated incessantly — arguably "the most monotonous song of all the vireos" (Kaufmann, 1996).

OTHER BEHAVIORS. They are very similar in appearance to the Ruby-crowned Kinglet, but they are relatively tame and much less active, twitching their wings less when perched or moving about. Around the nest they are surprisingly fearless.

Warbling Vireo, *Vireo gilvus.* Princ. distrib. Texas: Uncommon and local SR in n.c. Texas, most of e. Texas, and mountains of Trans-Pecos Texas, in deciduous forests and thickets; they generally avoid unbroken mature forests. In migration they are found throughout the state in most woodland habitats.

FEEDING BEHAVIOR. Warbling Vireos generally stay high in the tree-tops, where they glean, hover, or fly out for insects. Insects make up almost their entire diet. When feeding, they sometimes hang upside down like chickadees.

COURTSHIP BEHAVIOR. The male's courtship song elicits wing quivering in the female. He also spreads his wings, fans his tail, and hops around the female.

NESTING. Nest: a deep, compact, cup-shaped nest built by both parents from grass, leaves, bark strips, and other plant materials. Both parents incubate the eggs and feed the nestlings. They are fearless around the nest, but they are frequently victimized by Brown-headed Cowbirds.

VOICE. Call: a harsh, tense, ascending *kweeee,* suggesting the Gray Catbird's mewing. Song: a languid, flowing, unhurried warble, rendered as *brig adier, brig adier, brigate,* sung an estimated 4,000 times a day. The male sometimes sings even while he incubates.

Philadelphia Vireo, *Vireo philadelphicus.* Princ. distrib. Texas: Uncommon spring M in e. half of Texas, locally common on the coast, in woodlands and urban areas.

FEEDING BEHAVIOR. They hop deliberately from branch to branch, frequently lifting their wings slightly.

VOICE. Call: a soft *churr,* often uttered when disturbed. Song: suggests a Red-eyed Vireo, but higher, slower (about half the speed), and with weaker phrases. Philadelphia Vireos sing only infrequently during migration.

Red-eyed Vireo, *Vireo olivaceus.* Princ. distrib. Texas: Uncommon to common M and SR in e. half of Texas, much less common to the w., in open deciduous and mixed woodlands, including orchards and parks, and frequently near rivers.

FEEDING BEHAVIOR. They search methodically and deliberately in treetops, gleaning for insects, but sometimes hovering to pick insects from the foliage. In late summer and fall they feed on berries.

COURTSHIP BEHAVIOR. The male sleeks down his feathers. As he faces the female, whose feathers are fluffed up, he sways his head and body from side to side, then both birds simultaneously vibrate their wings.

NESTING. Unlike most of our vireos, the female contributes considerably more than the male to the process of building the nest and raising the young. Nest: a compact, dainty, cup-shaped structure of grass, rootlets, and bark that is attached by its rim to a horizontal, forked branch. Eggs: incubated by the female. Both parents bring food to the nestlings; curiously, the male sometimes sways back and forth when feeding them.

They respond to cowbird parasitism by building a second floor over the cowbird's eggs.

VOICE. Call: a harsh, mewing *tschay.* Song: a series of short, monotonous phrases that rise and fall to give the impression of asking a question, then answering it (*You see it, you know it, do you hear me, do you believe it?*). The song has been compared to the caroling of American Robins, but it is softer. Birds sing incessantly and throughout the day as they feed, continuing through the hottest parts of August. This species probably holds the record for singing the most songs during the day: 22,197 recorded during one 10-hour stretch. About 40 song types have been documented, and the same song type is rarely sung in succession.

CROWS AND JAYS: FAMILY CORVIDAE

Corvids, especially crows and ravens, give the impression of being intelligent, temperamental, and emotionally complex — traits we ordinarily associate with a highly developed or specialized brain, such as our own. (*See* "Other behaviors" under American Crow.) Anatomically, most corvids are generalists rather than specialists: their bills are stout but do not equip them to crack seeds as efficiently as parrots and most finches, and their unspecialized wings, feet, and bills render them far inferior to hawks and owls when chasing down and capturing animals.

Members of this family tend to be sociable. Some jays have complex breeding behaviors. Courtship involves a number of vocalizations, as well as unique postures, display flights, and courtship feeding.

Steller's Jay, *Cyanocitta stelleri.* Princ. distrib. Texas: Locally common PR in Guadalupe and Davis mountains, in pine-oak forests; they favor yellow pine and Douglas fir at higher elevations in the Guadalupes.

FEEDING BEHAVIOR. Although they typically confine themselves to the upper levels of trees, they sometimes come to the ground, especially in picnic areas. They eat many acorns and pine nuts, a fair number of insects, bird eggs, berries, fruits, and table scraps from picnic tables, and they occasionally capture small lizards and rodents. They crack open acorns by pounding them with their bill, and when feeding, they often flick their wings and tail, or whack their bill against a limb (possibly displacement behavior).

Steller's Jays cache food to be consumed later. They also steal food from the caches of Acorn Woodpeckers.

COURTSHIP BEHAVIOR. The male circles the female, alights nearby, then jumps and turns 180 degrees. He also engages in courtship feeding.

NESTING. In a coniferous tree. Nest: a bulky cup of twigs, leaves, and moss, built by both sexes. Eggs: incubated by the female. Both parents bring food to the nestlings. During the nesting season the young are quiet and secretive unless approached; then they become aggressive and noisy.

VOICE. Call: a distinctive *shook shook,* often given in flight, and a scream that resembles the Red-tailed Hawk's call. Song: soft and sweet, but rarely heard; it has been compared to the Ruby-crowned Kinglet's song.

OTHER BEHAVIORS. In many ways (especially behaviorally) Steller's Jays can be regarded as the western counterparts of the Blue Jay. They are bold and aggressive around campgrounds, yet shy and elusive in open forests. They are slightly less gregarious than other jays, though they do live in small flocks during the nonbreeding season. They are often seen flying single file across clearings in the forest.

They defend a territory immediately around the nest, but they join other birds, even during the breeding season, to feed together at other sites.

Blue Jay, *Cyanocitta cristata.* Princ. distrib. Texas: Common to abundant PR in e. two-thirds of Texas, becoming less common to the w. (although common in many w. Texas cities and towns); they occur in forests as well as rural and urban areas.

FEEDING BEHAVIOR. These omnivorous birds are consummate opportunists, feeding on acorns, seeds, grain, fruits, insects, snails, bird eggs, frogs, baby birds, mice, and even carrion. About 75 percent of the diet is plant material.

They forage in trees and on the ground, and they readily come to feeders. They pound hard nuts with their bill to break them open, and they sometimes store nuts in holes in the ground (often with the inadvertent result of planting trees). When carrying acorns, they may hold one in the throat and one in the mouth. They seem to utilize all of the usual passerine foraging tactics except flycatching.

COURTSHIP BEHAVIOR. Mainly aerial chases, courtship feeding, and up and down bobbing.

NESTING. In a vertical crotch or horizontal branch of a tree. Nest: a bulky cup of grass, bark strips, twigs, and other plant materials, built by both sexes and decorated with debris such as rags, strings, and paper. The male rarely incubates the eggs, but he feeds the female as she incubates. Both parents feed the nestlings.

VOICE. Call: a loud, raucous *jay jay*, as well as many varied notes, including a rusty-gate call. Blue Jays produce a call that is remarkably similar to the call of the Red-shouldered Hawk (*kee-yeeer*). The resemblance between the two calls might be coincidental and may not represent true mimicry. (Strictly speaking, mimics — vocal as well as visual — must benefit in some way from imitating another animal; *see* "Voice" under Northern Mockingbird.) On the other hand, Blue Jays have flushed birds from birdfeeders by producing the Red-shouldered Hawk's call (Clench, 1991).

The female makes a rapid clicking call of unknown function. Calls and postures are sometimes coordinated. For example, they raise and lower the head as they produce a *tull ull* call.

Flocks in fall and winter are especially noisy, frequently shrieking in chorus while flying from tree to tree. Song: low, soft whistles and chipperings (surprising for such a garrulous bird). Sung from a perch concealed in dense vegetation, it is unknown to most observers in spite of the bird's abundance and generally high profile.

OTHER BEHAVIORS. Jays are intelligent, cunning, inquisitive, and mercurial in temperament, and often difficult to ignore. Yet they become secretive and inconspicuous when returning to their nest, or when seeking another bird's eggs to rob. They are usually wary and flee readily when approached, yet they can be taught to eat out of one's hand. They commonly mob owls and hawks, presumably because these raptors pose a threat to them — yet they also mob herons (perhaps because herons resemble large hawks).

Blue Jays have been known to guard and feed old, blind, or otherwise disabled jays, a behavior they normally direct toward young birds (where the releaser for such behavior is more obvious). They sunbathe by facing the sun with spread wings, either on a perch or by spreading out on the ground. They also engage in anting, a curious behavior in which a bird takes an ant in its bill and rubs it over its plumage.

Prior to the breeding season, females become more aggressive and males less aggressive at feeders.

Green Jay, *Cyanocorax yncas.* Princ. distrib. Texas: Common PR in extreme s. Texas, becoming less common to the n., in mesquite brush, parks, and other wooded areas, especially trees and shrubs along the Rio Grande.

FEEDING BEHAVIOR. They consume insects, spiders, seeds, and other items as they move actively through the vegetation. They often drop to the ground for food items, occasionally flycatch for passing insects, and actively search for bird eggs. They sometimes hoard what they collect.

COURTSHIP BEHAVIOR. The female bobs a few times, and the male responds by assuming a sleek posture; this display is followed by mutual bill caressing.

NESTING. Concealed in dense shrubs or trees. Nest: a bulky structure made of sticks, built by both parents. Only the female incubates the eggs, and at this time she is fed by her mate. She is difficult to flush from the nest. Both parents feed and care for the young.

In California and some tropical areas, young adult birds help the parents care for the brood. In Texas, young birds remain in the family flock for one year, but they are forced from their natal territory by the breeding male after the young from the following year fledge. During that first year the yearlings provide a significant amount of territorial defense that spares the parents much of the energy costs demanded "by the stringent territoriality required in south Texas habitat" (Gayou, 1986).

VOICE. Call: a rapid, energetic *chick chick chick,* or slower *cleep cleep cleep,* as well as a rattle that has been compared to the sound of a cricket frog. These and other calls are heard year round.

OTHER BEHAVIORS. The feeding territory is defended by the pair or by the social group throughout the year. These noisy, lively birds can be very elusive, yet they can become tame and bold around gardens and public picnic tables, where they commonly eat leftover junk food.

They are not without a keen sense of curiosity. When a person enters a brushy area occupied by Green Jays, the birds approach the intruder and scream, caw, and coot—then they retreat. Oberholser and Kincaid (1974) observed that this jay "is skillful at keeping itself concealed when stalked, but if the stalker turns to walk away, several birds are quite likely to set up an outcry; some may even come to the edge of the brush to look the intruder over."

One jay in south Texas was observed smoke-bathing, anointing its plumage with smoke produced by a smoldering log.

Brown Jay, *Cyanocorax morio.* Princ. distrib. Texas: Locally uncommon PR in parts of extreme s. Texas, in relatively tall, dense vegetation along the Rio Grande. Very few nests have been reported from Texas; farther s. they nest communally.

FEEDING BEHAVIOR. Our largest jay, the Brown Jay is large enough to feed on small rodents and lizards. They also eat berries, insects, seeds, nuts (which they pound to open), and fruits. They feed by hopping around on the ground or by moving about in dense cover. Normally they occur in flocks.

VOICE. Call: a shrill, abrupt *peow peow,* often given at dawn and audible a quarter of a mile away; also, a popping of the throat, produced by an air sac connected to the lungs, that is often heard while the bird is calling. Other calls suggest the calls of the Blue Jay.

Western Scrub-Jay, *Aphelocoma californica.* Princ. distrib. Texas: Common PR in c. Texas, the Edwards Plateau, and many parts of the Trans-Pecos; locally common in the Panhandle and Rolling Plains. They occur in juniper and mesquite scrub, urban areas, and canyons. Juniper brakes seem to provide an especially important habitat on the Edwards Plateau.

FEEDING BEHAVIOR. They forage on the ground, where small groups search for insects, seeds, nuts, and small animals. They steal acorns from Acorn Woodpecker caches.

Sometimes they bury acorns, which presumably they later retrieve. Curiously, they also bury shiny objects such as pieces of glass. They swing their bill from side to side when retrieving buried objects.

COURTSHIP BEHAVIOR. The male drags his spread tail, erects his head feathers, and hops around the female. The pair bond is long-term.

NESTING. In a shrub or tree. Nest: a thick-walled cup made from grass, twigs, and moss, built by both parents, who construct it more carefully than do most jays. Incubation is by the female, who is brought food by the male while she is on the nest. Both parents feed the nestlings.

The very closely related Florida Scrub-Jay (*Aphelocoma coerulescens*) breeds in cooperative family groups, but apparently all Texas birds nest in isolated pairs.

VOICE. Call: a loud, harsh, and nasal *greenk greenk,* which rises in inflection; also, *quay quay quay.* Other distinctive calls include low throat rattles and a noise that sounds to some listeners like the sudden ripping of heavy canvas. Song: infrequently heard. It consists of soft, musical trills and coos, the so-called whisper song.

OTHER BEHAVIORS. Scrub jays are regarded as intelligent, alert, bold, quarrelsome, and sometimes playful. In remote areas they seem secretive and elusive, but in picnic areas they become bold and tame and learn to take peanuts from a person's hand. Their shifting moods seem to be revealed by their long, expressive tail.

They mob many animals, including snakes—but not all snakes. They eat small ones, ignore medium-sized ones, and mob only large ones (about 2 feet or more in length). Breeding males mob more intensely than breeding females and helpers.

Although in general they prefer the ground and the lower parts of trees, often they can be seen diving swiftly and gracefully from the top of a tree to a concealed perch in a nearby bush.

Mexican Jay, *Aphelocoma ultramarina.* Princ. distrib. Texas: Common PR in juniper-pine-oak forests of Chisos Mountains, to about 7,000 feet. Populations in Texas and Arizona are disjunct; they differ in several ways that include egg color, voice, and nesting behavior.

FEEDING BEHAVIOR. These jays are very dependent on a good supply of acorns. They open acorns by pounding them with their bill, or they bury them for later consumption. Usually several birds forage together in trees, less frequently on the ground. In addition to acorns, they eat seeds, insects, lizards, and other items.

COURTSHIP BEHAVIOR. The male circles the female with his wings and tail tilted. He aligns himself sideways to her, then quickly reverses his direction by jumping 180 degrees.

NESTING. In an oak, juniper, or pine tree, and usually well concealed. Nest: a bulky cup of twigs, sticks, and other plant parts, built by both parents. Eggs: incubated by the female. Both parents, as well as other members of the flock, bring food to the nestlings.

The nesting behavior of our Texas birds needs further study. Their mating system is probably less complex than that of the Arizona birds, but Texas birds have not been studied in the same detail. Arizona birds form nesting flocks of up to 22 males and females (in different proportions);

Texas birds tend to nest in smaller groups or in isolated pairs. The complex social system of Arizona birds (delayed maturation, plural breeding, helping, and loss of the rattle call) is thought to have arisen from a simpler state that had occasional nonbreeding associates (Brown and Li, 1995).

Interestingly, inbreeding in Mexican Jays is rare, despite their strong tendency to remain in the nesting area and associate with close relatives.

VOICE. Call: a rasping *wait wait wait,* and a querulous *jee ee eenk,* repeated several times. They do not seem to have a clear territorial call, perhaps because they live throughout the year in flocks, making territorial defense largely unnecessary.

OTHER BEHAVIORS. These are generally wary birds, but they are inquisitive enough to respond to squeaking sounds. They fly through trees in a leapfrog manner: those in the rear fly over birds in the front, which then fly over the birds that just passed over them.

Flocks, consisting primarily of close relatives, defend permanent territories that are passed down through several generations. These birds scream noisily at hawks, sleeping bobcats, foxes, and other potential predators, and there are reports that they have surrounded and killed a snake (Freeman, 2000).

American Crow, *Corvus brachyrhynchos.* Princ. distrib. Texas. Common to abundant PR in e. half of Texas, frequenting farmlands, river groves, and urban areas; in some areas, Fish Crows exclude them from shore habitats. In the parts of w. Texas where the ranges of American Crows and Chihuahuan Ravens overlap, crows prefer river bottoms and ravens occur mainly in the arid lands between watercourses.

FEEDING BEHAVIOR. To reach good foraging sites, these industrious birds often fly 50 or more miles from their roosting areas, returning at night to roost. They are omnivorous and notably opportunistic birds that feed on the ground, taking a wide variety of food items that include seeds, grains, insects, frogs, carrion, garbage, and bird eggs. They have been observed opening hard-shelled mollusks by carrying them high into the air and then dropping them on rocks, but this behavior has not been observed in Texas.

COURTSHIP BEHAVIOR. The male faces the female, fluffs up his body feathers, spreads his wings and tail, and utters a short rattling song. He repeatedly bows as he does this. He also pursues the female in flight, wheeling and diving during the chase.

The pair reinforce the pair bond by perching near each other, touching bills, and preening each other's feathers.

NESTING. Usually solitary; in a tree or shrub, at least 10 feet above ground. Nest: a large, bulky structure of twigs, bark, weeds, and mud, built by both males and females. The female usually incubates the eggs. Helpers — usually offspring from previous generations — sometimes aid the parents in raising the young.

VOICE. Call: an emphatic, far-carrying *caw caw caw*. It has the same intensity as the higher-pitched *jay jay jay* calls of the related Blue Jay. American Crows produce at least 23 different calls; they also can be taught to reproduce the human voice and other sounds.

OTHER BEHAVIORS. These are exceptionally gregarious birds; sometimes thousands roost communally in the same grove. Cooperative groups maintain shared territories throughout the year, even though they travel considerable distances to roost together at night. Groups readily assemble to mob owls or large hawks. In nature this is a very wary bird, but it can be tamed as a pet when young.

Flight is steady and direct, as if rowing through the air. In contrast, both of our ravens flap and glide, often along an irregular course, and sometimes soar like vultures.

Crows and ravens quickly solve problems that humans associate with intelligence (for example, adding or subtracting); they readily learn how to deal with novel situations (for example, taking advantage of an unconventional food source); and they respond to subtle environmental stimuli to which less intelligent birds are oblivious (for example, distinguishing an aggressive from a nonaggressive gait in a predator). Not surprisingly, anecdotal evidence abounds regarding the American Crow's savvy. One observer reported that American Crows announce the approach of a person in different ways, depending on whether the intruder is carrying a shotgun or not.

Tamaulipas Crow, *Corvus imparatus.* Princ. distrib. Texas. Locally common WR in Cameron County.

FEEDING BEHAVIOR. In Texas, Tamaulipas Crows scavenge refuse areas in Brownsville, alongside gulls and Chihuahuan Ravens. In Mexico they eat a number of different food items, including carrion, seeds, and insects.

VOICE. Call: Similar to that of the American Crow, but much lower than would be expected for a crow this size (it is the smallest crow in Texas).

Fish Crow, *Corvus ossifragus.* Princ. distrib. Texas. Locally uncommon PR in parts of e. Texas, especially the Sabine River north to Newton County;

they inhabit cypress swamps and other primarily freshwater habitats, as well as urban areas, farmlands, and wooded areas that are near water.

FEEDING BEHAVIOR. They are clearly omnivorous, in spite of their name, although they do fish in shallow water by grabbing minnows with their claws. They normally feed in flocks, walking along the shore or in very shallow water, where they not only search out animals such as crayfish, insects, and crabs but also eat carrion and garbage. They sometimes hover over shallow water looking for food items.

Fish Crows are known to carry mollusks into the air and drop them on rocks to break the shells; on the coast, they dive at gulls and terns and force them to give up their catches. Near heronries and other birds' nests, they sometimes eat large quantities of eggs after frightening the parent birds from the nest.

COURTSHIP BEHAVIOR. Birds fly near each other in a gliding display, approaching each other closely and touching wings and heads.

NESTING. They often nest in loose colonies. Nest: a bulky platform of sticks and strips of bark, built by both parents and lined with grass, rootlets, feathers, and miscellaneous items (including cow manure). Eggs: generally incubated by the female. Both parents bring food to the nestlings.

VOICE. Call: similar to the American Crow's, but more nasal; it has been compared to the Black-crowned Night-Heron's *quowk*. They also utter a two-note *oh oh.*

OTHER BEHAVIORS. They are more social and more gregarious than American Crows. Roosting flocks sometimes contain several thousand birds that include American Crows. Compared with American Crows, Fish Crows fly with quicker wingbeats, and they sail more frequently (flapping and gliding more like ravens). They are able to drink while skimming the water.

Chihuahuan Raven, *Corvus cryptoleucus.* Princ. distrib. Texas: Common to abundant PR in much of w. and s. Texas, preferring arid and semiarid grasslands to mountains and forests.

FEEDING BEHAVIOR. They consume a wide variety of food items, including seeds, grains, insects, frogs, bird eggs, and small animals. Frequently they scavenge dead livestock or consume garbage. They have learned, like vultures, to patrol highways for animals killed by automobiles (operant conditioning). They sometimes temporarily store certain foods in caches.

COURTSHIP BEHAVIOR. The male's acrobatic aerial displays include wheeling, soaring, and tumbling; on the ground, and in the presence of the female, his courtship displays include fluffing his neck feathers.

S. F. Blake (1957) alluded to a possible communicatory role for the white throat patch: "Every now and then the breast, and at times also the throat, of one bird would flash out brilliantly white, especially, it seemed, when the other bird caressed its beak with its own."

In some areas they breed later in the season if summer rains provide more food resources at this time, but apparently this behavior has not been observed in Texas.

NESTING. In a tree, yucca, utility pole, or shrub, occasionally on buildings or other artificial structures; they sometimes nest in loose colonies. Nest: probably constructed by the female, from sticks, thorny twigs, and stems, and lined with materials that include rags and pieces of paper. Both parents probably incubate the eggs and feed the nestlings. For such a common bird, it is surprising that we know so little about its breeding biology.

VOICE. Call: guttural croaks lower in pitch than the American Crow's call, but not as low as the Common Raven's.

OTHER BEHAVIORS. They are more sociable and gregarious than Common Ravens, and in this way are more like American Crows. They are not shy, and they often visit parks and other urban areas to feed.

Birds feeding together on the ground have been compared to a moving mass, in which birds at the rear constantly replace those in front. They often soar high on thermals on hot summer days.

Perhaps it is a form of play when they plunge into small whirlwinds (dust devils), then let the funnel carry them to great heights. Groups also perform spectacular aerial acrobatics, including tumbling and (according to some reports) sailing short distances on their backs.

Common Raven, *Corvus corax.* Princ. distrib. Texas: Uncommon PR in mountains of Trans-Pecos Texas, e. to the Edwards Plateau, in most habitats, including desert regions and mountain forests.

FEEDING BEHAVIOR. Common Ravens consume a wide variety of food items that include seeds, grains, insects, frogs, garbage, and bird eggs. They seem especially drawn to carrion, and in this respect they compete with vultures. When traveling to and from feeding areas, they often fly in a more or less straight line.

At carcasses, especially animals killed by automobiles, immatures and adults behave differently. Immatures aggregate into groups to feed on the same carcass, whereas breeding adults are territorial and defend a particular carcass. They cache food by burying it.

COURTSHIP BEHAVIOR. In the male's aerial display, he soars, wheels, and tumbles. Often the male and female soar together, the male above. Occasionally they touch wingtips. They also perch together, preen, and touch bills. Apparently most pairs mate for life.

NESTING. In a tall tree or on a cliff or ledge. Nest: a bulky and deep structure, contributed to by both parents, using large sticks, bark strips, and animal hair and often adding material to the previous year's nest. They never retrieve nest materials that fall from the nest, so large numbers of sticks accumulate below. The female is primarily responsible for incubating the eggs, and while she incubates, her mate brings her food. Both parents feed the nestlings.

VOICE. Call: a deep, guttural croak (lower than that of the Chihuahuan Raven), as well as bell-like gurgles and other sounds.

OTHER BEHAVIORS. The largest of our passerines, they are sagacious, resourceful, and quick to profit from experience. Their shaggy throat feathers are probably incorporated into both courtship and territorial displays. On the ground they walk in a stately manner, but they can also hop forward or sideways.

They perform spectacular aerial dives and other acrobatics that seem to be a form of play. They soar and hover like hawks, and occasionally they engage in mock combat with hawks. Interestingly, American Crows sometimes mob Common Ravens as intensely as they mob hawks.

These are not particularly social birds, but in winter they sometimes roost in small flocks. These flocks are much smaller than those of wintering American Crows or Chihuahuan Ravens. Apparently most birds roost and disperse independently, without significant flock cohesiveness.

Nocturnal roosts appear to serve as information centers, where birds that know the location of food pass this information along to other birds, possibly with vocal signals.

LARKS: FAMILY ALAUDIDAE

Horned Lark, *Eremophila alpestris.* Princ. distrib. Texas: Locally common PR on the High Plains and on the coast, in short-grass prairies, airports, shorelines, and other open habitats, especially those that offer a fair amount of bare ground. They seem to be increasing in numbers, perhaps because they adapt so well to overgrazed or abused land. In winter they visit freshly manured fields.

FEEDING BEHAVIOR. Outside the nesting season Horned Larks are usually seen in small flocks, walking or running in a manner reminiscent of mice as they glean for insects and seeds. Their generalized bill (neither as thin as an insectivorous warbler's nor as stout as a granivorous sparrow's) allows them to eat both insects and seeds.

COURTSHIP BEHAVIOR. The male's extraordinary flight-song display begins with a silent ascent to 800 feet or more above the ground. Once there, he flies in a large circle while singing a tinkling flight song sometimes rendered as *pit wit wee pi pit wee wee pit.* Suddenly he folds his wings and dives, quickly arresting his fall before reaching the ground. Shelley based his poem "To a Skylark" on the similar flight song of the Eurasian Skylark (*Alauda arvensis*).

The male also displays on the ground: he struts in front of the female with drooped wings and erect horns (the small black feathers on each side of the head).

NESTING. In a shallow depression next to a dirt clod or dried cow manure. Nest: The female lines the scrape with grass, rootlets, feathers, and plant down, often laying a flat doorstep of pebbles on one side of the nest. Eggs: incubated by the female. Both parents feed the nestlings.

When a person or animal approaches the nest, the female flies low over the ground for a short distance; if the intruder approaches too closely, she flutters away in a distraction display.

VOICE. *See* "Courtship behavior." Call: a thin *zeep.* Often a clod of dirt is a sufficient perch from which to call or sing.

Horned Larks have two types of songs, each delivered from the ground or in flight. The intermittent songs last about 2 seconds, but the recitative songs sometimes extend more than a minute. Ground songs advertise territory, and flight songs function in courtship.

OTHER BEHAVIORS. When disturbed, they fly away swiftly and erratically, pressing their wings tightly against the body between flaps to produce an undulating flight path. Migrating flocks are sometimes enormous and in some areas join flocks of Lapland Longspurs.

The elongated hind claw, or spur (the "larkspur" that lent its name to flowers with elongated, spurlike corollas), helps define the family to which all larks belong. How such an extreme structure functions in birds that habitually walk and run is not clear, since most cursorial birds such as quail and plovers increase efficiency in locomotion in quite the opposite way: the hind toe is elevated, reduced, or even lost. Possibly the long claw in larks helps the bird walk on soft substrates.

SWALLOWS: FAMILY HIRUNDINIDAE

Thanks to their long, pointed wings and streamlined bodies, swallows are capable of flying swiftly and gracefully (although not as rapidly as swifts). Their bills are small but their gapes are wide, an obvious advantage when capturing insects in flight. Most swallows forage in flocks.

Their forays often take them low over water; in this habitat they compete with very few other birds for the often high concentrations of insects that occur on or just above the surface of the water.

At rest they almost always sit on a utility line or other exposed perch, spacing themselves more or less evenly along the line. Sometimes they sunbathe, possibly to control ectoparasites. (They sunbathe less frequently when experimentally treated with pesticides.)

Many swallows are notably gregarious, feeding in flocks and nesting in colonies. The nest is usually the only territory that is defended in colonial species. Thus, territorial displays (at least the spectacular postures and stereotyped movements typical of most passerines) are essentially absent. Also, except for courtship flights, courtship displays in swallows seem relatively undeveloped.

The nest may be a cup inside a cavity (including birdhouses), a burrow in a vertical dirt bank, or a cup that is secured to a vertical surface like a cliff or wall.

Purple Martin, *Progne subis.* Princ. distrib. Texas: Common SR throughout most of Texas (except parts of the Panhandle), and common M e. to the Trans-Pecos, in virtually all semiopen habitats that provide nesting sites, including urban areas, especially if water is nearby. They are inexplicably absent from many apparently suitable areas in the state.

FEEDING BEHAVIOR. These seemingly indefatigable birds sweep gracefully over open water as they pursue beetles, bugs, flies, wasps, and a few mosquitoes (not the 2,000 mosquitoes a day, as previously claimed). They often ascend high into the air, and in harsh weather they may come to the ground to search for insects. Their fondness for bits of eggshell placed around martin boxes probably indicates a nutritional need for calcium.

COURTSHIP BEHAVIOR. The first birds to return to the nesting areas in spring are males (called scouts), which usually arrive in February to establish territories, in most cases a man-made martin house. Apparently the courtship displays have not been recorded.

A male may have one or several mates, but in any case he secures a room in a nesting box, defends it, then mates with the female that selects the room.

Floaters (nonbreeding, nonterritorial individuals that wander widely in search of an opportunity to breed) can be seen flying tirelessly around martin houses in an attempt to gain control over one of the several cavities defended by a resident male. Although floaters rarely win fights with resident males, their persistence often rewards them with a cavity.

NESTING. In Texas, most Purple Martins nest in colonial nest boxes, some quite large; one contained more than 350 rooms. In many areas, especially in the western United States, they utilize woodpecker holes and other natural cavities. There is a strong tendency for parents to return to the same nest site each year, but very few of their young return (whether by choice or necessity). Nest: a cup of grass, mud, twigs, and other materials constructed in the box by the parents, who occasionally place a dirt rim on the outside that prevents eggs from rolling out. Eggs: incubated by the female. Both parents feed the nestlings.

As long ago as the late 1800s it was noted that House Sparrows and European Starlings compete with Purple Martins by invading their nesting cavities and displacing them.

VOICE. A rich chirruping, with many low-pitched gurgles and a variety of other sounds; these are given in flight and when perched, especially during the predawn hours. Brown (1984) described 10 types of vocalizations from colonial-nesting Texas birds and solitary-nesting Arizona birds.

OTHER BEHAVIORS. Around their nesting boxes they turn, soar, and dart about noisily, alternating rapid wingbeats with glides and sweeps; over open water they skim low to drink and to bathe.

Before and during migration, Purple Martins form enormous roosting flocks of 100,000 or more individuals.

Tree Swallow, *Tachycineta bicolor.* Princ. distrib. Texas: Common M throughout most of Texas, in most open habitats, wintering locally on the coast.

FEEDING BEHAVIOR. They fly low over water or fields for insects, perch in bushes for berries, and occasionally come to the ground for food. They eat more berries (20 percent of the diet) than any other swallow, especially in cold weather, but the importance of berries for Texas migrants and winter residents is not known.

VOICE. Call: a thin, liquid *chileet,* or *chee weet.* Song: a variety of chatterings and twitterings that are infrequently heard in Texas.

OTHER BEHAVIORS. They fly with strong, quick, but slightly flickering wingbeats. They sail more than most swallows, carrying their wingtips down and shoulders forward.

Violet-green Swallow, *Tachycineta thalassina.* Princ. distrib. Texas: Uncommon SR in mountains of Trans-Pecos Texas, where they are also a common M. They are found in semiopen habitats, including prairies, canyons, and mountain slopes.

FEEDING BEHAVIOR. These handsome birds consume many insects, usually while flying high in flocks but also by skimming low over prairies and water. They only rarely forage on the ground. They fly in company with other swallows and with White-throated Swifts.

COURTSHIP BEHAVIOR. Essentially unknown, but possibly includes predawn song flights.

NESTING. Nest: a cup-shaped nest in a natural cavity or nesting box, built by both parents. Eggs: apparently incubated almost entirely by the female. Both parents feed the nestlings. There are reports of pairs that have assisted Western Bluebirds in raising their fledglings, then appropriated the bluebirds' cavities for their own nests.

VOICE. Call: a repetitious *chip cheep chip,* as well as rapid twitters.

OTHER BEHAVIORS. They soar less and fly with more rapid wingbeats than Tree Swallows.

Northern Rough-winged Swallow, *Stelgidopteryx serripennis.* Princ. distrib. Texas: Locally uncommon SR throughout most of Texas, common M in most of the state; locally uncommon WR on the coast and in extreme s. Texas. They occur in most open habitats, especially near water. Their distribution is probably limited by the availability of nesting sites.

FEEDING BEHAVIOR. They frequently fly low and alone, and visit fields, ponds, and rivers in their search for insects. They only rarely feed on the ground.

COURTSHIP BEHAVIOR. As the male pursues the female, he displays the white feathers at the base of the tail.

NESTING. Usually solitary. Nest: They typically dig a burrow in a sand or dirt bank, but unlike Bank Swallows, they also utilize drain pipes, holes in buildings, and culverts. Eggs: most likely incubated only by the female. Young birds are fed by both parents. *See* Bank Swallow.

VOICE. Generally silent. Call: a rasping, squeaky *kee zeep,* lower and harsher than the Bank Swallow's.

OTHER BEHAVIORS. Their flight is swift and powerful. Compared with the Bank Swallow, they fly in a straighter line, with fewer twists and turns, more gliding and sailing, and deeper and slower wingbeats. They do not gather in large, premigratory flocks as do Bank Swallows.

Apparently recurved hooks along the outer primaries produce a sound in flight that functions in communication. ("Rough-winged" refers to these hooks, not to the species' manner of flying.)

Bank Swallow, *Riparia riparia.* Princ. distrib. Texas: Locally common SR along the Rio Grande; common M throughout most of the state.

FEEDING BEHAVIOR. Groups fly low over fields, rivers, ponds, and roadsides, where they capture insects, especially flies. Apparently they feed on the ground only in bad weather.

COURTSHIP BEHAVIOR. During the courtship flight, the male passes a white feather to the female. Copulation probably occurs inside the burrow.

NESTING. Pairs in the colony usually breed synchronously. Nest: in a burrow 2–5 feet long near the top of a vertical bank. They nest in dense colonies, the entrances to the burrows being about a foot apart. Several hundred birds sometimes nest on the same dirt bank.

The burrow is dug by both sexes. First, they cling to the side of the bank and dig an opening with their bill. Once the burrow is deep enough, they enter it and kick out more dirt with their feet. The nest cavity at the end of the burrow is lined with grass, weeds, and feathers. Both parents incubate the eggs and feed the young.

Bank Swallows, being colonial nesters, can learn the calls of their own young—a useful ability because it reduces erroneous feeding in the colony. Rough-winged swallows do not have this ability, since, being solitary nesters, they never have an opportunity to feed the wrong nestling. Their inability to discriminate between nestlings was demonstrated when young Bank Swallows placed in the burrows of Northern Rough-winged Swallows were completely accepted by the latter species.

VOICE. Call: a gritty *spee dzeet.*

OTHER BEHAVIORS. Premigratory communal roosts, in tall-grass marshes, trees, or bushes, are sometimes enormous. It has been suggested that these colonies serve as information centers that allow birds to follow each other to patchy food resources.

The smallest of our swallows, they fly with quick, fluttering wingbeats (like a butterfly), twisting and turning along an erratic course. When gliding, they tend to hold their wings close to the body.

Cliff Swallow, *Petrochelidon pyrrhonota.* Princ. distrib. Texas: Generally common to locally abundant SR and M throughout most of Texas, in semiopen habitats where nesting sites and mud for nests are available. Curiously, they are absent in parts of their range where nesting and feeding conditions appear suitable.

FEEDING BEHAVIOR. They eat insects captured while flying low over water and high over land, often while in association with Barn Swallows. They occasionally gorge on berries. Individuals that have been unsuccessful while foraging sometimes watch successful parents as they feed their young at the nest, then follow the successful birds to good foraging sites. In this way colonies seem to serve as information centers.

Careful listeners report a squeak call that they hear only in birds that are feeding. Since nonfeeding birds can be recruited to playbacks of this call, possibly the squeak functions to alert other birds to a food source.

Why would a bird share knowledge about the location of a food source by making a squeak or other signal? Perhaps because several birds are able to track swarms of insects more efficiently than a single bird (which might lose the swarm), thereby benefiting the individual that behaves altruistically by sharing insects with the group.

COURTSHIP BEHAVIOR. Courtship flights are followed by copulation on the ground.

NESTING. Nest: a gourd-shaped mud structure stuck securely to a sheltered, vertical surface such as a cliff or, more frequently, the underside of a bridge or other man-made structure. Both parents bring mud, bit by bit, from the bank of a lake or river. Several hundred families often nest alongside each other, placing their nests just a few inches apart.

Cliff Swallows often parasitize nests of other Cliff Swallows. When they do this, they make a fine-scale assessment of the quality of the nest they parasitize. For example, earlier in the season they tend to lay eggs in nests that are protected from the bad weather so often experienced in spring; later, they choose the nests of birds that have successfully produced fledglings. At all times they parasitize nests with the lowest infestations of ectoparasites.

They sometimes lay eggs in another bird's nest (brood parasitism), and occasionally they carry eggs in their bill to another nest.

Both parents incubate the eggs and feed the nestlings. Parents and young (including foster fledglings) recognize each other's voices, thereby reducing brooding errors that could easily occur in a closely packed colony. Colonies often alternate between nesting sites, probably to avoid parasites that infest the nests. Young birds return to the natal area to nest.

VOICE. Call: "an incessant chatter of husky creaking notes" (Pough, 1946).

OTHER BEHAVIORS. Even during the last century their nests were usurped by House Sparrows, a problem that becomes severe in many urban areas. Prior to migration, large flocks assemble that include other swallows.

Although relatively steady in direction, their flight follows a roller-coaster path: they fly up rapidly on quick wingbeats, then glide down.

One benefit of living in large colonies was experimentally demonstrated in Cliff Swallows. When small and large colonies were exposed to an artificial snake at varying distances, the large colonies detected the snake farther away and mobbed it. Curiously, mobbing Cliff Swallows do not swoop down close to predators, as do the solitary-living Barn Swallows.

Cave Swallow, *Petrochelidon fulva.* Princ. distrib. Texas: Locally common SR in parts of c., s., and w. Texas, and in scattered localities elsewhere in the state, in semiopen habitats, especially sites near open water. They are currently expanding their range to the n. and e.

FEEDING BEHAVIOR. Flocks fly low over water or higher over land and consume numerous insects.

COURTSHIP BEHAVIOR. Evidently their courtship displays have not been reported.

NESTING. Colonies often contain hundreds of nesting pairs. Formerly they located almost exclusively in the dimly lit interior of caves or sinkholes; however, during the last few decades they have increasingly utilized artificial structures such as bridges and culverts, where they compete for suitable nesting surfaces with Cliff Swallows and Barn Swallows. This recent emergence from their restricted nesting habitat is assumed to be a learned response to the increasing number of such artificial sites in Texas. Nest: a mud cup the pair build and attach to a vertical surface. In natural caves it contains bat guano and mud picked up from the floor of the cave. It is lined with plant down, feathers, and grass. Both parents incubate the eggs and feed the nestlings.

VOICE. Call: a clear *wheet,* given in flight. Song: "squeaks which merge into a melodic warble" (Oberholser and Kincaid, 1974).

OTHER BEHAVIORS. When returning to the nest, they close their wings and plunge into the cave at high speed, then circle around inside to lose momentum.

Barn Swallow, *Hirundo rustica.* Princ. distrib. Texas: Locally common SR and common to abundant M in most of Texas; they occupy virtually any semiopen habitat, including farms, marshes, lakeshores, and towns, but usually only when water is nearby. They generally avoid deserts, which supply no water for mud, as well as deep forests, where flight is restricted.

FEEDING BEHAVIOR. Barn Swallows capture and eat most of their prey during tireless flights that take them over lakes, fields, and other open areas; they sometimes drop to the ground for insects, and occasionally consume berries.

COURTSHIP BEHAVIOR. They sit side by side on a perch, preen each other's feathers, rub their heads together, and touch bills. The male chases the female as part of a long, graceful courtship flight.

In the European subspecies, males with the longest tail more readily acquire mates. This observation suggests that tail length is important in mate acquisition and thus is under the influence of sexual selection, an idea supported by the observation that during courtship the male flies in front of the female with his tail fanned out.

NESTING. Several pairs often nest near each other. Nest: a cup of mud and grass built by both parents, who line it with feathers. They cement it to a vertical surface in open buildings, under eaves, or in other sheltered sites.

Barn Swallows originally nested in sheltered crevices such as cave entrances, and even now they sometimes nest under bridges with Cliff Swallows. The male helps with incubation. Both parents feed the nestlings, and previous offspring sometimes help raise them. Young birds do not return to the natal area to breed.

VOICE. Call: a gentle *wit wit*. Song: liquid, energetic chattering at different pitches.

OTHER BEHAVIORS. Their flight is direct and flowing. It is given a graceful appearance by the skillful maneuvering of the forked tail while pulling the wingtips back quickly at the end of each stroke.

Birds both drink and bathe on the wing. During migration they form large, mixed flocks.

In male Barn Swallows, song rate is linked to the concentration of white blood cells and to the ratio of gamma-globulins to plasma proteins, both indicators of health status; thus, females may be assessing a male's phenotypic and genetic quality when they listen to his song.

CHICKADEES AND TITMICE: FAMILY PARIDAE

Petite busybodies that clamber about in trees and shrubs in search of insects and seeds. In the process they engage in amusing acrobatics such as hanging upside down from twigs. They are generally sociable and sometimes become trusting enough to eat out of a person's hand.

They nest in natural cavities, woodpecker holes, and nest boxes. When approached while in their nesting cavities, many species raise their head feathers. Many members of this family also show conspicuous postures and movements when threatened outside the cavity. These displays include injury-feigning and distraction displays. The name "waving" has been given to the primarily visual display in which birds slowly move the body and wings.

The volume of the hippocampal formation (a part of the brain associated with memory) is relatively larger in members of this family that store more food or store food for longer periods of time. Moreover, in Black-capped Chickadees, the volume of the hippocampal formation is greater in October, when they hoard food most intensely. This relationship suggests that seasonal differences in the use of spatial memory in birds influence hippocampal volume.

Carolina Chickadee, *Poecile carolinensis.* Princ. distrib. Texas: Common PR in e. two-thirds of the state, in mixed and deciduous woodlands, including heavily wooded urban areas. In c. Texas they seem particularly fond of live oaks.

FEEDING BEHAVIOR. They join small, mixed-species feeding groups (called guilds) that include titmice and nuthatches. Guilds are especially common in fall and winter. Chickadees glean for caterpillars, moths, and other insects, hang upside down under leaves and stems, hover for food on bushes and trees, and sally out for flying insects. They also eat berries, seeds, and other plant parts, sometimes stuffing items into bark crevices for later retrieval.

COURTSHIP BEHAVIOR. The male vigorously pursues the female; he also engages in courtship feeding. Apparently many pairs mate for life.

NESTING. Nest: in a natural cavity of a tree, constructed of bark strips and other plant materials. Eggs: probably incubated only by the female. Both parents bring food to the nestlings. When disturbed at the nest, the female produces an explosive, snakelike hissing sound.

VOICE. A soft but husky *chicka dee dee dee dee;* also, a measured, carefully enunciated, whistled *dee dee dee dee.* The first and third notes are higher in pitch to produce a simple melody.

Birders usually separate this species and the Black-capped Chickadee by their different songs, but occasionally the two species imitate each other's songs.

OTHER BEHAVIORS. These are active and industrious birds. Their curiosity compels them to approach whistled imitations of their own song, as well as the Eastern Screech-Owl's. They also respond to squeaking.

Mountain Chickadee, *Poecile gambeli.* Princ. distrib. Texas: Common PR in Guadalupe and Davis mountains, in healthy coniferous forests at high elevations.

FEEDING BEHAVIOR. Essentially as in the Carolina Chickadee. They forage in trees at both high and low levels. In winter they descend from higher altitudes in the mountains to feed in mixed woodlands and sometimes desert scrub.

VOICE. A repeated *chick a dee a dee a dee,* more extended than the Carolina Chickadee's song; also, a sweet, whistled *fee bee bay,* descending in pitch and suggesting the song "Three Blind Mice."

OTHER BEHAVIORS. In other parts of the United States, Mountain Chickadees apparently adjust their breeding season to the growth of the white fir so as to insure that expanded new shoots are available for their young.

Juniper Titmouse, *Baeolophus ridgwayi.* Princ. distrib. Texas: Locally uncommon PR in Delaware and Guadalupe mountains, in deciduous, coniferous, and mixed woodlands.

FEEDING BEHAVIOR. Essentially like that of the Tufted Titmouse, except that apparently they eat more plant than animal matter.

VOICE. A clear *whit ee whit ee whit ee,* as well as a *tchick a deer.*

Tufted Titmouse, *Baeolophus bicolor.* Princ. distrib. Texas: Common to abundant PR throughout most of the state (although absent from most of the High Plains), in open woodlands, parks, and gardens. The black-crested race occurs in w. half of the state (often in mesquite or juniper scrub), and the tufted race in e. half.

FEEDING BEHAVIOR. In their lively searches for insects (especially caterpillars), titmice cling to tree trunks and examine crevices in the bark; they also hang upside down under leaves and twigs, flit about limbs and branches, or drop to the ground. They eat plants as well as animals, sometimes cracking open acorns by holding them with their feet and rapidly pounding them with their bill; they also store seeds (including sunflower seeds they take from feeders) in crevices or in the ground.

COURTSHIP BEHAVIOR. Since most birds mate for life, exactly how they select their mates (a process that occurs only once in life) is apparently unknown. What is usually observed is simply the pursuit of the female by the male.

NESTING. Nest: in a cavity, often a woodpecker hole. Unlike their close relatives the chickadees, they apparently do not enlarge the cavity but simply line it with moss, leaves, and animal hairs (often plucked from live animals, in one instance, a person). Eggs: incubated by the female, who is a close sitter and not easily flushed. Both parents bring food to the nestlings; they are sometimes assisted by helpers, which generally are offspring from the previous year.

VOICE. A loud, ringing *peter peter peter,* sung persistently throughout the year; also, a harsh *day day day,* sometimes described as peevish.

OTHER BEHAVIORS. These restless, vivacious little birds are notably tame, even to the point of becoming quite bold; they readily avail themselves of birdfeeders in winter, and even learn to take food from a person's hand. They are hopelessly inquisitive and invariably respond to squeaking, sometimes approaching to within an arm's length.

Their flight is bounding and irregular, but they rarely fly farther than to the next tree. They often spread their tail when they fly.

Although they mate for life, the pair separate in winter to join guilds of chickadees and other birds. At night they roost in cavities.

PENDULINE TITS AND VERDINS: FAMILY REMIZIDAE

Verdin, *Auriparus flaviceps.* Princ. distrib. Texas: Uncommon to common PR in most of w. and s. Texas. They are well adapted to the harshest of desert habitats, including cactus and creosote-bush scrub, but they also frequent more mesic areas like mesquite woodlands and densely vegetated gullies. This is a tough little bird with a loud voice and a big nest.

FEEDING BEHAVIOR. They consume many insects, but they also take berries, fruits, nectar, and, less frequently, seeds. They flit around small branches and leaves, often hanging upside down like chickadees. They also come to the ground, as well as fly out for passing insects. Their fondness for nectar frequently leads them to hummingbird feeders.

Evidently they are capable of surviving long periods of time without drinking water, as their nests are sometimes more than 10 miles from water.

COURTSHIP BEHAVIOR. Apparently unrecorded.

NESTING. The male builds several nests, of which his mate selects one for raising the young. Nest: a large, conspicuous, oval or spherical structure made of thorny twigs, in a tree, thorny shrub, or cholla. It gives the appearance of being too large for a bird this size. Nests built early in the season have the entrance oriented away from the prevailing cold winds; later nests are oriented toward the wind. In winter, Verdins construct additional nests for sleeping. Eggs: incubated by the female. Fledglings are brought regurgitated insects by both parents.

VOICE. A loud, penetrating *tswee tswee tswee tsweet;* also a rapid *tzit tzit tzit.* When foraging, they sometimes utter a *tsee tyu tyu,* like a chickadee.

OTHER BEHAVIORS. They are shy during the nesting season, yet rather bold at other times of the year. They often travel in pairs, flying from bush to bush with rapid, jerky wingbeats. In winter they may forage with other species.

Although active at all ambient temperatures, they greatly reduce their foraging time at temperatures above 90 degrees F. Also, above this temperature they reduce exposure to sun and wind by moving from exposed to shaded microhabitats. This behavior reduces their rate of evaporative water loss by a factor of four or more in summer. In winter, they seek areas that protect them from wind.

LONG-TAILED TITS AND BUSHTITS:
FAMILY AEGITHALIDAE

Bushtit, *Psaltriparus minimus.* Princ. distrib. Texas: Uncommon to common PR in much of w. Texas, in pine, juniper, and oak woodlands.

FEEDING BEHAVIOR. Flocks (usually 6–30) of these exceedingly active birds seem constantly on the move. As they flit nimbly from tree to tree searching for insects, they utter soft ticking and lisping notes. They occasionally eat berries and seeds.

COURTSHIP BEHAVIOR. Their courtship involves mainly posturing and vocalizations.

NESTING. In a shrub or low tree. Nest: a tightly woven gourd-shaped pocket, often a foot long, that hangs like a pendulum from a limb of a tree or shrub. The entrance to the nest chamber is a small hole near the top. Inside, the nest proper is constructed of grass, moss, leaves, spiderwebs, and other materials. When pairs are disturbed while building their nests or incubating their young, they usually abandon them and begin a new nest, sometimes with a new mate.

Up to 40 percent of Bushtit nests in an area may be attended by one to four supernumerary birds. These extra birds include other females that lay eggs in the nest, or other males that mate with the female that built the nest. Both parents incubate the eggs and feed the nestlings.

VOICE. Call (heard throughout the year): a high-pitched *tsit tsit tsit;* also, an alarm note sometimes rendered as *sre e e e e e.* Apparently, Bushtits have no song.

OTHER BEHAVIORS. In winter they join mixed flocks of kinglets and wrens. Although these flocks break up into pairs before spring, the Bushtit's gregarious nature evidently does not weaken as much in spring as that of most bird species, which typically reject all other birds at that time. Bushtits only weakly defend a territory, and they tolerate other Bushtits nearby.

Pairs generally sleep in the nest at night, even before the eggs have hatched. On winter nights, groups of individuals huddle together at the roost to keep warm.

The significance of the light eye in females (as opposed to the dark eye in males and in both sexes of most birds) is not known. Since both parents feed the nestlings, the female's light eye would not seem to function as a releaser for begging behavior. Possibly it plays a role in courtship.

Flight is short, weak, and undulating.

NUTHATCHES: FAMILY SITTIDAE

Nuthatches are compact, short-tailed birds with strong toes and claws that allow them to climb headfirst up and down tree trunks without using the tail for support (as do woodpeckers and creepers). Presumably their ability to go down a tree headfirst exposes them to insects not seen by woodpeckers and other climbing birds, which always ascend the trees.

To open a nut, they wedge it in a crevice, then hack on it to break it open. They do not hold seeds with their feet, as do chickadees and titmice. They store seeds in crevices for later retrieval. Nuthatches also dart out to capture insects.

They often forage in mixed groups that include chickadees, titmice, Downy Woodpeckers, and other small birds.

Red-breasted Nuthatch, *Sitta canadensis.* Princ. distrib. Texas: Uncommon M and WR in most of Texas, in both coniferous and deciduous forests, but probably a pine habitat specialist (Herb and Burt, 2000).

FIGURE 9. Courtship feeding in the White-breasted Nuthatch.
(L. Kilham, 1972, *Auk* 89: 115–129. With permission.)

FEEDING BEHAVIOR. Typical for the family. They are more energetic than White-breasted Nuthatches, moving rapidly over the tree and often creeping out to the ends of small branches. After the nesting season they unite to form small flocks. Apparently their periodic winter flights southward are correlated with a scarcity of seeds in northern conifers.

COURTSHIP BEHAVIOR. The male turns his back to the female, droops his wings, raises his head, and sways from side to side.

VOICE. Call: a quiet *aank aank,* higher pitched and more nasal than the White-breasted; also, a *hit hit hit hit.* Song: soft and musical, sung by the male during courtship, but very rarely heard in Texas.

White-breasted Nuthatch, *Sitta carolinensis.* Princ. distrib. Texas: Locally uncommon PR in the Trans-Pecos and e. third of Texas, s. to the coast; locally uncommon M in most parts of the state. They are more common in deciduous or mixed forests than in purely coniferous woodlands.

FEEDING BEHAVIOR. Typical for the family.

COURTSHIP BEHAVIOR. Mating appears to be for life. The male spreads his tail feathers, droops his wings, raises his head, sways back and forth, and bows. He also brings food to the female (Fig. 9).

NESTING. In a large natural cavity, frequently a woodpecker hole, rarely in nesting boxes. Nest: constructed by the female from grasses, bark fibers, hair, and other materials. Interestingly, both parents sweep the inside and outside of the nest with a crushed insect that they hold in their bill, apparently utilizing chemical secretions from the insect as an insect repellent. The male feeds the female when she is incubating the eggs, and both parents feed the young.

VOICE. Call: a nasal *yank,* stronger, lower, and louder than the Red-breasted's call.

OTHER BEHAVIORS. These nuthatches become quite tame and have been taught to take food from a person's hand. Their agility is extraordinary: they can catch a seed in midair, hang upside down on a small, swinging branch, and run down a swaying rope.

In winter they join mixed groups of chickadees, titmice, and other small birds. At that time they usually roost singly in tree cavities.

Pygmy Nuthatch, *Sitta pygmaea.* Princ. distrib. Texas: Locally common PR in Guadalupe and Davis mountains, predominantly in pines along the 8,000-foot contour line. Ecologically, this is the w. equivalent of the Brown-headed Nuthatch of e. North America.

FEEDING BEHAVIOR. Typical for the family. They drift noisily and in a disconnected manner through the outermost and highest limbs of trees, less frequently on the main branches and trunks. They glean small twigs, leaves, and terminal foliage and probably concentrate more than our other nuthatches on these parts of the tree.

COURTSHIP BEHAVIOR. Apparently not recorded. Pair bonds are usually long-term.

NESTING. Nest: a cavity in a dead limb, excavated by both parents. Interestingly, they tolerate as neighbors such hole-nesting species as bluebirds and swallows but not chickadees or other nuthatches. The cavity is lined with plant down, feathers, and other materials and is used as a roost at night, even before eggs are laid. The female incubates the eggs and is fed by the male and occasionally helpers (usually previous offspring). Pairs that are aided by helpers have a higher reproductive success than those that nest alone.

VOICE. Call: *ti di ti di ti di,* a soft *kit kit kit,* and other twitterings, incessantly delivered while foraging.

OTHER BEHAVIORS. The Pygmy is sociable, alert, and active: "an astonishing sight to see . . . the crown of a Douglas fir hopping with these little birds—like animated Christmas tree ornaments" (Oberholser and Kincaid, 1974). A group of these nuthatches suggests roving Bushtits. Sometimes in winter 5–15 individuals may roost together in the same tree cavity.

Brown-headed Nuthatch, *Sitta pusilla.* Princ. distrib. Texas: Locally common PR in e. Texas, normally in pine forests, especially burns and clearings that support healthy stands of young saplings. They are the e. equivalent of the Pygmy Nuthatch.

FEEDING BEHAVIOR. Typical for the family. They usually search the main trunk by ascending to the uppermost parts. Males forage lower than females and also sometimes drop to the ground for food. Tool use has been reported: birds use pieces of bark to expose insects. They also cache food.

COURTSHIP BEHAVIOR. Apparently not recorded.

NESTING. Nest: a cavity excavated by both parents, in a dead tree or fence post. The male selects the site, and several excavations are sometimes attempted before the final one is completed. They also utilize nesting boxes and woodpecker holes, often competing with Eastern Bluebirds for these sites. The male brings food to the incubating female; at night he roosts with her in the nest. Both parents feed the nestlings. They are sometimes assisted by helpers, which are generally unmated males.

VOICE. Call: most are reedier and harsher than the calls of our other nuthatches; one variant is a *pit pit pit*. Song: a musical *ki dee ki dee*. They also constantly twitter, chirp, and hiss, probably to maintain contact with each other.

OTHER BEHAVIORS. These are quick and restless birds. After nesting, they join groups of 5–20 chickadees, titmice, kinglets, and other small birds to roam noisily through the forest in search of food.

CREEPERS: FAMILY CERTHIIDAE

Brown Creeper, *Certhia americana.* Princ. distrib. Texas: Locally common M and WR in most of Texas e. of the Pecos River, and SR in Guadalupe Mountains, in almost any wooded habitat that supports large trees, including urban areas.

FEEDING BEHAVIOR. Presumably by slowly and methodically scaling tree trunks, they locate insects overlooked by the less meticulous nuthatches, titmice, and wrens. Using their tail as a prop, they follow a spiraling course up a tree trunk. When they reach the top they flutter to the bottom of a nearby tree and repeat the process, occasionally making short hops backward to reinvestigate parts of the bark.

Their diet consists almost exclusively of insects, but they occasionally supplement it with seeds. Brown Creepers usually feed alone, but at times they associate with flocks of chickadees, warblers, and other small birds. They sometimes cache food.

voice. Call: a thin, reedy *tseeee*, and a high-pitched, rolling trill. Song: high-pitched and musical, but it is rarely heard in Texas.

other behaviors. They are relatively tame; in one case a bird alighted on a man's leg. When pursued, they take advantage of their camouflaged plumage by flattening their body, spreading their wings, and remaining motionless.

WRENS: FAMILY TROGLODYTIDAE

Wrens are spirited, noisy birds that appear indefatigable as they hop around looking for insects. They use their relatively long, slender, and often curved bills to probe crevices, holes in trees and rocks, and other places that conceal insects. When scolding at intruders, they characteristically hold their tail above their back.

They often nest in tree cavities, nest boxes, and other enclosed places. Many males build dummy nests that are never used, and some species actively enter the nests of unrelated species to destroy their eggs, even when the victimized birds pose no obvious threat.

Courting males display vocally and by spreading their wings and hopping around the female; however, the courtship behavior of most Texas wrens is essentially unknown.

Cactus Wren, *Campylorhynchus brunneicapillus.* Princ. distrib. Texas: Uncommon to locally common PR in w. and s. part of the state, generally in arid or semiarid habitats.

feeding behavior. They forage singly, in pairs, or in family groups, probing leaf litter and bark crevices in search of insects, and occasionally capturing a small lizard or frog. They sometimes lift a rock or leaf with their bill, or take smashed insects from car radiators. They eat more vegetal matter than our other wrens and appear especially fond of berries and seeds.

courtship behavior. Apparently their initial pair-bonding displays have not been recorded. These displays are probably brief and occur very few times during a bird's lifetime, as Cactus Wrens usually mate for life. Possibly pair-bonding behavior resembles the greeting display described under "Other behaviors."

nesting. In a cactus or in mesquites, acacias, or other thorny trees or shrubs. Nest: a bulky, football-shaped mass of twigs, grass, and weeds, built by both parents. A narrow tubular passage terminates in a nest chamber lined with soft materials, including hair and feathers. Although conspicuous, the nest is usually so entangled in thorns that it is inaccessible to almost all predators except snakes. When the female is incubating the eggs, the male may build one or more dummy (supernumerary) nests that possibly function to distract predators from the real nest. Even after the nesting season, birds often sleep in the nest at night.

VOICE. Call: harsh, scolding notes. Song: a harsh *rawh rawh rawh rawh rawh rawh rawh,* often compared to an automobile's starter on a cold morning. Birds sing all months of the year, frequently from a conspicuous perch.

OTHER BEHAVIORS. When the pair greet each other, they perch upright, partially spread their wings and tail, and call harshly. They run swiftly on the ground, and fly with short, jerky wingbeats. Curiously, they jerk their tail like flycatchers.

Although usually shy, they become inquisitive when not molested. For no apparent reason, Cactus Wrens sometimes destroy eggs of other birds.

Rock Wren, *Salpinctes obsoletus.* Princ. distrib. Texas: Common SR and locally common WR in w. half of Texas. Rock Wrens favor piles of boulders, rock fences, rock dams, and bare ground, whereas Canyon Wrens prefer rock cliffs, shallow caves, and deep canyons.

FEEDING BEHAVIOR. As they creep, bob, and jump among rocks and boulders, these nondescript birds locate and capture insects and spiders by probing their long, curved bill into crevices and spaces between rocks.

COURTSHIP BEHAVIOR. Apparently not recorded.

NESTING. In a cavity in a dirt bank, under a rock ledge, in a crevice among boulders, or in a similar site. Nest: a cup built, probably by both parents, from weeds, grass, and twigs and lined with hair, spiderwebs, feathers, and other soft materials. They place stones, bits of bone, and other objects about 10 inches out from the nest. One nest had more than 1,500 such items, recalling comparable behavior in the bowerbirds of Australia and New Guinea. Why birds expend so much energy to do this is puzzling. In some species (the unrelated wheatears), stone-carrying appears to enable the female to assess the quality of her mate.

Probably the female does most, if not all, of the incubation, but both parents feed the nestlings.

VOICE. Call: a metallic, ringing *ti keer,* as well as numerous other sounds. Song: an accelerating series of three or four double notes that are sometimes musical, sometimes harsh. The rambling song suggests to some listeners a mockingbird's song, but without sounds appropriated from other birds.

A male may have 100 or more songs, and neighboring males may countersing with similar song types.

OTHER BEHAVIORS. They frequently cock their tail as they hop over rocks and boulders.

Canyon Wren, *Catherpes mexicanus.* Princ. distrib. Texas: Uncommon to locally common PR in w. half of Texas (*see* Rock Wren).

FEEDING BEHAVIOR. Essentially the same as the Rock Wren's.

COURTSHIP BEHAVIOR. Apparently unrecorded.

NESTING. In a crevice on a rocky cliff, rock pile, or similar site. Nest: built by both parents using twigs, bark chips, grass, and other plant materials and lined with finer materials, including plant down, feathers, and spiderwebs. Eggs: incubated by the female. During this time her mate brings her food. Both parents feed the nestlings.

VOICE. Call: a low *peupp.* Song: a series of silvery, bell-like, whistled notes that descend in pitch and end in a buzz. Some listeners hear laughter in this song; others, whimsy. Surely it is one of the most unmistakable and unforgettable of all passerine songs.

OTHER BEHAVIORS. Sometimes they become quite fearless and nest in houses and buildings. In the 1950s one pair (possibly more) nested in the dome of the State Capitol in Austin, entering the building through a broken window. Their facility in navigating through corridors, houses, and other human structures is described by Oberholser and Kincaid (1974).

Carolina Wren, *Thryothorus ludovicianus.* Princ. distrib. Texas: Common PR in e. two-thirds of the state, in forests, dense shrubbery, urban areas, and especially river bottoms (their primary habitat in w. Texas).

FEEDING BEHAVIOR. Pairs search for insects along limbs, trunks, fallen trees, tangled vegetation, and densely vegetated shorelines. They also consume seeds and berries.

COURTSHIP BEHAVIOR. Apparently unrecorded.

NESTING. In cavities in stumps and trees, woodpecker holes, nest boxes, nooks and crannies in human habitations, and other enclosed spaces. Nest: a bulky structure of leaves, weeds, and other materials (often including a snake skin), built by both parents. Eggs: incubated by the female. She is brought food by the male at this time. Both parents feed the nestlings, and the male may continue feeding them alone if his mate begins a new nest. Carolina Wrens mate for life and maintain their pair bond throughout the year, sometimes reinforcing it by singing in duet.

VOICE. Call: buzzes and scolds, including a long, drawn-out *djerrrrrrr*. Song: a rich, rollicking *tea kettle tea kettle tea kettle*, including several variations that may appropriate the sounds of other birds (including Pine Warblers, Eastern Bluebirds, and meadowlarks). The song is usually delivered as birds feed rather than from a perch.

Each male has 27–41 song types in his repertoire, and after repeating one song several times, he switches to another. Males often match the song type they hear their neighbors sing.

They sing all year long, at night and during the day, and in all types of weather. Birds also have dialects that apparently are adapted to their region, as recorded songs were found to carry farthest in the habitat where the individual lived.

OTHER BEHAVIORS. Like most of our wrens, they are active, alert, and inquisitive, hopping or running about nervously with the tail cocked and turning the head back and forth. They readily respond to squeaking and swishing.

Although many birds sunbathe, evidently this species and the Greater Roadrunner are the only Texas species that expose their rumps to the sun's rays. They do this by partially spreading their wings and tail and parting their back and rump feathers to allow sunlight to reach the bare skin. They often remain in this position for several minutes.

Bewick's Wren, *Thryomanes bewickii.* Princ. distrib. Texas: Uncommon to common PR in all but e. third of Texas; uncommon to common M and WR throughout most of the state in scattered woodlands and urban areas. It would be a rare barn or corral in Texas that had never been colonized by a pair of Bewick's Wrens.

FEEDING BEHAVIOR. They forage primarily for insects, hopping and climbing in brush piles, on fences, under eaves of houses and barns, and among pieces of junk such as tires and scrap lumber. On the ground they flip over leaves with their bill. They also use their bill to probe into litter.

COURTSHIP BEHAVIOR. Apparently unrecorded.

NESTING. Like the better-known House Wren, Bewick's Wrens utilize a number of unconventional nesting sites, including flowerpots, shoes, cans, and drain pipes. On a ranch near Junction, Texas, a pair built a nest in the bed of a pickup and continued to attend to the nest, even though the pickup was taken to town for an hour or so each day.

Frequently the male builds several incomplete dummy nests. One is selected by the female, who then completes it. Nest materials include twigs, bark, leaves, debris, and sometimes pieces of snake skin. Apparently only the female incubates the eggs, but both parents bring food to the nestlings.

VOICE. Songs vary considerably throughout the United States. Regional variations also occur within Texas, but they have not been well documented. Call: a single or double *tchitt*. Song (often from a conspicuous perch): a clear, melodious, jumbled series of notes that usually begins with an upward slurred *jreet jreet* and concludes with a buzzy trill. It is more rigidly structured than the somewhat similar song of the Lark Sparrow, which is faster, longer, and has more buzzlike notes. They throw back their head when they sing.

Bewick's Wrens must perfect the crude song they acquire as juveniles. They do this by countersinging with a neighboring adult, a process that can often be heard in summer.

OTHER BEHAVIORS. They hop around actively, cocking or waving their tail, and responding to the least provocation by scolding. They sometimes puncture the eggs of birds nesting nearby.

House Wren, *Troglodytes aedon.* Princ. distrib. Texas: Locally common M and locally uncommon WR throughout Texas, as well as a SR in the Panhandle and a rare breeder in Davis Mountains. They generally favor dense vegetation.

FEEDING BEHAVIOR. They are lively, bouncy, and inquisitive; with tail cocked, they explore nooks and crannies in trees and logs, often pausing briefly to sing. They stay low in the vegetation, often along fencerows and in woodland undergrowth. Their diet is primarily insects.

VOICE. Call: a deep, grating scold, and a rapid *churr churr churr*. Song: an energetic but not particularly musical series of notes that are executed too rapidly to distinguish clearly. The flutter of notes rises, then falls toward the end.

Winter Wren, *Troglodytes troglodytes.* Princ. distrib. Texas: Rare to uncommon M and WR throughout most of Texas, in dense thickets near water, brush piles, and similar habitats.

FEEDING BEHAVIOR. They creep like mice along logs and rock fences, or down low in shrubbery, bobbing their head as they glean insects from twigs, trunks, and foliage. They frequently drop to the ground. When next to a river or pond, they sometimes take insects from the water's surface.

VOICE. Call: a sharp *tsick,* and an abrupt *chirr.* Song: a series of rich, loud, high-pitched notes, about 110 of them packed into 7 seconds and woven into "a fine silver thread of music" (Terres, 1980).

Sedge Wren, *Cistothorus platensis.* Princ. distrib. Texas: Common M and WR on the coast, less common in e. part of the state, in marshes, damp grassy areas, reedy meadows, and coastal prairies, especially those with scattered shrubs. They typically avoid the dense stands of cattails that Marsh Wrens frequent.

FEEDING BEHAVIOR. They creep among reeds and grasses searching for insects and spiders, generally staying low in the vegetation and out of sight. They sidle up and down stems with agility and occasionally sally out for a passing insect.

VOICE. Call: a short, high-pitched *tzick.* Song: a chattering trill, not frequently heard in Texas.

OTHER BEHAVIORS. When flushed, they flee a short distance with an awkward, weak, and fluttering flight, then drop back into the vegetation. They appear more shy and retiring than the Marsh Wren.

Marsh Wren, *Cistothorus palustris.* Princ. distrib. Texas: Common M and WR on the coast, and PR in upper and c. coastal marshes, less common in e. part of the state; they inhabit marshes with cattails and tall grasses and brushy edges of lakes and ponds. (*See* Sedge Wren.)

FEEDING BEHAVIOR. Essentially as in the Sedge Wren.

VOICE. Call: a series of grating notes running into a chatter, often produced when disturbed. Song (frequently heard at night): scraping notes that evolve into a loud, rapid rattle that ends in a whistle. Males have 30–300 different songs, depending on where they breed. Anatomical studies show that several parts of the brain that are active when these birds sing are larger in males than in females.

OTHER BEHAVIORS. They behave basically like Sedge Wrens. Compared with other wrens, the eggs are rounder (therefore more difficult to break) and have a thicker shell. The tendency for Marsh Wrens to destroy the eggs of other Marsh Wrens may have led to the evolution of unusually strong eggs in this species.

KINGLETS: FAMILY REGULIDAE

The tiny kinglets are usually recognized at once by their movements: they flit about nervously and often flick their wings. They eat mainly insects.

Golden-crowned Kinglet, *Regulus satrapa.* Princ. distrib. Texas: Uncommon to locally common M and WR in all parts of Texas, occurring in mixed woodlands but favoring conifers.

FEEDING BEHAVIOR. Essentially like the Ruby-crowned Kinglet.

VOICE. Call: a high, hissing *zee zee zee.* Song (rarely heard in Texas): a series of high notes that ascend in pitch, then descend to end in a chatter.

OTHER BEHAVIORS. *See* Ruby-crowned Kinglet. This species is "astonishingly fearless of people, comes into open cabins, allows itself to be stroked at times and even picked up" (Terres, 1980).

Ruby-crowned Kinglet, *Regulus calendula.* Princ. distrib. Texas: Common to abundant M and WR throughout the state, in scattered forests, bushes, parks, and urban gardens.

FEEDING BEHAVIOR. Intensely active, they are seemingly indefatigable as they glean for insects in trees and shrubbery, restlessly hopping among limbs and twigs and occasionally flicking their wings open and shut. They sometimes hover by a bush or sally out like a flycatcher. Compared with Golden-crowned Kinglets, they hover and flycatch more and hang on twigs less. They sometimes eat sap that oozes from trees.

VOICE. Call: an abrupt, husky *ji dit.* Song (variable): a series of high *tees* followed by several lower notes and terminating in a *tee tee tee tee tee tee tee tee . . . tee dadee tee dadee tee dadee,* or *liberty liberty liberty.* The song is surprisingly loud for a bird this size.

OTHER BEHAVIORS. In winter they associate with titmice, nuthatches, and other small birds. They are exceptionally inquisitive and are among the first birds to respond to squeaking. Their flight is jerky, irregular, and rarely prolonged.

OLD WORLD WARBLERS AND GNATCATCHERS: FAMILY SYLVIIDAE

Small, active birds that pump their tail as they forage for insects and spiders, generally in the outer parts of trees and bushes.

Blue-gray Gnatcatcher, *Polioptila caerulea.* Princ. distrib. Texas: Rare to common M and SR throughout most of Texas; uncommon WR in s. part of the state. They frequent both dense and scattered woodlands, including junipers and oaks, and often (but not necessarily) habitats near water. (*See* Black-tailed Gnatcatcher.)

FEEDING BEHAVIOR. Energetic to the point of being fidgety, Blue-gray Gnatcatchers hop through foliage searching for insects and spiders, every so often raising their wings and twitching their tail up and down or from side to side. They hover when picking insects from foliage, or fly out to capture insects on the wing. Large insects are beaten against a branch before being eaten.

COURTSHIP BEHAVIOR. Apparently no specific displays or postures are associated with courtship.

NESTING. In a tree. Nest: a compact open cup of weeds, grass, plant fibers, and bark that is camouflaged with spiderwebs and lichens and built by both parents. Eggs: incubated by both parents. The male brings food to his mate while she broods the nestlings; later, both parents feed the young.

VOICE. Call: a thin, twanging *jeeeee,* sometimes compared to a plucked banjo string; when agitated, they scold at intruders with rapid, buzzy squeaks. Song: a soft, musical warble that has a lisping quality.

OTHER BEHAVIORS. These curious birds readily respond to squeaking. They sometimes form small groups in winter.

Black-tailed Gnatcatcher, *Polioptila melanura.* Princ. distrib. Texas: Rare to common PR in Trans-Pecos Texas, becoming less common to the s., where they are uncommon in winter; they are restricted to arid habitats. Where Bluegrays and Blacktails overlap in range, the Bluegrays favor pinyon-oak mountain forested areas (although often occurring in other habitats), whereas the Blacktails occur in desert scrub, the brushy banks of the Rio Grande, and occasionally on creosote-bush desert flats.

FEEDING BEHAVIOR. Pairs remain together all year, actively searching for insects and occasionally berries in creosote bushes, mesquites, acacias, and other desert trees and shrubs. They feed much like Blue-gray Gnatcatchers, except that they rarely fly out to capture passing insects.

COURTSHIP BEHAVIOR. Apparently unrecorded.

NESTING. In a shrub, usually less than 5 feet above the ground. Nest: built by both parents using plant fibers, strips of bark, and spiderwebs, and lined with softer materials. Both parents incubate the eggs and bring food to the nestlings.

VOICE. Call: a harsh, mewing *jee jee jee,* unlike the Bluegray's clear *jeeeee;* most notes have a querulous or plaintive quality. The calls serve to maintain contact between individuals. A few sounds suggest Verdins and Black-throated Sparrows. Song: an insectlike *tsee dee dee dee dee.*

OTHER BEHAVIORS. This is the smallest passerine in Texas, weighing less than a fifth of an ounce.

THRUSHES: FAMILY TURDIDAE

Most of our thrushes have a thin bill and strong legs. Many forage for terrestrial insects by hopping or running; some climb into foliage for insects or berries, generally eating berries during fall and winter when insects are scarce.

Courtship displays are varied and include postures in which birds point their bill forward or upward. In some species the male chases the female around the territory. Their extraordinarily fine songs are important for courtship as well as for advertising territory.

When excited, thrushes perform displacement behaviors such as raising the head feathers or quickly opening and shutting their wings.

Eastern Bluebird, *Sialia sialis.* Princ. distrib. Texas: Common SR in e. half of Texas; uncommon to locally common M and WR in most parts of the state; they inhabit open woodlands, farmlands, and roadsides, in some areas showing a penchant for orchards.

FEEDING BEHAVIOR. They fly out from a utility line or tree branch to capture insects in midair. They also pick insects from the foliage or from the ground, or pick berries while hovering next to a tree or bush (especially in winter).

COURTSHIP BEHAVIOR. While singing, the male ascends to about 50 feet and floats down, fluttering in front of the female with his wings half open and his tail spread. He also preens his mate and offers her food.

NESTING. In a woodpecker hole or other cavity, including abandoned metal mailboxes (where young birds often die from the heat). Fortunately, breeding pairs readily utilize nesting boxes, and they are usually successful in defending them against sparrows and swallows (but not European Starlings). Local populations often increase dramatically when boxes are made available. Nest: constructed by the female, using twigs, dry grass, and weeds. Eggs: incubated by the female. Both parents bring food to the nestlings. Helpers, chiefly young birds from a previous brood, occasionally assist in caring for the young.

VOICE. Call: a melancholy *chirrr eee.* Song: a distant, warbled *tru a lee tru a lee.*

OTHER BEHAVIORS. Broods may represent offspring from more than one male or one female; this phenomenon is attributable to egg dumping and promiscuity. Outside the breeding season they form flocks that contain 100 or more individuals.

Western Bluebird, *Sialia mexicana.* Princ. distrib. Texas: Uncommon and local PR in Davis and Guadalupe mountains, locally common M and WR in w. third of the state; they prefer slopes and canyons in juniper-pine-oak woodlands.

FEEDING BEHAVIOR. Essentially like that of the Eastern Bluebird.

COURTSHIP BEHAVIOR. Reported to be similar to the Eastern Bluebird's.

NESTING. Nest: largely like that of the Eastern Bluebird, although helpers at the nest have not been reported.

VOICE. Call: a soft *kyew,* and a chattering *cut cut cut.* Song: a repeated *kyew,* given in chorus in early spring during the predawn darkness, while they fly above the treetops.

OTHER BEHAVIORS. In winter they associate with Mountain Bluebirds, American Robins, and Yellow-rumped Warblers. Compared with Eastern Bluebirds, they are quieter, a little less active, and slightly more deliberate in their movements.

Mountain Bluebird, *Sialia currucoides.* Princ. distrib. Texas: Uncommon to locally common (and irregular) M and WR in w. half of Texas, in open terrain with scattered trees or bushes; they also inhabit pinyon-juniper woodlands.

FEEDING BEHAVIOR. Mountain Bluebirds hover low over the ground, then drop to the ground to capture insects such as beetles, caterpillars, and grasshoppers. They also fly out from low perches such as bushes or rocks to capture insects on the wing. In winter they consume berries from junipers, mistletoe, and other plants.

VOICE. Notably quiet. Call: a soft *curuuuu,* given in flight. Song (rarely heard in Texas): a clear, short warble, usually sung only in the predawn darkness; it has been compared to the American Robin's song.

OTHER BEHAVIORS. Besides their cerulean plumage, these birds are set apart from the other two bluebirds by their behavior. They are lighter on the wing, almost swallowlike, and when small flocks course over the cold prairie in winter, they do so in a quiet, gentle, and ghostly manner. At times they assemble in flocks of several hundred birds, often mingling with Western Bluebirds.

Townsend's Solitaire, *Myadestes townsendi.* Princ. distrib. Texas: Uncommon to common (but irregular) M and WR in w. third of Texas, in semi-open wooded areas, especially near thickets, canyons, and cedar brakes.

FEEDING BEHAVIOR. If berry supplies are low in winter, they defend a feeding territory that comprises a specific group of junipers. They often hover when picking berries. In summer they feed mainly on insects, which they locate from a perch; they drop to the ground or sally out like a flycatcher to capture insects in midair.

VOICE. Call: "a monotonously repeated short ventriloquial creaking" (Pough, 1946). Song: a thrushlike, prolonged series of trills and warbles that rise and fall in pitch; this song ranks among the most beautiful of our birdsongs, along with that of the Wood Thrush and Hermit Thrush. Townsend's Solitaires also sing a special song in flight.

OTHER BEHAVIORS. Generally shy and retiring, usually seen alone or in pairs. They fly with slow and irregular wingbeats. On the ground they run gracefully, like a robin. When they do this they sometimes move their wings slowly up and down.

Veery, *Catharus fuscescens.* Princ. distrib. Texas: Locally uncommon M in e. half of Texas, common spring M on the Upper Coast, in deciduous woodlands and sometimes parks and gardens.

FEEDING BEHAVIOR. They hop along the ground in shaded areas as they search for insects and berries, sometimes hovering to pick insects from leaves, at other times flipping dead leaves with their bill. They also search for insects in trees and bushes or fly out to capture them in midair.

VOICE. Call: a soft *yew* that drops in pitch; imitations readily draw birds to an observer. Song (not often heard in Texas): four or five downward-spiraling *whree uus*, each weaker and lower than the one before, to create an eerie and remote effect. The song only vaguely suggests the bird's name.

Gray-cheeked Thrush, *Catharus minimus.* Princ. distrib. Texas: Locally uncommon M in e. half of the state, in almost any wooded or brushy habitat, including roadsides, gardens, and parks.

FEEDING BEHAVIOR. For the most part they feed like Swainson's Thrushes, except that they hover less frequently and spend more time on the ground.

VOICE. Call: a harsh, downward-slurred *queeep,* recalling the Common Nighthawk's call. Song: a musical *whee wheeo teedee wheee* that rises quickly at the end; it is heard less often in Texas than the song of the Swainson's or Hermit Thrush.

OTHER BEHAVIORS. Very shy, they are probably more elusive than the other brown-backed thrushes. They fly with quick and strong wingbeats.

Swainson's Thrush, *Catharus ustulatus.* Princ. distrib. Texas: Uncommon to locally common M in e. half of Texas, becoming less common to the w., in woodlands and urban areas.

FEEDING BEHAVIOR. They forage mainly in trees, sometimes hovering at trees and shrubs for insects and berries. Although they also feed on the ground, they are the least terrestrial of the brown-backed thrushes.

VOICE. Call: a sudden *pwit,* or a high-pitched *peep,* suggesting the piping of a spring peeper. Song: an upward-spiraling *whib o willow willow zee zee zee,* which fizzles out at the end; in Texas it is heard in spring, very frequently near mulberry trees.

OTHER BEHAVIORS. In migration they sometimes feed in company with juncos and wintering sparrows.

Hermit Thrush, *Catharus guttatus.* Princ. distrib. Texas: Common M and WR in most parts of Texas (except the Panhandle); common SR in Guadalupe and Davis mountains. They inhabit woodlands and seem more habitat-tolerant than most of the other brown-backed thrushes. In the Guadalupes they favor remote pine, Douglas fir, and oak forests above 7,800 feet.

FEEDING BEHAVIOR. They feed essentially like Swainson's Thrushes, but they spend more time on the ground, hopping about and lifting their tail when they stop.

COURTSHIP BEHAVIOR. Displays include flicking the wings, raising the crest, and assuming a sleeked erect posture while pointing the bill upward.

NESTING. In a tree. Nest: a bulky but nicely woven cup of weeds, twigs, moss, bark, and other plant materials, built by the female. Eggs: incubated by the female. Both parents bring food to the young.

VOICE. Call: a low *chuck,* a sound generally not expected from a thrush. They produce a variety of other sounds and song types, including mews (Rivers and Kroodsma, 2000). Song: a deliberate, flutelike note followed by a series of bell tones that rise and fall and fade away at a dizzying high pitch. This song (regarded by many listeners as the finest of our birdsongs) is serene, relaxed, and rarely forgotten once heard. Among the thrushes, it competes only with the songs of the Wood Thrush and Townsend's Solitaire.

OTHER BEHAVIORS. Regarded as slightly hardier than the other brown-backed thrushes, they are able to remain in Texas throughout the winter.

In summer, Hermit Thrushes are shy and retiring, but they seem less so at other times of the year. When startled, they often fly to a nearby perch, flick their wings, raise their tail quickly, lower it slowly, and stare at the intruder.

In this and the previous three thrushes, foot-quivering seems to function both as a hostile (intraspecific) display and, more commonly, to flush prey. The male sings to advertise his territory.

Wood Thrush, *Hylocichla mustelina.* Princ. distrib. Texas: Uncommon to common M and locally common SR in e. half of Texas, in heavily forested areas, especially along streams, as well as in urban areas. They generally avoid pure stands of pines.

FEEDING BEHAVIOR. Wood Thrushes hop around on the ground scratching and probing for insects, sometimes ascending into trees and bushes to harvest berries.

COURTSHIP BEHAVIOR. The female raises her wings and fluffs her feathers, then leads the male in swift, circling flights; afterward they feed together.

NESTING. Nest: an open cup of moss, leaves, weeds, grass, and other plant materials, constructed by the female. Rather atypically for thrushes, she often mixes these materials with mud and adds paper and other trash. Eggs: incubated by the female. Both parents bring food to the nestlings.

VOICE. Call: a liquid, guttural *quoit,* as well as other sounds that include an abrupt *pit.* Song: a series of liquid, flutelike notes interspersed with pauses, guttural sounds, and *ee o lays.* These notes are calm, unhurried, and delivered at a lower pitch than the Hermit Thrush's song.

The Wood Thrush seems to offer the only serious challenge to the Hermit Thrush's title as the most beautiful songster in the United States. Experimental analysis of the singing process suggests that the sequence of notes is regulated by a feedback mechanism in which the bird constantly attends to the rapidly produced notes it hears.

OTHER BEHAVIORS. Flight is quick and graceful. They seem to be less shy than the other brown-backed thrushes, but less bold than American Robins. When alarmed, they raise the feathers on the head to form a slight crest.

American Robin, *Turdus migratorius.* Princ. distrib. Texas: Common SR in n. half of Texas and common to locally abundant M and WR in most parts of the state, in clearings, lawns, parks, and other sparsely wooded areas.

FEEDING BEHAVIOR. Robins run and hop in an upright posture, stopping abruptly to feed on seeds, fruits, insects, and several kinds of invertebrates that live in the soil, especially earthworms. Early experimenters suggested that robins do not locate subterranean prey by sound, but later studies demonstrated that they do. Careful observations show that American Robins are capable of locating buried meal worms in the absence of visual, olfactory, and vibrotactile cues; furthermore, they are less successful when auditory cues are obscured by white noise.

Earthworms make up as much as 20 percent of the American Robin's diet, and capture rates of 20 earthworms per hour have been reported. Oberholser and Kincaid (1974) report that if a bird is pulling an earthworm from the ground and "the worm suddenly gives way, the bird may fall backward with such force that it virtually turns a complete somersault."

Flocks that feed on chinaberries sometimes become intoxicated to the point of being immobilized.

COURTSHIP BEHAVIOR. Males sing and fight among themselves, or peck at their reflected images in a window or a shiny part of an automobile. In his courtship display, the male spreads his tail, shakes his wings, inflates his throat, and struts around the female. Early in the season several courting males may pursue the same female.

NESTING. Generally in a tree or shrub, but occasionally in buildings, under bridges, on porches, or on the ground. Nest: built primarily by the female, a cup constructed of twigs, grasses, and debris that is worked into a mud foundation. The female incubates the eggs and does the larger share of feeding the nestlings.

Parents generally allocate more food to the nestlings that occupy the central part of the nest, and nestlings learn to jockey for access to this location. Food-deprived young birds beg more intensely, but those that are not deprived of food beg more intensely in the presence of hungrier siblings that are begging (social facilitation).

Both parents aggressively defend the nest. When humans approach, parents with exposed nests attack more vigorously than those with concealed nests.

Sometimes females initiate a second nest before the fledglings from the first clutch have departed, leaving the male to care for the fledglings while she incubates the second clutch of eggs. Curiously, although his commitment to the fledglings obviously limits his ability to guard his mate, she now shows less willingness to engage in extrapair copulations than when she was being guarded by the male at the first nest.

VOICE. Call: a variety of scolding notes, some loud and piercing, including a *chuck chuck*. Song (frequently referred to as caroling): a series of melancholy notes and double notes that rise and fall in a slow rhythm, heard during predawn darkness as well as throughout the day.

OTHER BEHAVIORS. After the breeding season they congregate in enormous flocks that persist together throughout the winter, but they rarely mix with blackbirds, European Starlings, and other flocking birds.

Flight is rapid, steady, and direct, with intermittent wingbeats. In temperament, American Robins are bold and sometimes boisterous; they habituate readily to humans and seemingly prefer to nest near houses when given a choice.

MOCKINGBIRDS AND THRASHERS: FAMILY MIMIDAE

Most thrashers have strong legs and a heavy bill that help them forage for insects and seeds in ground litter; some species actually dig in the ground with their stout bill.

Thrashers sing to announce territories and to attract mates. Some songs contain skillful imitations of other species (*see* "Voice" under Northern Mockingbird). Courtship displays vary among the species, as would be expected for a family that includes birds as elusive as the Sage Thrasher and as bold as the Northern Mockingbird. Among the displays commonly observed are courtship feeding, presentation of sticks to the other member of a pair, and posturing.

Gray Catbird, *Dumetella carolinensis.* Princ. distrib. Texas: Locally common SR in e. half of the state, in dense, low growth (including thickets and gardens). They generally avoid unbroken woodlands and coniferous forests. They are an uncommon to common M throughout most of Texas, in virtually any wooded or brushy habitat, especially brushy fencerows, parks, and orchards.

FEEDING BEHAVIOR. They forage for insects and berries, mainly on the ground. They flip leaves aside with their bill in typical thrasher fashion, but they are probably not as effective as the thrashers that possess a long curved bill. They ascend to trees and shrubs for berries.

COURTSHIP BEHAVIOR. The male returns to the breeding grounds before the female. He immediately begins singing while holding his body low to the perch and depressing his tail. When courting the female, he struts with his wings low and his tail erect, then turns to display his chestnut undertail coverts, occasionally pausing to sing.

NESTING. In dense thickets, shrubs, brier patches, or low trees. Nest: a bulky cup of weeds, leaves, twigs, and bits of trash that is built by the female. Eggs: incubated by the female. Both parents bring food to the nestlings. Gray Catbirds puncture and eject eggs that cowbirds deposit in their nest.

VOICE. Call: a catlike mewing. Song: similar to the Northern Mockingbird's and the Brown Thrasher's, but less musical, less repetitive, and with frequent harsh notes; to some ears it has a complaining quality (Oberholser and Kincaid, 1974). Gray Catbirds often appropriate songs of other birds, including quail, jays, and hawks, but not with the mockingbird's facility. They sing more often from concealed than from exposed perches.

They sing at night as well as in autumn, when their barely audible whisper song can be heard.

OTHER BEHAVIORS. Compared with Brown Thrashers, they are more conspicuous, more inquisitive, and bolder (sometimes fearless); they adapt well to houses and gardens and respond readily to squeaking. Their frequent tail-jerking and tail-spreading behavior may be a displacement behavior derived from an intention movement to flee.

Northern Mockingbird, *Mimus polyglottos.* Princ. distrib. Texas: Common PR throughout Texas, in open areas that support scattered trees and bushes; they are very much at home in urban areas. In Trans-Pecos Texas they inhabit brushy streamsides, arroyos, and thickets.

FEEDING BEHAVIOR. They run along the ground searching for insects and seeds, ascend to trees and bushes for berries, and sometimes flycatch for insects on the wing.

COURTSHIP BEHAVIOR. In their mating dance, the male and female face each other with their heads and tails held high, dart at each other, but then quickly retreat. This behavior is also part of their territorial display and is seen at territorial boundaries.

Frequently, during an extended bout of singing, the male flies up vertically from his perch, then glides back without missing a note of his song. Unmated males sing more than mated males, and they more frequently fly up into the air as they sing. It is said that only unmated males sing at night.

The pair bond sometimes lasts for life. Although both members of a pair sometimes deviate from monogamy, opportunistic breeding in males is limited by the demands of territory maintenance.

NESTING. In a tree or shrub, often hidden in dense foliage. Nest: a cup of leaves, weeds, grass, and other plant materials, lined with softer materials such as moss, plant down, and animal hair. Eggs: incubated by the female. Normally only the female brings food to the nestlings. Generally birds cannot fly until about a week after they leave the nest. The parents boldly defend their nesting area, diving without hesitation at cats, dogs, and people.

Males often begin building a new nest before the fledglings from the first brood are independent of parental care (clutch overlap). This leaves the female to care for the fledglings. Although relatively uncommon in birds, clutch overlap normally increases the number of offspring produced in a season. In the case of the Northern Mockingbird, up to four broods can be produced each year this way.

A few days prior to building the new nest, the male resumes singing and continues throughout the nest-building period. Although generally birdsong functions to strengthen the pair bond and to allow individuals to recognize each other, in this case it seems to be a stimulus for renesting.

VOICE. The male's tireless outpouring of trills, warbles, squawks, and scolds recalls the verve of a Rossini overture. He may sing uninterruptedly for more than an hour. He may produce 30 or more different sounds during a 10-minute time span, including virtually any sound heard in nature and human-produced sounds like sirens and whistles. He often repeats a sound three or more times before going on to the next one.

Strictly speaking, Northern Mockingbirds do not mimic sounds, in spite of their scientific name (L. *Mimus*, mimic; Gr. *polyglottos*, many tongued). In the context of biology, mimicry implies deception as well as a benefit to the mimic that results from the deception. Because mockingbirds do not seem to be deceiving other animals when they sing, it is preferable to say that they appropriate sounds into their songs.

The reason mockingbirds engage so abundantly in singing is not clear. Most likely the extraordinary variety of sounds produced by the male influences, through physiological mechanisms, both his own sexual readiness and that of his mate. The chatburst, heard mainly during the period of fall territoriality, may be a response to demands posed by the arrival of strangers in fall.

Call: harsh and grating scolds, including two rendered as *shack* and *tchair*. Refer to Oberholser and Kincaid's (1974) interesting discussion of singing in this species.

OTHER BEHAVIORS. Males sparring at the border of their territories often hop sideways. Wing flashing (running along the ground, then pausing to lift the wings briefly) is a common behavior, as is anting. Wing flashing probably stirs up insects and possibly distracts snakes and other predators.

Males and females often defend separate feeding territories in winter.

Sage Thrasher, *Oreoscoptes montanus.* Princ. distrib. Texas: Locally uncommon M and WR in w. half of Texas, tending toward open scrubby areas rather than dense thickets along streams (a favorite habitat of many thrashers).

FEEDING BEHAVIOR. Sage Thrashers run rapidly along the ground searching for insects, sometimes ascending into shrubs and low trees to feed on berries.

VOICE. Call: a guttural *tchuck,* like a blackbird's. Song: a continuous series of clear, rapid, warbled, and rarely repeated phrases, very few of which are appropriated from songs of other birds. The song is infrequently heard in Texas. When singing, birds sometimes turn the head from side to side.

OTHER BEHAVIORS. Elusive by nature; when disturbed, they usually fly some distance rather than to the next bush.

In appearance, these are the most thrushlike of the thrashers. When running, Sage Thrashers suggest an American Robin, but unlike robins and other thrushes, they sometimes run with their tail elevated.

Brown Thrasher, *Toxostoma rufum.* Princ. distrib. Texas: Uncommon to locally common PR in e. part of Texas; uncommon to common WR throughout the state except the most s. parts. They occur in thickets, shrubby edges of swamps, and urban areas as long as dense vegetation is nearby.

In s. Texas, wintering Brown, Long-billed, and Curve-billed Thrashers compete very little for resources because they tend to occupy different habitats. Brown Thrashers favor riparian woodlands; Long-billed Thrashers, shrub cover of chaparral; and Curve-billed Thrashers, open mesquites.

FEEDING BEHAVIOR. When searching for terrestrial insects, they thrash the ground with their bill (hence the name), tossing leaves aside and digging up soil. The curved bill aids them in doing this. Occasionally, they flycatch.

Although they spend much of their time foraging on the ground, they climb about in trees and shrubs to pick berries and fruits. They crack open acorns by abruptly striking them with their bill.

Both Brown and Long-billed Thrashers feed almost entirely in dense vegetation, where they sweep debris aside with their bill. Curve-billed Thrashers, which rely on digging, frequently forage outside the shrub-tree habitat. Apparently Brown Thrashers exclude most Long-billed Thrashers from riparian habitats rich in arthropods and gastropods.

COURTSHIP BEHAVIOR. Their courtship is rather dignified. The male sings softly as he approaches the female; this prompts her to pick up a stick or leaf and hop to him, chirping and fluttering her wings. He may also pick up a leaf and present it to her.

NESTING. In a tangled vine, low tree, or shrub. Nest: built by both parents, who use sticks for the foundation, and twigs, weeds, bark, and other plant materials for the cup. Both parents incubate the eggs and bring food to the nestlings.

VOICE. Call: hisses, clicks, and whistles, including a loud *spack.* Song: somewhat like the Northern Mockingbird's, but throaty and with fewer appropriated sounds. Compared with the mockingbird, the phrases are generally more abrupt and in twos instead of threes or more; they sound to some listeners like a telephone conversation (Terres, 1980): *Hello, hello, yes, yes, who is this? who is this? I should say, I should say, how's that? how's that?*

They usually sing from a conspicuous perch, holding their head high and drooping their long tail. Their repertoire contains more than 1,100 song types (Boughey and Thompson, 1981).

OTHER BEHAVIORS. They can be shy and elusive in thickets, yet at other times bold enough to venture out on lawns to feed. On the ground they walk, run, or hop.

Long-billed Thrasher, *Toxostoma longirostre.* Princ. distrib. Texas: Uncommon to locally common PR in s. Texas, in mesquites and brushy undergrowth, especially if water is nearby. Where their ranges overlap in s. Texas, Long-billed Thrashers favor denser vegetation (bottomland mesquites, willows, huisaches, and other bushes), whereas Curve-billed Thrashers prefer more open habitats such as low mesquites and prickly pears in well-drained soil. *See* Brown Thrasher.

FEEDING BEHAVIOR. Basically like the Brown Thrasher, although they sometimes dig deeper (as much as an inch) in the soil for food items.

COURTSHIP BEHAVIOR. Apparently not recorded. Most pairs probably remain together throughout the year.

NESTING. In dense, shaded vegetation, amid spiny shrubs and trees, making the nest generally inaccessible to predators. Nest: a cup of sticks, leaves, twigs, weeds, and other plant parts, probably constructed by both parents. Unlike most thrashers, hatching is synchronous.

Both parents incubate the eggs and bring food to the nestlings. *See* Curve-billed Thrasher.

VOICE. Call: sharper and higher pitched than the Brown Thrasher's.

Song: very similar to the Brown Thrasher's, with fewer repetitions of phrases. The male sings from the top of a tree or bush, or from a hidden perch in dense cover.

OTHER BEHAVIORS. Where their wintering ranges overlap in south Texas, Long-billed Thrashers and Brown Thrashers maintain separate feeding territories. Long-billed Thrashers are shy and wary and not likely to be observed in open areas (as are Brown Thrashers, which occasionally become quite bold).

Curve-billed Thrasher, *Toxostoma curvirostre.* Princ. distrib. Texas: Uncommon to common PR in w. half of the state; they inhabit most arid and semiarid habitats, including mesquite brush and roadsides, as long as there are sufficient shrubs and small trees in which to hide. *See* Brown Thrasher.

FEEDING BEHAVIOR. They vigorously dig for insects by pounding on the ground with their bill. Sometimes they use their tail to brace themselves when the ground is hard. They also eat berries, cactus seeds, and other plant items. *See* Brown Thrasher.

COURTSHIP BEHAVIOR. The male sings softly as he pursues the female.

NESTING. In the desert, often in the fork of a cholla; elsewhere, in habitats that include prickly pear, mesquite shrub, and other spiny plants. In south Texas, Curve-billed Thrashers nest and forage in open habitats; they usually nest in dense chaparral only if clearings are nearby. Long-billed Thrashers always nest in sheltered habitats. Nest: probably built by both parents, who utilize thorny twigs, grass, feathers, and animal hair to shape a bulky, loose cup. Compared with Long-billed Thrashers, Curve-billed Thrashers build a deeper nest and incubate their eggs for shorter periods of time each day. Both parents incubate the eggs, and both tend to the young.

VOICE. Call: an abrupt and arresting *wheet whit*, like someone whistling for another person's attention. It is frequently and repeatedly uttered at dusk. Song: similar to Northern Mockingbird's, but a little lower, slower, and with fewer appropriated sounds. The male sings from the top of a barn, utility pole, tree, or other conspicuous place.

OTHER BEHAVIORS. They become tame around ranch houses, corrals, and other human structures. If they locate a Cactus Wren's nest in their territory, they often destroy it. They occasionally build a winter roosting platform that may be used as their nest site the following spring.

Crissal Thrasher, *Toxostoma crissale.* Princ. distrib. Texas: Locally common PR in Trans-Pecos Texas and w. Edwards Plateau. They occur in either the dense vegetation near rivers (in particular, the Rio Grande) or, less frequently, in the juniper-pinyon-oak associations of the Chisos.

FEEDING BEHAVIOR. Essentially like the Curve-billed Thrasher.

COURTSHIP BEHAVIOR. Apparently unrecorded.

NESTING. In dense vegetation, including mesquites and a variety of shrubs, and usually well concealed. Nest: built by both parents, a bulky structure of thorny twigs. Both parents incubate the eggs and feed the nestlings.

VOICE. Call: a rolling *toit toit,* or *pichoo ree,* usually at dawn or dusk. Song (from the ground or an elevated perch): rich, full, and sweet. It is thrasherlike but has few appropriated sounds.

OTHER BEHAVIORS. These highly terrestrial birds usually perch on structures only when singing to advertise their territories. They walk quickly, with long, graceful strides, and raise their tail when they stop. They typically avoid intruders by running through thick brush instead of flying.

STARLINGS: FAMILY STURNIDAE

European Starling, *Sturnus vulgaris.* Princ. distrib. Texas: Uncommon to abundant PR in almost all parts of Texas, especially urban areas, becoming more common in winter in many parts of the state. They occur in most habitats except deep forests.

FEEDING BEHAVIOR. Powerful jaw muscles allow them to open their bill after they jam it into the ground; this action forces open a hole that exposes insects and other food items. The same mechanism is used to pry open vegetation for insects inaccessible to most birds. As the bill opens, the eyes move forward toward each other, providing the bird with binocular vision that helps them detect insects uncovered by the bill action.

European Starlings usually forage in flocks. Individuals ascend to trees for fruit, or sally out from a perch to capture flying insects.

COURTSHIP BEHAVIOR. Males sometimes have more than one mate. Among their several courtship postures and displays is a hunchback stance accompanied by a flailing of the wings. Apparently female European Starlings use both aggressive behavior (including song) and the solicitation of copulations to prevent their mates from pursuing new mates. Attacking

FIGURE 10. Attack behavior in the European Starling. Birds attack with their feet rather than their beak, which is analogous to a bird pouncing on prey. (C. R. Ellis, Jr., 1966, *Wilson Bulletin* 78: 208–223. With permission.)

with the feet rather than the beak is considered analogous to a predator pouncing on prey (Fig. 10).

NESTING. In virtually any hole, cavity, or crevice, both natural and artificial, including human structures; indeed, the species' reproductive success is partly attributable to its tolerance of so many nesting sites. Nest: a very shallow cup of loosely connected grass, leaves, twigs, feathers, and bits of trash. It is initiated by the male, but his mate proceeds to build the rest of it and sometimes discards some of the nesting material that he provides. The green material is often fresh and rich in chemical substances that repel parasites. Both parents incubate the eggs and feed the nestlings. Like many birds, they keep the nest clean by removing the fecal sacs (membranous sacs that contain the nestlings' feces). Some females lay eggs in the nests of other European Starlings. Bachelor males sometimes help in nesting activities.

VOICE. Call: various and numerous sputterings, wheezing sounds, and mechanical squeaks, often with skillful imitations of other birds. These garrulous birds vocalize constantly throughout the year. They are capable of memorizing and later recognizing vocalizations from particular individuals.

OTHER BEHAVIORS. Tough, adaptable, spirited, intelligent, and extremely gregarious, they sometimes form flocks (mixed with various blackbirds) of a million or more individuals. Flocks move about in concert, like shorebirds, wheeling and twisting together with astonishing precision, even though no single bird takes the lead.

Their flight is strong and direct; on the ground, they waddle with short, jerky steps, projecting an image of cockiness. Their territory hardly extends beyond the nest hole.

WAGTAILS AND PIPITS: FAMILY MOTACILLIDAE

Pipits are small, plain-colored birds that walk, rather than hop (like sparrows). In open areas they forage for insects and, to a lesser extent, seeds.

American Pipit, *Anthus rubescens.* Princ. distrib. Texas: Common M and WR throughout most of Texas, in plowed fields, prairies, and along muddy shores.

FEEDING BEHAVIOR. As they walk over bare lakeshores or sparsely covered ground, they daintily pick up insects. Sometimes they venture into shallow water.

VOICE. Call: a lispy *tsee-tseep*, only remotely suggesting *pipit*. They call in flight. The courtship flight song (*chee weee*) is rarely heard in Texas.

OTHER BEHAVIORS. Wintering birds often forage in small flocks that include Horned Larks and longspurs. Pipits walk or run with ease, habitually wagging their tail (either up and down or from side to side) and bobbing their head like a dove.

Their flight is swift, buoyant, erratic, and undulating. Flushed birds ascend high into the air, hover, then return to the ground. Flocks in winter are sometimes quite large.

Sprague's Pipit, *Anthus spragueii.* Princ. distrib. Texas: Rare to uncommon WR throughout Texas, except the Panhandle, and locally common WR on the coast; they inhabit short-grass prairies in winter (Freeman, 1999).

FEEDING BEHAVIOR. They walk along in short grass while foraging for insects and seeds.

VOICE. Call: a sharp *tsip,* harsher than the American Pipit's. The spectacular courtship flight song consists of a series of seven or eight descending notes delivered for half an hour or so while floating in circles high in the air. The song is rarely heard in Texas.

OTHER BEHAVIORS. When flushed from the grass, birds fly high, circle, then dive quickly back to the ground. They form flocks in winter.

WAXWINGS: FAMILY BOMBYCILLIDAE

Cedar Waxwing, *Bombycilla cedrorum.* Princ. distrib. Texas: Common to abundant WR throughout Texas, less common in w. parts; they occur in open woodlands, orchards, and virtually all cities and towns within their range.

FEEDING BEHAVIOR. Waxwings take berries while perched in trees, but they also capture insects by hovering next to foliage or by sallying out like a flycatcher. They do not select berries randomly but instead choose those they can more efficiently handle or those that produce the highest rate of sugar intake. Thus, the following observation by King (1996) probably represents anomalous behavior: "The Cedar Waxwing paused briefly after landing, then swallowed the [unidentified] nestling whole in approximately three successive swallowing motions."

VOICE. Call: a very high, thin *seeeee,* uttered both in flight and while perched.

OTHER BEHAVIORS. Wintering birds almost always associate in flocks. They fly swiftly and purposefully from tree to tree, and when they alight, they sit near each other in an upright posture and usually face the same direction. They are unusually sociable: individuals sitting side by side on a branch sometimes pass a berry back and forth with their bill before one of them swallows it. Their droppings (usually containing highly colored, undigested parts of berries) are deposited below their roosting sites, frequently in such places as parking lots and sidewalks.

The waxwing (the small, red waxy substance exuded by the feather shafts of the secondaries) appears to signal age or social status. Second-year birds usually lack these waxy tips, and older birds usually mate with each other rather than younger birds, suggesting that the red spot plays a role in mate selection.

SILKY FLYCATCHERS: FAMILY PTILOGONATIDAE

Phainopepla, *Phainopepla nitens.* Princ. distrib. Texas: Locally common PR in the arid regions of Trans-Pecos Texas, notably in cottonwoods, pecans, and other tall trees that are near isolated ranch houses (Oberholser and Kincaid, 1974); also in mesquite groves, juniper-oak hillsides, and desert washes.

FEEDING BEHAVIOR. They consume berries (especially mistletoe berries) and insects. They feed while perched, but they also hover; males seem especially prone to sally out to capture insects flying past.

COURTSHIP BEHAVIOR. The male follows a circular or zigzag path high (300 feet) over the nesting territory, often somersaulting while holding his wings in a V. In Arizona, males from adjoining territories display together in this manner. The male also feeds the female and pursues her in a courtship flight.

NESTING. Well concealed in dense vegetation, often in a clump of mistletoe. Nest: a shallow cup of weeds, twigs, leaves, and other plant parts, constructed primarily by the male. Both parents incubate the eggs, and both parents bring crushed insects and berries to the nestlings.

Some Phainopeplas are thought to nest first in one area, then leave that area to nest a second time elsewhere, probably following areas of rainfall. This behavior has not been documented in Texas.

VOICE. Call: a querulous *ka rak,* given when alarmed; also, a low, liquid *wuurp,* as well as several other calls. Song: a disconnected sweet gargle, often sung by males in flight.

OTHER BEHAVIORS. Phainopeplas are restless, shy, and suspicious. When flushed from their perches (typically the top of a tree or cactus), they fly quickly into the brush. When alarmed, they raise their crest and jerk their tail (displacement behaviors). Their white wing patches, which are displayed in flight, evidently help maintain contact between individuals.

WOOD WARBLERS: FAMILY PARULIDAE

As a rule, wood warblers are small, lively, tree-dwelling birds that are often brightly colored and have pleasing (though not extraordinary) songs. Few of their songs come across as "warbles." Many species constantly flit about trees or shrubs looking for insects and spiders, often darting out like flycatchers to capture passing insects. Others forage like creepers, meticulously working their way along trunks and limbs as they extract insects, pupae, and eggs from crevices in the bark; still others forage on the ground, where they move about teetering like Spotted Sandpipers.

The males of many species sing two separate songs, one to advertise their territory and another to attract and communicate with a mate. Species that sing from treetops generally produce higher-pitched songs than those that sing from lower levels. These differences are correlated with the acoustical properties of specific bird sounds: lower-pitched sounds efficiently penetrate the dense foliage typical of the understory, and higher-pitched sounds carry farther in the unobstructed space at treetop level.

It is remarkable that in a country teeming with millions of birdwatchers so little is known about the courtship behavior of our warblers. Nonetheless, a number of courtship behaviors are commonly observed. Many males spread their wings and tail to exhibit parts of their brightly colored plumage; these displays serve to maintain contact between mates, to advertise their territory, and to attract a mate. Also, being agile and energetic by nature, many males perform elaborate courtship displays that may include spectacular aerial acrobatics.

When disturbed at the nest, some warblers feign injury or engage in other distraction displays, many of which call attention to markings on their wings and tail.

Many migrating warblers forage in mixed groups that contain other warbler species and other small passerines.

Blue-winged Warbler, *Vermivora pinus.* Princ. distrib. Texas: Locally common M in e. Texas and on the coast, especially in shrubby woodland edges.

FEEDING BEHAVIOR. Blue-winged Warblers generally forage quietly in shrubs and trees at low to middle levels, gleaning for insects and spiders with deliberate, vireolike movements, often with their head pointed downward. They probe clumps of dead leaves with their long, spikelike bill, which appears ideally adapted for extracting insects from hidden places. They also search for food items by hovering or hanging up-side down.

VOICE. Call: often given in flight, a sharp *chip* and a thin *zeeee*. Song: a buzzy *beee bzzzz*, the first note ascending and the second descending, "suggesting inhalation and exhalation" (Oberholser and Kincaid, 1974).

Golden-winged Warbler, *Vermivora chrysoptera.* Princ. distrib. Texas: Rare M in e. Texas, in woodland edges and old pastures; they are fairly common on the coast, where they favor salt cedars and live oaks.

FEEDING BEHAVIOR. Golden-winged Warblers carefully inspect the forest floor for insects and spiders; they also creep along trunks and stumps or probe dead leaves with their long, spikelike bill. They are more acrobatic than their close relative the Blue-winged Warbler, and like that species, they feed with their head down or hang upside down like a chicka-dee. Their foraging flights are erratic and usually over short distances.

VOICE. The male throws his head back, points his bill to the sky, and utters a buzzy *djeee,* followed by a lower-pitched *dzz dzz dzz.* Apparently the songs of this species do not differ appreciably from those of the Blue-winged Warbler.

Tennessee Warbler, *Vermivora peregrina.* Princ. distrib. Texas: Uncom-mon to locally abundant M in e. half of Texas, in trees and bushes. They are more common in spring, when they seem to prefer oaks and pecan trees.

FEEDING BEHAVIOR. Tennessee Warblers favor the outer parts of trees. They generally forage high in the canopy during spring and at all levels during fall. When searching for insects, spiders, and occasionally berries, they sometimes suspend themselves from twigs like a titmouse or hang their head downward. They move about nervously, and when they fly from perch to perch, they do so quickly and erratically.

VOICE. Call: a soft *chit.* Song: seemingly too loud for a bird this size, it comes across as a strident, staccato, unevenly spaced *zidit zidit zidit zidit zit zit zit zit zit zit,* the last notes becoming more rapid and suggesting a Chipping Sparrow.

Orange-crowned Warbler, *Vermivora celata.* Princ. distrib. Texas: Com-mon M throughout the state, and uncommon WR in most parts except the Panhandle. They inhabit woody and shrubby habitats, especially dense thickets near water. In Guadalupe and Davis mountains they nest at high altitudes.

FEEDING BEHAVIOR. They forage nervously in the lower parts of trees and shrubs, flitting rapidly and jerkily from perch to perch, or flying out like a flycatcher to capture passing insects. They also descend to the ground to search for food items among dead leaves.

Their diet consists primarily of insects and spiders; they also eat berries and nectar, especially in winter.

COURTSHIP BEHAVIOR. Apparently not recorded.

NESTING. Usually on the ground under overhanging vegetation. Nest: a small cup built by the female from grass, leaves, and bark. Eggs: incubated by the female. Both parents feed the nestlings.

VOICE. Call: a thin, metallic *dzeep,* or *dzee,* often given in flight. Song: a weak, high-pitched trill that drops in pitch and volume at the end, rarely heard away from the breeding site.

OTHER BEHAVIORS. When birds engage in their song flight, their wings flutter in a manner unlike normal direct flight. Interestingly, flight songs and song flights apparently are not associated with special circumstances such as the presence of a female or a competing male.

Nashville Warbler, *Vermivora ruficapilla.* Princ. distrib. Texas: Common to abundant M in all parts of Texas except the extreme w.; uncommon WR in extreme s. Texas, in open, mixed forests. This is one of the most abundant spring migrants in c. Texas.

FEEDING BEHAVIOR. Dainty, agile Nashville Warblers forage actively in trees and shrubs at all levels, flicking their tail as they move about nervously. They favor low trees and thickets, and occasionally descend to the ground. They feed mainly on insects, taking them from the undersides of leaves, the tips of twigs, and other parts of trees.

VOICE. Call: a soft *chip.* Song: loud, musical, and separable into two parts: *suweet suweet suweet suweet,* suggesting the Black-and-white Warbler's song, and a lower-pitched twittering *ju ju ju ju,* dropping in pitch at the end. The song is heard frequently during spring migration.

OTHER BEHAVIORS. Their flight is quick and erratic. During migration, they join mixed species of warblers.

Virginia's Warbler, *Vermivora virginiae.* Princ. distrib. Texas: Rare and local SR in Guadalupe and Davis mountains, usually above 6,000 feet. They inhabit dry mountainsides, ravines, and rocky slopes that are covered with dense scrub oak thickets. This species is an ecological counterpart of the closely related Colima Warbler of the Chisos.

FEEDING BEHAVIOR. Not well known. They flick their tail as they actively forage for insects in thick brush, usually at low levels, or on the ground; they sometimes flycatch.

VOICE. Call: a dry, abrupt *tchick*. Song: a series of 10–15 cheerful notes, sometimes rendered as *chewee chewee chewee chewee*. The first 7–10 are high-pitched and are followed by 3–4 lower-pitched notes. According to Oberholser and Kincaid (1974), it is "much like the Colima's, although louder, more variable, and usually rising in pitch at the end."

OTHER BEHAVIORS. Generally, they are more shy and elusive than the Colima Warbler and stay closer to the ground.

Colima Warbler, *Vermivora crissalis.* Princ. distrib. Texas: Uncommon SR at Boot Spring (6,500+ feet in altitude) in the Chisos Mountains of Big Bend National Park, primarily in maple and oak thickets, oak-pine canyons, and other undisturbed woodlands with dense understory (Wauer, 2000).

FEEDING BEHAVIOR. Apparently they eat only insects, sometimes capturing them on the wing, but more often moving sluggishly and deliberately, like vireos, among the vegetation. They generally glean at the lower levels of undergrowth.

COURTSHIP BEHAVIOR. Apparently unrecorded.

NESTING. On the ground among rocks, more or less concealed in vegetation or leaf litter, generally where trees are shorter and shrubs are taller (Lanning et al., 1990). Nest: built by both parents from grass, bark strips, dead leaves, and other plant material; often lined with animal hair. Both parents incubate the eggs and feed the nestlings. Interestingly, young birds become independent a few days after fledging, sooner than most warblers.

VOICE. Call: a sharp *chit*, suggesting a Nashville Warbler. Song: a trill that resembles the songs of the Chipping Sparrow or Pine Warbler, but shorter and more musical. Also, it varies more in pitch and terminates with two slightly lower notes. Birds sing even as they forage.

OTHER BEHAVIORS. In Texas, they occur alone or in pairs. They defend their territories by singing and by chasing away intruders.

Lucy's Warbler, *Vermivora luciae.* Princ. distrib. Texas: Rare, local SR in s.w. Trans-Pecos (the extreme e. edge of its range), in mesquites, willows, and brushy habitats bordering watercourses. This is the smallest North American warbler and the only one that nests in the hot s.w. deserts.

FEEDING BEHAVIOR. With the industry of a gnatcatcher, they hop around the lower levels of mesquite and other plants, quickly snapping up insects from twigs and leaves and occasionally flying out to capture prey in midair. They often flick their tail while foraging.

COURTSHIP BEHAVIOR. In his courtship display, the male fluffs his plumage, raises his crown feathers, and spreads his wings and tail.

NESTING. Nest: cuplike, compact, and built by both parents. Unlike all other North American warblers except the Prothonotary Warbler, Lucy's Warblers nest in natural cavities such as woodpecker holes and spaces under loose bark. Eggs: usually incubated by the female. Both parents bring food to the nestlings.

VOICE. Call: a loud *tsip,* which has been compared to the call notes of the White-crowned Sparrow and the Nashville Warbler. Song: similar to the Yellow Warbler's, a series of chirps (*weetee weetee weetee weetee*) followed by four to eight slow, slurred notes on a lower pitch.

OTHER BEHAVIORS. They are generally shy birds and not easily approached.

Northern Parula, *Parula americana.* Princ. distrib. Texas: Common SR and M in e. half of the state. Although favoring humid woodlands for nesting, they occur in many woody habitats during migration.

FEEDING BEHAVIOR. They forage high in the canopy of mature trees, especially in swampy areas, where they are more likely to be heard than seen. These relatively sedate birds search for insects by creeping along branches like a nuthatch, but they also hover at leaf clusters, suspend themselves upside down like chickadees, or flycatch. Occasionally they drop to the forest floor to feed.

COURTSHIP BEHAVIOR. Apparently not recorded.

NESTING. In open coniferous and deciduous forests. Nest: well hidden and hollowed out in Spanish moss or hanging lichens. Eggs: usually incubated by the female. Both parents feed the nestlings.

VOICE. Call: an abrupt but musical *tzip;* in flight, a high *tzif* that drops in pitch. Song: an ascending (trickling up), insectlike trill, the last note dropping abruptly as a *tzip.* Birds sing during migration as well as on the breeding grounds. Two song types have been studied in east Texas (Bay, 1999).

OTHER BEHAVIORS. They are generally fearless and easily observed.

Tropical Parula, *Parula pitiayumi.* Princ. distrib. Texas: Rare resident in extreme s. Texas, mainly in live oak woodlands.

The behavior and vocalizations of this species apparently do not differ significantly from those of the closely related Northern Parula. See

Brush (1999) and Dunn and Garrett (1997) for more details, and Ober-holser and Kincaid (1974) for their views on how changing agricultural practices have influenced the distribution and abundance of this species in south Texas.

Yellow Warbler, *Dendroica petechia.* Princ. distrib. Texas: Common to abundant M throughout the state, in gardens, parks, thickets, and other woody or shrubby habitats.

FEEDING BEHAVIOR. These active and restless birds normally glean for insects; occasionally they hover at the tip of a limb or sally out like a fly-catcher. In winter they eat berries.

VOICE. Call: an emphatic, musical *chip;* in flight, a high, insectlike *zeee.* Song: bright and sweet, traditionally rendered as *sweet sweet sweet I'm so sweet.* This series of loud and energetic notes ascends at the end in a stac-cato warble like a goldfinch's. It is sung persistently in spring and summer. The song varies according to specific situations. For example, birds sing one song if near their mate, another if they are away from their mate.

OTHER BEHAVIORS. The flight is quick and erratic. On their breed-ing grounds, males forage higher in trees and in less dense foliage than females, supposedly to advertise the territory and to gain their mate's attention.

Yellow Warblers cover cowbird eggs with a new nest floor, and in some cases they eventually produce a multilayered nest. Another anticowbird tactic is to nest near colonies of Red-winged Blackbirds, possibly because Redwings tend to intimidate cowbirds (although these blackbirds them-selves are sometimes parasitized). Also, female Yellow Warblers probably are more vigilant when cowbirds are in the vicinity, as studies show they can recognize cowbirds on the basis of bill shape and vocalizations.

OTHER BEHAVIORS. They are generally unwary and easily approached and observed.

Chestnut-sided Warbler, *Dendroica pensylvanica.* Princ. distrib. Texas: Uncommon M in e. Texas and locally common M on the coast, in brushy thickets, woodland borders, and cutover woods, as well as deciduous and coniferous forests. Apparently migrants favor deciduous forests. On their breeding grounds (outside Texas), they have benefited from the clearing of forests and are much more common now than 150 years ago.

FEEDING BEHAVIOR. They are primarily insectivorous, eating seeds and berries when insects are scarce. They hop among the lower and mid-level branches of trees, puffing their breasts, drooping their wings, and flicking their tail to expose the white undertail coverts, most likely a display that reinforces the pair bond. They occasionally flycatch.

VOICE. Call: a husky *chip;* in flight, a soft *tzeeet.* Song: loud, clear, and distinctive. It has inspired transcriptions that convey contrasting sentiments: (1) *I wish, I wish, to see, Miss Beecher,* (2) *pleased pleased pleased to meet you,* and (3) *sweet sweet sweet I'll switch you.* All renditions emphasize the accented next-to-last syllable and the slurred final note. The song suggests the Yellow Warbler's song and is heard during spring migration.

Magnolia Warbler, *Dendroica magnolia.* Princ. distrib. Texas: Uncommon M in e. half of Texas, sometimes very common on the coast, in gardens and both evergreen and deciduous woodlands, including mesquites.

FEEDING BEHAVIOR. These relatively plump (yet quite energetic) birds pursue insects at all levels of vegetation but favor the center part of the tree. They move about with wings and tail partially spread, exposing the white wing and tail patches that probably help maintain contact with other warblers (including their mate). In winter they eat fruit, although probably less than most warblers.

VOICE. Call: a soft *tlip;* in flight, an insectlike *dzeee.* Song: a loud, clear, ascending *weeta weeta weeta wichee,* recalling the Yellow Warbler's song but generally considered richer in quality.

OTHER BEHAVIORS. Their flight is quick and erratic.

Cape May Warbler, *Dendroica tigrina.* Princ. distrib. Texas: Rare spring M in e. Texas (even rarer in fall), in open forests and along forest edges.

FEEDING BEHAVIOR. These energetic warblers forage in the canopy for insects and other invertebrates, sometimes moving deliberately, at other times quickly and erratically. Frequently they dart out for insects like a flycatcher. Berries form a lesser part of their diet.

VOICE. Call: a thin *dzeep* or *dzee dzee.* Song: a series of thin, high-pitched, unhurried notes (*see see see see*) that are sometimes doubled. To some listeners it suggests a Black-and-white Warbler's song.

Black-throated Blue Warbler, *Dendroica caerulescens.* Princ. distrib. Texas: Rare M in e. half of Texas, more common in fall, in pastures, woodland understory, tall brush, fencerows, and gardens.

FEEDING BEHAVIOR. Sometimes they are quite active, moving about with a quick, jerky flight, but they generally feed quietly and deliberately (like a vireo). They glean insects and berries from the inner parts of shrubs and trees, and sometimes flycatch.

VOICE. Song: variable but usually with a series of three to five or more slow, insectlike *zweee*s, the last slurred upward.

OTHER BEHAVIORS. They often hold their wings partly open while moving about on limbs or when perched, perhaps as a signal to maintain the pair bond or to advertise the territory. Normally they do not associate with other warblers.

Yellow-rumped Warbler, *Dendroica coronata.* Princ. distrib. Texas: Common to abundant M and WR throughout Texas, in forests, gardens, and brushy habitats. One subspecies (formerly called the Myrtle Warbler) occurs mainly to the e., and two other subspecies (both formerly called Audubon's Warbler) to the w. One of the latter two subspecies is a common SR at higher elevations in the Guadalupe Mountains and an uncommon SR in the Davis Mountains.

FEEDING BEHAVIOR. Yellow-rumped Warblers forage in trees and bushes at all levels, sometimes dropping to the ground. They glean for insects from trunks and foliage, hover at tips of limbs, sally out like flycatchers, or skim over water like swallows. On their breeding grounds, males generally feed at a higher level in trees than females.

Insects (an important food item when available) make up a greater portion of the diet in the Audubon races (about 85 percent, even in winter). Berries, especially bayberries and poison ivy, are very important in winter.

COURTSHIP BEHAVIOR. The male's courtship posture emphasizes his yellow rump and breast patches. When pursuing the female he fluffs his side feathers, raises his wings, erects his crown feathers, and calls repeatedly.

NESTING. On a horizontal limb, 10–25 feet high in a tree, usually a conifer. Nest: an open cup of twigs, grasses, feathers, and other plant materials, built by the female. Feathers are incorporated into some nests, and these curl over to shield the eggs. Eggs: generally incubated by the female. Both parents feed the nestlings.

VOICE. Call: a sharp *chuck,* suggesting the Northern Cardinal's call, although it is less musical in quality. It is softer and less emphatic in the Audubon races. Song: variable, but basically a weak, rambling *whee see see see see,* sometimes repeated at a higher pitch. It is delivered more slowly and deliberately in the Audubon races.

OTHER BEHAVIORS. Flight is "ordinarily easy and fluttering and performed with a minimum of wingbeats" (Oberholser and Kincaid, 1974). Birds flick their tail after alighting; this is more likely a displacement behavior than a balancing mechanism. Wintering Yellow-rumped Warblers often roost communally and may be joined by other species.

Black-throated Gray Warbler, *Dendroica nigrescens.* Princ. distrib. Texas: Rare to uncommon M in much of w. and c. Texas, possibly nesting in Guadalupe Mountains. They occur in dry juniper-pinyon-oak scrub.

FEEDING BEHAVIOR. They forage for insects in bushes or the lower levels of trees. Oberholser and Kincaid (1974) described this species as "retiring and where foliage is thick may remain inconspicuous. Its movements are not as rapid or as nervous as those of most warblers."

VOICE. Call: *tchup;* or in flight, a high-pitched *tzee* or *tzeep.* Song: a soft, *weezee weezee weezee weezee weezee tsee,* the last note lower. The sound quality has been compared to that of the Black-throated Green Warbler and the Yellow-rumped Warbler.

Golden-cheeked Warbler, *Dendroica chrysoparia.* Princ. distrib. Texas: Rare to locally common SR on the Edwards Plateau and adjacent parts of c. Texas. They require stands of Ashe juniper–oak woodlands with diverse canopies, where mature oaks and other hardwoods serve as foraging sites during spring and early summer.

Near Austin, they are incompatible with Blue Jays, a species historically new to the area that adapts readily to urbanization. More important, habitat destruction continually threatens these warblers.

FEEDING BEHAVIOR. They glean and occasionally sally out for insects at medium to high levels in juniper-oak stands. The shy female is not often seen, but the male is conspicuous as he flies from tree to tree.

COURTSHIP BEHAVIOR. Not notably spectacular: the male fluffs his feathers and utters a *chip,* sometimes spreading his wings while facing the female. Although the male transports nesting materials as part of nest construction, this behavior also plays a role in courtship.

NESTING. The nesting site, well concealed in the fork of a tree, is selected by the female. Nest: constructed by the female from grass, feathers, hair, bark strips, spiderwebs, lichens, and leaves and camouflaged with bark strips that are almost always from junipers. Eggs: incubated by the female. When threatened at the nest, the female performs a unique distraction display: she spreads her tail and rapidly flutters among the tree branches. Sometimes the male also engages in this display. (See Mark Lockwood's informative 1996 account of the nesting behavior of this species.)

VOICE. Call: a sharp, high-pitched *tzip.* Song (one variant; see Bolsinger, 2000): a wheezy *zee zuu zeedee zee,* suggesting the Black-throated Green Warbler but lower pitched. Parts of the song have also been compared to the songs of the Field Sparrow and Olive Sparrow.

Interestingly, "many Texas Hill Country birders, straining to hear the first Goldencheek of spring, have been fooled, at least momentarily, by the resident [Bewick's] wren" (Oberholser and Kincaid, 1974).

OTHER BEHAVIORS. Their flight is "quick and jerky" (Oberholser and Kincaid, 1974). Both males and females show a strong tendency to return the next year to the same breeding territory.

Black-throated Green Warbler, *Dendroica virens.* Princ. distrib. Texas: Common M in e. half of Texas, in shrubby thickets, river bottoms, and wooded urban areas. Uncommon WR in extreme s. Texas.

FEEDING BEHAVIOR. They forage briskly for insects, mainly high in trees, less frequently at lower levels. They glean, hover, and flycatch. In winter and during migration they add poison ivy and other berries to their diet.

VOICE. Call: a high-pitched *tzeek* that resembles the Townsend's Warbler's call. Song: a slow, deliberate, and lisping *zee zee zee zuu zee,* the next to last note slightly lower.

"The bird is a persistent singer and often utters an incredible number of songs within a short period: up to 466 repetitions have been counted in a single hour!" (Oberholser and Kincaid, 1974).

OTHER BEHAVIORS. These restless birds fly with swift and erratic movements. Although migrating and wintering with mixed-species flocks of warblers, they seem less gregarious than the closely related Townsend's Warbler. They are not especially shy.

Townsend's Warbler, *Dendroica townsendi.* Princ. distrib. Texas: Uncommon M in the High Plains and Trans-Pecos, in habitats that include conifers, canyons, oak-covered slopes, and scrubland.

FEEDING BEHAVIOR. Townsend's Warblers flit about in the tops of conifers, nervously gleaning or flycatching for insects. Migrants descend to lower branches and shrubs to feed.

VOICE. Call: a repeated, high-pitched *tzeek;* in flight, a thin *zeee.* Song: a brief, high-pitched *zee zee zee zee zee* followed by one or two clear notes at a slightly higher pitch. It has been compared to the songs of the Black-throated Green and Black-throated Gray Warblers.

OTHER BEHAVIORS. They normally occur singly or in pairs, but during migration and in winter they join small groups of other warblers. They fly with quick and irregular wing movements.

Blackburnian Warbler, *Dendroica fusca.* Princ. distrib. Texas: Rare to uncommon M in e. half of Texas, frequenting shrubby growth, upland and bottomland thickets, and other woody habitats; they are sometimes common on the coast, at least in spring.

FEEDING BEHAVIOR. They glean for insects mainly in treetops, hopping from limb to limb or making short, jerky flights. They also fly out for passing insects and eat berries when insects are scarce.

VOICE. Call: a very high *tsip,* and a thin *zeet,* given in flight. Song: high, thin, and insectlike, ascending to an even higher pitch before ending in a very high trill (*tzip tzip tzip titi tzeee*).

Yellow-throated Warbler, *Dendroica dominica.* Princ. distrib. Texas: Locally common SR in open pine forests, woodlands, river bottoms, and bayous in e. part of the state.

FEEDING BEHAVIOR. Moving slowly and deliberately, these attractive birds forage meticulously for insects concealed in holes and crevices in bark, often at the highest levels of the trees. They have the longest bill of any warbler, enabling them to extract insects from crevices too deep for other warblers.

Like Pine Warblers, they inch along both horizontal and vertical branches and trunks; however, sometimes they are quite wrenlike and investigate buildings, lumber piles, old machinery, and other human structures. They also fly out to capture passing insects.

COURTSHIP BEHAVIOR. Apparently not recorded.

NESTING. Often concealed in a clump of Spanish moss. Nest: a cup of grass, bark strips, plant down, and other plant materials constructed by the female (and to a lesser extent the male). Eggs: incubated mainly by the female. Probably both parents feed the nestlings.

VOICE. Call: a loud *tsip;* in flight, a clear, high *tseee.* Song: a clear and robust *teeyew teeyew teeyew TWEE* that descends in pitch and ends emphatically with a rising inflection, suggesting the Indigo Bunting's song.

OTHER BEHAVIORS. Their quick and nervous flight is typical of warblers. They are said to bathe more frequently than most warblers.

Grace's Warbler, *Dendroica graciae.* Princ. distrib. Texas: Uncommon SR in Davis and Guadalupe mountains, in pine-oak forests above 6,000 feet. This is the s.w. U.S. counterpart of the Yellow-throated Warbler.

FEEDING BEHAVIOR. These retiring birds prefer to forage at high levels in pine trees, where they creep along branches gleaning for insects. They also flycatch. They are not as inclined to investigate human structures as the Yellow-throated Warbler.

VOICE. Call: a soft, sweet *tsip* and a thin *tsee* (given in flight), similar to but softer than the Yellow-throated Warbler's calls. Song: a series of musical, slightly down-slurred *tzip tzip tzip tzip* notes, suggesting a Chipping Sparrow; the notes accelerate in pitch toward the end. Each male warbler possesses two distinct subsets of preferred songs (Staicer, 1989).

OTHER BEHAVIORS. Flight is quick, erratic, and slightly undulating (Oberholser and Kincaid, 1974).

Pine Warbler, *Dendroica pinus.* Princ. distrib. Texas: Common PR in the pine forests of e. Texas; in winter they are less particular about their habitat and frequent deciduous and mixed forests.

FEEDING BEHAVIOR. Like Yellow-throated Warblers, they have a long bill that enables them to extract insects from crevices inaccessible to most other warblers. They creep leisurely along the middle to high branches in trees. They also sally out for insects like a flycatcher, hang upside down in search of insects hidden beneath leaves, or descend to the ground to walk around looking for prey or seeds. Their movements are sluggish compared with most warblers. In winter they eat pine seeds, fruits, and berries, and unlike most warblers, they visit feeders for suet.

COURTSHIP BEHAVIOR. Unknown.

NESTING. Well concealed on a horizontal limb in a pine tree. Nest: constructed by the female, who utilizes grass, weeds, bark strips, and other plant materials, as well as feathers and spiderwebs. Both parents incubate the eggs and feed the nestlings.

VOICE. Call: a weak *tzip.* Song: a liquid, musical trill that suggests a Chipping Sparrow, only slower and softer.

Singing appears to be an exceptionally important element in the life of this species. They sing to advertise their territory, as do most birds, but they also sing while feeding and throughout the year. Since there is nothing striking about the Pine Warbler's plumage, perhaps singing plays a larger role than usual in signaling the bird's presence to other birds.

OTHER BEHAVIORS. They are generally unwary and easy to approach. Their flight is quick and erratic. Winter flocks may contain more than 50 individuals.

Prairie Warbler, *Dendroica discolor.* Princ. distrib. Texas: Uncommon local PR in parts of e. Texas, in secondary pine and deciduous forests, thickets, and abandoned fields. They do not occur on our prairies. (The species was named after its winter habitat in s. United States: grassy areas, with scattered trees, which there are called prairies.)

FEEDING BEHAVIOR. They fly quickly and erratically through branches as they search for insects and spiders, twitching (or bobbing) their tail and fluttering their wings as they move. They also flycatch and hover for insects at low to middle levels of trees, in shrubs, and occasionally on the ground. They sometimes feed on nectar.

COURTSHIP BEHAVIOR. The male's mothlike display flights, performed with highly exaggerated wingbeats, alternate with energetic courtship chases.

NESTING. Sometimes several birds nest near each other in a loose colony, each pair defending a territory. Nest: a compact cup of down, grass, and leaves, built by the female at a site she selects (usually low in a tree). Eggs: incubated by the female. Both parents care for the brood and feed the nestlings.

VOICE. Call: a weak but musical *tzip;* in flight, a buzzy *tzeep.* The song is distinctive: a fast or slow buzzy *zee zee zee zee* that ascends a chromatic scale and accelerates slightly. It is sometimes sung by the female.

OTHER BEHAVIORS. When alarmed, they vigorously bob their tail (displacement behavior).

Palm Warbler, *Dendroica palmarum.* Princ. distrib. Texas: Uncommon M in e. Texas and uncommon WR on the Upper Coast, in weedy fields, pastures, cottonfield stubble, and other open habitats. (In winter they occur in low-growing palmettos in Florida and the Bahamas; hence their name.)

FEEDING BEHAVIOR. These ground-dwelling birds wag their tail as they forage for insects, possibly to maintain contact with other birds; they also fly up after insects or hover while picking insects from leaves and twigs. In winter they eat seeds and berries as well.

Pine Warblers could be mistaken for pipits (which also wag their tail and occur in similar habitats) except that pipits walk, whereas most warblers (including this one) hop.

VOICE. Call: a sharp *tzip;* in flight, a slightly husky *tzeep.* Song: a buzzy trill of 7–10 notes that usually rises and accelerates; it has been rendered as *tzee tzee tzee tzee tzee tzee tzee tzee.* The song has invited comparison to the songs of a number of birds, including the Chipping Sparrow, Dark-eyed Junco, Yellow-rumped (Myrtle) Warbler, and Pine Warbler.

OTHER BEHAVIORS. Their flight is "rather flitting and performed with moderately rapid wingbeats" (Oberholser and Kincaid, 1974). Birds feed close together in loose flocks, sometimes in association with Yellow-rumped and other warblers.

Bay-breasted Warbler, *Dendroica castanea.* Princ. distrib. Texas: Locally common to abundant M in spring in virtually all wooded habitats in e. Texas and on the coast; less common in fall.

FEEDING BEHAVIOR. They forage in a deliberate (some say sluggish) manner at all levels in trees and shrubs, wagging their tail as they move along limbs gleaning for insects. They also hop and flit in a jerky manner while moving from branch to branch, and they occasionally flycatch. Their diet consists primarily of insects, with some berries taken during migration.

VOICE. Call: a loud, high-pitched *tzipp,* not frequently heard, and a buzzy *tzee,* uttered when birds are in flight or are feeding. Song: a high-pitched, sibilant *teetzee teetzee teetzee teetzee, teetzee,* on one pitch but louder in the middle, suggesting the Black-and-white Warbler's song.

OTHER BEHAVIORS. They form flocks that often contain other species.

Blackpoll Warbler, *Dendroica striata.* Princ. distrib. Texas: Uncommon spring M in e. Texas, especially on the coast, in trees and bushes.

FEEDING BEHAVIOR. Blackpoll Warblers glean mainly for insects and spiders, less often for seeds and berries. They creep along lower limbs in the interior foliage and occasionally sally out for passing insects.

VOICE. Call (infrequently heard): a muffled, guttural *tzipp,* not as strong or emphatic as the Yellow-rumped Warbler's, and generally indistinguishable from that of the Bay-breasted Warbler. Song: a series of 6–18 deliberate, high-pitched, and thin *tzees,* staccato, evenly spaced, and on the same pitch, although slightly louder in the middle.

OTHER BEHAVIORS. They associate with other warblers during migration. This species is noted for being the warbler with the longest migratory route, from Alaska to Bolivia.

Cerulean Warbler, *Dendroica cerulea.* Princ. distrib. Texas: Uncommon M in e. Texas, in virtually any woody or shrubby habitat that has tall trees.

FEEDING BEHAVIOR. These warblers are not easily located, as they occupy the highest levels of treetops. They move quickly from branch to branch, gleaning for insects and spiders, and they also fly out for passing insects.

VOICE. Call: a weak, musical *tzip;* in flight, a loud *dzeeee.* Song: a loud, high-pitched *zree zree zree zree bzzzz,* ending in a high, buzzy trill. The middle notes accelerate and rise in pitch.

OTHER BEHAVIORS. They are almost always seen singly or in pairs, but occasionally migrants form small, loose flocks.

Black-and-white Warbler, *Mniotilta varia.* Princ. distrib. Texas: Locally common SR primarily in e. Texas, but also the Edwards Plateau, in deciduous and mixed forests; common M throughout most of Texas (except the far w.), in most woody habitats.

FEEDING BEHAVIOR. They forage somewhat like nuthatches, creeping at all levels up and down tree trunks and along main branches. Their short legs equip them to move efficiently in this manner, and their long bill helps them extract insects and their eggs and pupae from cracks and crevices in the bark. Because they can take advantage of this food supply, they arrive earlier in spring than most warblers, which generally must wait for flying and leaf-dwelling insects to hatch and develop. They also take insects while hovering by leaves and twigs.

COURTSHIP BEHAVIOR. The male follows the female and assumes various courtship postures as he utters a rapid, loud *chit chit chit.* The female responds with a begging posture.

NESTING. On the ground, often at the base of a tree or concealed under branches or leaves, and usually in the vicinity of water. Nest: an open cup of leaves, grass, pine needles, hair, and other materials that is constructed by the female. Eggs: incubated by the female. Both parents bring food to the nestlings. The female engages in a distraction display when threatened at the nest.

VOICE. Call: a soft *dzeet,* and a louder *tzink* that accelerates into a *tzek tzek tzek* when birds become alarmed. Song: 6–12 high, thin, buzzy, whistled notes, *see weesy weesy weesy weesy,* the second note of each couplet lower. Birds sing while foraging.

American Redstart, *Setophaga ruticilla.* Princ. distrib. Texas: Uncommon SR and M in the thickets and forests of e. Texas; migrants occur in most wooded habitats.

FEEDING BEHAVIOR. This species is the most flycatcherlike of our warblers. Although the general body proportions are those of a warbler, they have acquired through convergent evolution the long, broad, boat-shaped bill and the long rictal bristles that adapt tyrant flycatchers so well for aerial feeding. Also, like flycatchers, they have a long tail that aids them in flight.

Active and agile like a butterfly in flight, American Redstarts sometimes hover as they pick insects from leaves, and they often leap into the air after prey. They less frequently glean for insects by creeping along branches, and occasionally they eat seeds and berries.

VOICE. Call: a sharp *tzi*; in flight, an ascending *sweet*. Song (both sexes): a short and variable series of high, thin notes, rendered as *dsee dsee dsee dsee* (the last note abrupt and lower), suggesting a Black-and-white Warbler.

OTHER BEHAVIORS. Redstarts repeatedly exhibit their bright markings by drooping their wings, spreading their tail, and flicking the tail from side to side, most likely a display for maintaining contact with their mate.

Prothonotary Warbler, *Protonotaria citrea.* Princ. distrib. Texas: Uncommon to common SR and M in e. third of the state, strongly favoring swamps, river bottoms, and other wet woodlands, even during migration (hence its older names "willow warbler" and "golden swamp warbler").

FEEDING BEHAVIOR. Their foraging behavior has been compared to that of the Black-and-white Warbler: they creep along branches in the understory, often over water and sometimes on the ground, gleaning bark, crevices, and comparable sites for insects. Occasionally they pick up seeds and berries.

COURTSHIP BEHAVIOR. The male displays around the female by fluffing his plumage. According to Oberholser and Kincaid (1974), "An aerial nuptial song, delivered while the bird hovers butterfly-like, sometimes at considerable height, is longer, softer, and more complexly melodious" than its usual song.

NESTING. In a woodpecker cavity or other cavity. They show a strong tendency to nest in trees surrounded by water, and older nestlings swim if they fall into the water. Early in the breeding season they simultaneously work on several nests but eventually complete only one of them. Curiously, Prothonotary Warblers have been observed selecting cardboard milk cartons over wooden nest boxes. Nest: built by the female, who fills the cavity with twigs, dry leaves, moss, and bark almost to the level of the entrance. These warblers are unique in that they select large quantities of bryophytes for their nests. While the female is building the nest, her mate sometimes makes dummy nests nearby, especially early in the nesting season. The function of dummy nests is not clear. One hypothesis is that snakes discontinue visiting nests to prey on eggs and young birds once they encounter several empty nests. Interestingly, their response to cowbird parasitism varies according to the availability of nest sites in the area; they are more likely to abandon their parasitized nest if there are other nest cavities in the area than if nesting sites are scarce.

Eggs: incubated by the female. Both parents bring food to the nestlings. Although relatively unwary in the presence of humans, breeding birds aggressively defend their territory against bluebirds, wrens, robins, woodpeckers, and other birds.

VOICE. Call: a loud *tzenk,* a sibilant *pseet,* and in flight, a softer *tzeet.* Song: loud and energetic, recalling the penetrating notes of the Spotted Sandpiper, *sweet sweet sweet sweet sweet sweet sweet,* all delivered on the same pitch. It is sung persistently throughout the nesting season.

OTHER BEHAVIORS. Their flight is "swift and strong, more direct and less erratic than that of most warblers" (Oberholser and Kincaid, 1974). On their Central and South American wintering grounds they roost communally at night, not a common behavior in warblers.

Worm-eating Warbler, *Helmitheros vermivorus.* Princ. distrib. Texas: Locally common M in e. third of Texas, on the coast, and in s. Texas. They rarely stray from forested areas.

FEEDING BEHAVIOR. Apparently they were named after the caterpillars in their diet (not a particularly useful name, since many warblers eat caterpillars). They locate hanging clusters of dead leaves from which they extract insects with their long bill. They hop, probe, and scratch in the undergrowth, all the while elevating their tail, bobbing their head, and performing acrobatic movements that are facilitated by their short legs.

In spite of their energetic feeding behavior, they have been described as lethargic by some observers. They also glean at most levels in trees and shrubs, sometimes while creeping over limbs.

COURTSHIP BEHAVIOR. Rarely observed in Texas; it includes a musical flight song.

VOICE. Call: an abrupt, pleasant *tzip;* in flight, a soft and slightly buzzy *dzeet dzeet.* Song: suggests a Chipping Sparrow (or Pine Warbler), rendered as a monotonous, bubbling *che e e e e e e,* on one pitch.

Swainson's Warbler, *Limnothlypis swainsonii.* Princ. distrib. Texas: Uncommon (or locally common) SR and M in the swamps and damp forested areas of e. Texas. They favor early pine growth.

FEEDING BEHAVIOR. They forage by walking along the ground with "deliberate thrushlike movements, sometimes drooping [their] wings but not pumping [their tail]" (Oberholser and Kincaid, 1974). As they do this they expose insects by tossing leaves with their long bill. They also forage in the understory.

COURTSHIP BEHAVIOR. The male crouches, flutters his wings, and chases his potential mate.

NESTING. Hidden in vines, canes, low bushes or trees. Nest: usually constructed of leaves and lined with finer materials. Eggs: unique among Texas warblers in being white and unmarked. The female is a close sitter on the nest, perhaps because her eggs are not camouflaged. In one case a person fed insects to an incubating bird. If flushed by humans or predators, she may feign injury.

VOICE. Call: *ztiiip,* somewhat like the Worm-eating Warbler's, but more extended and emphatic. Song: loud, delivered by the male with his bill pointed toward the sky. A rich cocktail of slurred whistles followed by a slow warble, it is somewhat ventriloquial in quality. It has been transcribed as *whee whee whee whip-poor-will,* suggesting at a distance the song of the Louisiana Waterthrush or Hooded Warbler. Many listeners consider it the finest of the warbler songs.

OTHER BEHAVIORS. When approached, these secretive birds remain several minutes in the same place without moving, yet when they fly away, they do so swiftly and with strong wingbeats. They very rarely associate with other warblers.

Ovenbird, *Seiurus aurocapillus.* Princ. distrib. Texas: Locally common M in the forested areas of e. Texas.

FEEDING BEHAVIOR. Ovenbirds walk deliberately along the forest floor searching for insects, snails, worms, and other invertebrates; while doing this they drop their wings slightly, bob their head, and flick their tail.

VOICE. Call: a soft *chip;* when alarmed, a loud *tzuck.* Song: well known on the bird's breeding grounds, but probably not sung by spring migrants in Texas. A popular transcription is *teacher teacher teacher teacher.*

OTHER BEHAVIORS. Their flight is swift and powerful for a warbler. They almost always occur either singly or in pairs.

Northern Waterthrush, *Seiurus noveboracensis.* Princ. distrib. Texas: Locally common M in e. Texas, in forests, thickets, bogs, lakeshores, streamsides, and gardens and parks.

FEEDING BEHAVIOR. They teeter (constantly bobbing their tail) like a Spotted Sandpiper as they walk along turning over leaves in search of insects and other invertebrates; they move about on wet logs and sometimes wade into shallow water. Because of their slightly smaller bill, they take, on the average, slightly smaller prey than do Louisiana Waterthrushes.

VOICE. Call: an abrupt *pennk*. Song (rarely heard in Texas): a loud, clear, ringing *whit whit twee twee twee chew chew chew*.

OTHER BEHAVIORS. They are normally observed alone or in pairs, but occasionally large numbers migrate together.

Louisiana Waterthrush, *Seiurus motacilla.* Princ. distrib. Texas: Uncommon SR and M in e. Texas, in deep, moist forests, especially near running water.

FEEDING BEHAVIOR. These handsome birds favor stream banks, shores of ponds, wet logs, and similar habitats. They walk along deliberately, teetering by bobbing their head and pumping the entire rear end of the body. When foraging for aquatic and terrestrial insects, as well as for other invertebrates, they flip over dead leaves or pull in floating leaves with their bill. They also wade and, when necessary, jump into the water for floating insects. *See* Northern Waterthrush.

COURTSHIP BEHAVIOR. Basically unrecorded, except for the flight song, which curiously is delivered as the male flies above the treetops. Such high aerial displays are common in prairie and tundra birds but understandably are rare in forest species.

NESTING. On the bank of a river next to logs, stumps, and other structures. Nest: a bulky structure of leaves, bark, moss, twigs, and other plant parts, often linked to the water by a pathway of leaves. Eggs: incubated by the female. Both parents bring food to the nestlings.

VOICE. Call: a sharp *pennk*, more emphatic than the Northern Waterthrush's; in flight, a *zeeet*. Song: differs sharply from the Northern Waterthrush's, beginning with three high, shrill, slurred notes (*pseur pseur pseur*), followed by a rapid, jumbled, descending *per see ser*. It is delivered with remarkable vigor from a rock or a low limb. The male sings regularly only until the eggs are laid.

OTHER BEHAVIORS. On short flights they fly with jerky wingbeats, but on longer flights their wingbeats are swift and powerful. These notably shy birds (almost always seen alone) can easily be missed in the field. When agitated, they rapidly pump their tail, apparently a displacement behavior; however, under other circumstances these movements may serve to maintain contact between individuals.

Kentucky Warbler, *Oporornis formosus.* Princ. distrib. Texas: Uncommon M and SR in the thick, damp woodlands of e. Texas; common M on the coast, in less forested habitats.

FEEDING BEHAVIOR. Kentucky Warblers search alone or in pairs. They move about deliberately and quietly as they turn over leaves looking for insects. They also forage in shrubs, undergrowth, and low tree limbs, frequently walking out to glean branches that project over water.

COURTSHIP BEHAVIOR. Apparently unrecorded, except for a nuptial flight song evidently not yet heard in Texas.

NESTING. On or near the ground. Nest: a bulky cup of weeds, stems, grasses, and other materials built by both parents. Eggs: incubated by the female, who also feeds the nestlings. She feigns injury when flushed from the nest.

VOICE. Call: a low, husky *chup.* Song: a melodious, whistled *ter wheeter wheeter wheeter,* suggesting a Carolina Wren. These warblers engage in an uncommon type of social behavior called countersinging: resident males alter the frequency characteristics of their songs to approximate songs of nearby males.

OTHER BEHAVIORS. These are shy, skulking birds. When agitated, they flick their tail and raise their crown feathers. Their flight is swift and strong.

Mourning Warbler, *Oporornis philadelphia.* Princ. distrib. Texas: Uncommon to common M in e. and c. Texas, but probably often overlooked, as these shy, secretive birds typically hide in undergrowth, thickets, hedgerows, and garden shrubbery.

FEEDING BEHAVIOR. They hop along on the ground looking for insects, spiders, and other invertebrates, but they also skulk in low, heavy underbrush, moving quietly from one bush to another.

VOICE. Call: a sharp *peen* and, in flight, an insectlike *zeee.* Song: soft and clear, a rapidly delivered series of slurred, rolling, liquid notes, sometimes rendered as *teryee teryee teryee wee see.* It is only rarely heard in Texas.

OTHER BEHAVIORS. They do not normally associate with other warblers.

MacGillivray's Warbler, *Oporornis tolmiei.* Princ. distrib. Texas: Uncommon M in w. Texas, in a variety of habitats, but favoring open woodland, underbrush, and other low vegetation. This species is the w. counterpart of the Mourning Warbler.

FEEDING BEHAVIOR. While skulking in dense cover, these timid and elusive birds hop along searching for insects and other invertebrates.

VOICE. Call: a sharp, colorless *tship.* Song: loud and variable, consisting of three or four buzzy *chuwee* notes on one pitch, followed by two or three more melodic *peechee* notes on a lower pitch.

OTHER BEHAVIORS. They are rarely seen other than singly or in pairs.

Common Yellowthroat, *Geothlypis trichas.* Princ. distrib. Texas: Common M throughout Texas in almost all habitats except prairies and deserts; locally uncommon SR in scattered localities throughout the state, in marshes, swamps, and cattail ponds; locally common WR in e. and s. Texas, as well as along the densely vegetated Rio Grande. This is one of the most abundant warblers in North America.

FEEDING BEHAVIOR. Their feeding movements have been compared to the Marsh Wren's. They flit among thick cattails, usually near or on the ground, and glean industriously for insects, spiders, and other invertebrates. Males tend to forage higher up in the vegetation than females (Kelly and Wood, 1996).

COURTSHIP BEHAVIOR. The male closely follows the female for a short time, then, from a low perch, he ascends to 25–100 feet above the ground and swoops dramatically to another perch. While doing this he gives his flight song, a series of descending notes.

NESTING. In marshes, swamps, or cattail ponds. Nest: a bulky, loose cup of grass, stems, bark, and other materials, with a partial roof attached to the rim; built by the female. Eggs: incubated by the female, who receives food from her mate at this time. When disturbed, she unobtrusively retreats from the nest.

Both parents feed and care for the nestlings, attending to them longer than do most warblers. Since the family generally stays together throughout summer, most families probably begin their fall migratory journey together. Males sometimes breed with more than one female (recalling the polygamous behavior of the Red-winged Blackbird, which nests in basically the same habitat).

VOICE. Call: a distinctive, dry, throaty *tchap;* in flight, an insectlike *tzeet.* Song (sung from a prominent perch): a loud and rollicking *witchity witchity witchity,* which, according to some listeners, occasionally incorporates poor imitations of other birds' songs.

OTHER BEHAVIORS. Common Yellowthroats are notably wrenlike in behavior. They respond to human presence by actively darting from limb to limb, flitting and scolding, then withdrawing to hide, only to reappear and scold again. At this time they appear irresolute and easily intimidated, but at other times they seem completely unresponsive to human presence and sing with abandon from a conspicuous perch.

Also, like many wrens, they readily respond to squeaking, at times perching with their tail half-cocked. They fly with rapid wingbeats, but in a labored manner. They are typically solitary in winter.

Hooded Warbler, *Wilsonia citrina.* Princ. distrib. Texas: Uncommon to locally common SR and M in e. Texas (although there are very few nesting records), frequenting forested areas that support abundant undergrowth.

FEEDING BEHAVIOR. The pair usually forage together in vegetation that is less than 15 feet from the ground, but they do not seem to compete with each other because they have different foraging tactics. In general, females glean for insects on the ground or in the lower dense shrub layers, whereas males typically flycatch (with considerable skill) from a low perch.

COURTSHIP BEHAVIOR. Apparently unrecorded. Although males spend less than 8 percent of their time intruding onto another male's territory, extrapair copulations with neighboring females are common.

NESTING. In dense understory and near water, close to the ground in a bush or a low limb; usually the female selects the site. Nest: constructed by the female, of dead leaves, spiderwebs, grass, and other materials. Eggs: normally incubated by the female. Both parents bring food to the nestlings, and sometimes they divide the brood between them. Interestingly, males often return to their previous year's territory, whereas females generally move to a different location. Cooperative breeding has been reported.

VOICE. Call: a loud, musical, metallic *cheep* or *chink,* and a soft, insect-like *zshrrt,* given in flight. Song: a series of loud, clear notes, the last pair slurred and accented, and transcribed as *toowee toowee toowee toowee CHOO.*

OTHER BEHAVIORS. These lively birds endlessly fan their tail, revealing white tail spots that probably function to maintain contact between individuals. Birds usually approach an observer who sits quietly.

On their Mexico wintering grounds, males and females apparently defend feeding territories in different habitats, males preferring landscapes with more tree trunks, and females the more shrubby areas. These preferences were demonstrated with captive birds. When given a choice, males orient toward vertical stripes on walls, which resemble tree trunks, and females orient toward oblique stripes, which appear more like shrubby habitats (Morton et al., 1993).

Wilson's Warbler, *Wilsonia pusilla.* Princ. distrib. Texas: Uncommon to common M throughout Texas, in thickets, fencerows, gardens, and other brushy habitats; common WR on the coast and in s. Texas. They usually avoid thick forests.

FEEDING BEHAVIOR. They actively examine all parts of the vegetation for insects, usually foraging less than 10 feet from the ground. They also commonly flycatch and occasionally eat berries.

VOICE. Call: a slightly harsh, flat *tjep* or *chip;* in flight, an abrupt *taLIK* or *tzip* (down slurred). Song: 15–20 sharp, rapidly produced staccato notes that drop in pitch toward the end, *tsee tsee tsee tsee tsee tsee;* also, *tchee tchee tchee,* followed by a trill.

OTHER BEHAVIORS. They sometimes flick their tail and wings, probably to maintain contact with conspecifics.

Canada Warbler, *Wilsonia canadensis.* Princ. distrib. Texas: Uncommon M (especially in late spring) in e. Texas and on the coast, generally in scrub or deciduous forests that support a thick understory; they are fond of thickets near water.

FEEDING BEHAVIOR. These highly active warblers cock their tail as they search for insects. They forage at low to middle levels, less frequently on the ground, often flushing and chasing insects from the foliage. They also dart out after insects, snapping them up with an audible click of the bill.

VOICE. Call: a sharp, harsh *tchik* and a softer *tsiip,* as well as a high-pitched flight call (*dzee*). Song: loud, variable, and animated, beginning with a *tzip* and continuing with a sputtering warble of rich, liquid notes.

OTHER BEHAVIORS. They are not particularly timid. Many pairs remain together during migration.

Painted Redstart, *Myioborus pictus* (also called Painted Whitestart). Princ. distrib. Texas: Chisos Mountains (usually 6,000–10,000 feet above sea level), where they are an irregular SR in canyons, frequenting deciduous and evergreen trees and thickets.

FEEDING BEHAVIOR. As foragers, these are among the most versatile of our warblers. They glean for insects by climbing over trunks, logs, and stumps, sometimes clinging like a Brown Creeper, at other times flitting briefly on a tree trunk in search of prey. They also fly out after passing insects.

VOICE. Call: an abrupt whistle, *chweep* or *chereep*, with the second syllable higher pitched, like the peep of a young chicken. Other calls include a soft and low *tzeeyoo*, an alarm call (*zreeeeet*) and, during courtship, a high-pitched *tee tee dee*. Song: a clear, rich, musical warble, commonly transcribed as *weeta weeta weeta wee*, and suggesting a Yellow-rumped Warbler. Females sing and sometimes duet with their mates.

OTHER BEHAVIORS. Flight is flitting and rarely prolonged. Although Painted Redstarts and American Redstarts are in different genera and are less closely related than their common names suggest, they both move in a jerky manner and share the habit of partially spreading their wings and tail. This display probably helps maintain contact between individuals.

Yellow-breasted Chat, *Icteria virens.* Princ. distrib. Texas: Locally common M and SR throughout most of the state, in dense thickets and woodland edges in both wet and dry habitats.

FEEDING BEHAVIOR. They forage in low, dense cover for insects, berries, and fruits, sometimes using their feet to hold food, a behavior rarely observed in passerines.

COURTSHIP BEHAVIOR. The male sways from side to side while pointing his bill upward. In his aerial display, he flies up from a conspicuous perch, momentarily hovers, then glides down to his perch, all the while singing. This display resembles the display of the unrelated Northern Mockingbird.

NESTING. Well concealed in vegetation. Nest: a large, bulky, open cup that the female constructs using weeds, grass, leaves, and other materials. Eggs: the largest eggs of any warbler, incubated by the female. Both parents bring food to the nestlings.

VOICE. Call: several, including a harsh *tzack,* and a catlike *mew.* Song: truly remarkable, and hardly recognizable as belonging to a warbler. It is a jumble of disjointed, unrelated sounds that suggests some sort of thrasher. The sounds are deliberately spaced and include squeals, whistles, caws, cackles, squawks, and rattles, some seemingly appropriated from other birds' songs. The song is delivered from both conspicuous and concealed perches, as well as while flying between perches. It is often heard at night.

OTHER BEHAVIORS. In his conspicuous flight from perch to perch (probably a territorial display) the male pumps his tail and dangles his feet, making his wings appear double-jointed and giving the whole performance a curious awkwardness. Otherwise, these birds are normally very secretive and elusive. They occur singly or in pairs, although they sometimes migrate in groups.

TANAGERS: FAMILY THRAUPIDAE

Our tanagers forage with slow and deliberate movements, searching trees and shrubs for insects and sometimes berries. The consumption of vertebrates by tanagers, especially tropical species but also the Summer Tanager, seems to be more frequent than previously suspected (Perez-Rivera, 1997).

Curiously, the courtship behaviors of our tanagers are not well known, even though they are conspicuous and well-known garden birds.

Hepatic Tanager, *Piranga flava.* Princ. distrib. Texas: Uncommon SR in Davis, Chisos, and Guadalupe mountains, in tall trees, in particular pines.

FEEDING BEHAVIOR. Hepatic Tanagers move slowly and deliberately in the treetops. They creep over branches, gleaning for insects and sometimes flying out for them; in autumn, they consume fruits and berries.

COURTSHIP BEHAVIOR. Apparently unrecorded.

NESTING. In a tall tree, 15–50 feet above ground, away from the trunk and on a horizontal limb. Nest: a structure made with grass and weed stems woven into a shallow cup, built mainly by the female. Much of their basic nesting behavior is still unrecorded.

VOICE. Call: an abrupt *chuck,* repeated several times. Song: loud, clear, and sometimes compared to the Black-headed Grosbeak's, but less varied. It is usually sung from a high perch.

OTHER BEHAVIORS. When disturbed, these shy birds fly restlessly from tree to tree.

Summer Tanager, *Piranga rubra.* Princ. distrib. Texas: Uncommon to locally common SR throughout most of Texas, except the Southern High Plains, Rolling Plains, and extreme s. Texas. Uncommon to common M throughout the state. They prefer deciduous forests, riparian woodlands, and urban areas, especially near open water.

FEEDING BEHAVIOR. Summer Tanagers forage in the tops of trees, moving about deliberately and frequently pausing to look around. They forage for insects, especially bees and wasps, sometimes raiding beehives and nests of paper wasps. They capture insects on the wing with more skill than would be expected from a bird that generally moves slowly. They also hover while picking insects from foliage.

Males and females sometimes forage differently with respect to height in the tree, foraging maneuvers, tree species, and location within the tree (Holmes, 1986).

COURTSHIP BEHAVIOR. Apparently unrecorded.

NESTING. On a horizontal branch in a deciduous or coniferous tree, away from the trunk, and usually 10–30 feet above ground. Nest: a shallow cup of weed stems, leaves, grass, Spanish moss, and other plant materials, apparently built only by the female. Eggs: evidently incubated only by the female. Both parents bring food to the nestlings.

VOICE. Call: a dry, mechanical *chick a tuck,* which descends in pitch. Song: a rich, melodious series of warbled phrases that suggests a robin, but more hurried and more closely connected. It is uniquely structured. Individual males combine several subunits to form larger units. Because these vocal configurations do not vary from region to region, they cannot be considered dialects (Shy, 1985).

OTHER BEHAVIORS. The birds are not always easy to locate. They are usually solitary, and they move about slowly while concealed in vegetation. In some respects, however, they are perhaps the least shy of our tanagers, at times becoming unwary enough to take food out of one's hand.

Scarlet Tanager, *Piranga olivacea.* Princ. distrib. Texas: Locally common spring M in e. half of the state, in wooded habitats.

FEEDING BEHAVIOR. They prefer to forage high up in oaks, where they pursue insects and take a few berries and fruits. They move about deliberately, hover next to leaves, or occasionally sally out to capture insects on the wing. On the breeding grounds, at least, females forage higher and fly out more for insects; this sexual difference in foraging behavior should be looked for in migrants.

VOICE. Call: a buzzy, low-pitched *TIP cher.* Song: like the American Robin's, but containing a hoarse *burr.*

OTHER BEHAVIORS. In spite of their bright coloration, Scarlet Tanagers are not easily observed. They stay high in trees, move about slowly, and sit for several minutes without moving. Moreover, during migration they are generally silent. Their flight is swift and graceful.

Western Tanager, *Piranga ludoviciana.* Princ. distrib. Texas: Uncommon SR in Guadalupe, Davis, and Chisos mountains, preferring Douglas fir–pine–oak forests such as those in the bowl of the Guadalupes; they are a locally common M in w. Texas, frequenting wooded habitats.

FEEDING BEHAVIOR. They spend considerable time foraging in the upper canopy of mature, open woodlands, in the deliberate manner characteristic of tanagers. However, they come to the ground more frequently than the other tanagers. They also sally out for passing insects and sometimes visit flowers, presumably for insects and nectar.

Like Scarlet Tanagers, they occasionally stop and pause for several minutes during their foraging bouts. They also feed on fruits and berries.

VOICE. Call: a dry, colorless *pit i tik*, uttered frequently; less often, a plaintive, purring *tu weep*. Song: like the American Robin's, but shorter, hoarser, and lower pitched. Some listeners report that the song suggests a mood of wildness and freedom.

OTHER BEHAVIORS. They fly swiftly and with powerful wingbeats.

EMBERIZIDS: FAMILY EMBERIZIDAE

The conical bills of sparrows and other members of this group clearly reflect their seed-eating habits. Finches (family Fringillidae) also have conical bills. Because sparrows consume such small seeds and because finches are much more efficient at handling large seeds, sparrows need to locate three or four times as many seeds during the day as do finches. As a result, sparrows spend much more time foraging, have less time to search for new food patches, and tend to be relatively sedentary, restricting themselves to sites with dense seed concentrations. In contrast, finches move over very large expanses searching for new feeding areas.

Furthermore, resource partitioning is more pronounced among finches: small and large finches eat seeds that vary considerably in size, whereas the diets of small and large sparrows overlap greatly (Benkman and Pulliam, 1988).

Most sparrows forage on the ground. Their courtship behavior does not seem to be particularly complex, and it often involves little more than singing, fluttering the wings, or chasing the female. Birds that live in open areas perform aerial displays.

Experiments show that singing behavior (during territorial defense) is linked to gene regulation in brain centers that control song perception and production (Jarvis et al., 1997).

White-collared Seedeater, *Sporophila torqueola.* Princ. distrib. Texas: Uncommon and local PR in s. Texas, formerly more common than now, in tall grassy areas, brushy open woodlands, as well as patches of ragweed, huisaches, and low bushes.

FEEDING BEHAVIOR. These tropical birds feed energetically on seeds and insects, usually down in the weeds and grass, occasionally higher in bushes.

VOICE. Call: a dry *tick tick.* Song: a loud and clear *sweet sweet sweet cheer cheer,* suggesting an Indigo Bunting, and notably loud for the size of the bird.

OTHER BEHAVIORS. They usually occur in flocks except when breeding.

Olive Sparrow, *Arremonops rufivirgatus.* Princ. distrib. Texas: Common PR in s. Texas, in clearings, brushy thickets, undergrowth, and tall grass, especially in low, moist areas.

FEEDING BEHAVIOR. They feed on the ground, usually in or next to dense thickets. Often one sees a pair feeding quietly in the undergrowth. They scratch in the leaves with both feet, like a towhee, as they search for insects, other invertebrates, and, less frequently, seeds and berries.

COURTSHIP BEHAVIOR. Apparently unrecorded.

NESTING. In dense vegetation, usually less than 3 feet above ground. Nest: bulky and appearing large for a bird this size, a cup of grass and leaves with a domed top and an entrance on the side.

VOICE. Call: a loud *clink,* suggesting a Northern Cardinal's call; also, soft ticking notes, probably to maintain contact between individuals. Song: a series of notes terminating in a trill, recalling the Field Sparrow's bouncing-ball song pattern, but louder and more metallic.

OTHER BEHAVIORS. These shy birds are not easy to observe, as they usually remain concealed in dense vegetation, sometimes singing from a low perch in the shrubbery. In south Texas they often occupy the same habitat as Long-billed Thrashers.

Green-tailed Towhee, *Pipilo chlorurus.* Princ. distrib. Texas: Rare SR in Trans-Pecos Texas, usually at higher elevations; locally common M and WR in s. and w. Texas, frequenting thickets, densely covered mountain slopes, and semiopen habitats.

FEEDING BEHAVIOR. They feed on the ground under thickets, exposing insects and seeds by scratching litter with both feet simultaneously.

They sometimes ascend into low bushes. In gardens, they are more likely to forage under a birdfeeder than on it.

VOICE. Call: a plaintive, catlike *pee yu eee,* and a softer *mew.* Song: rich and melodious, sometimes beginning with *wee churrr* and ending with a trill; they often appropriate songs from other species.

OTHER BEHAVIORS. Their method of distracting predators from a nest seems unique: they drop down from the nest, lift their tail, and run away in a mouselike manner.

Spotted Towhee, *Pipilo maculatus.* Princ. distrib. Texas: Common PR in mountains of Trans-Pecos Texas, in semiopen woodlands and thickets; common M and WR in w. two-thirds of the state, in open forests and shrubby areas.

FEEDING BEHAVIOR. These colorful, retiring birds forage alone under thick vegetation, sometimes moving up to low bushes. In typical towhee fashion, they scratch noisily amid the leaf litter by making a quick hop forward, then scratching backward with both feet. They eat primarily insects, berries, acorns, and seeds.

COURTSHIP BEHAVIOR. The male fluffs his body feathers and rapidly spreads his tail and wings to show the white spots. He also chases the female while singing a whispered version of his song.

NESTING. Usually on the ground, occasionally in a low bush. Nest: an open cup of grass, rootlets, weeds, and twigs, built by the female. Most or all incubation is done by the female, but both parents feed the nestlings. When molested at the nest, the female runs away in a mouselike manner or feigns injury.

VOICE. Call: a nasal, inquisitive *shreenk,* which at dusk on a cold winter day comes across as remote and haunting. Song: a *chup zur eeeeeeeee.* Females sometimes sing.

OTHER BEHAVIORS. When disturbed, they prefer fleeing on foot to flying. They sometimes freeze. Their flight is jerky and rarely sustained.

Family members tend to stay together throughout summer, then join other families to form loose winter flocks.

Eastern Towhee, *Pipilo erythrophthalmus.* Princ. distrib. Texas: Common M and WR in e. part of the state. Their behavior is essentially like the Spotted Towhee's. Until recently, the two species were regarded as races of the same species (Rufous-sided Towhee).

VOICE. Call: a distinctive *to whee* (hence its name). Song: an energetic *drink your teeeeee*, the final sound wavering. Evidently females occasionally sing.

Canyon Towhee, *Pipilo fuscus.* Princ. distrib. Texas: Locally common PR in w. third of Texas, in arid or semiarid habitats except open deserts.

FEEDING BEHAVIOR. These chipper little birds feed mainly on the ground, hopping around like sparrows under cars, bushes, and sheds. Apparently they do not scratch in dirt as much as the previous two species, but when they do, they scratch with both feet as is typical of towhees. Their diet consists of insects, seeds, and berries. Some of their water requirements are met by drinking morning dew.

COURTSHIP BEHAVIOR. The male approaches the female while drooping and quivering his wings. The pair stay together throughout the year, and apparently most birds mate for life.

NESTING. Usually less than 10 feet above ground, in a tree, cactus, or shrub. Nest: a solidly built, bulky cup of twigs, grass, and strips of bark. Eggs: incubated by the female. She is not easily flushed from her nest, but when disturbed, she runs away in a mouselike fashion. Both parents bring food to the nestlings.

VOICE. Call: a loud, emphatic *chink,* and a hoarse *shidapp.* Song: a drawn-out series of musical notes, sometimes rendered as *chili chili chili chili.* The three races of Canyon Towhee in Texas differ to some extent in their vocalizations.

OTHER BEHAVIORS. Both members of the pair perform a squeal duet and assume a characteristic posture when they meet after being apart. Where not molested, as in Big Bend National Park, they become quite unwary and forage near people in parking lots, on porches, and in gardens; however, away from human habitations they are usually quite shy and wary.

Interestingly, they are relatively nonaggressive when other towhees approach their territory. This relatively peaceful response toward intruders differs markedly from the highly territorial California Towhees, which, like robins, attack their image in windows and hubcaps. This difference in temperament is particularly interesting considering that until recently the two species were considered races of the same species.

Cassin's Sparrow, *Aimophila cassinii.* Princ. distrib. Texas: Locally common SR in w. half of Texas, in scattered mesquite scrub, brushy fields, and arid grasslands. *See* Botteri's Sparrow.

FEEDING BEHAVIOR. They hop on the ground in open areas in search of insects and seeds.

COURTSHIP BEHAVIOR. The male's distinctive flight-song is perhaps the best field mark for identifying this frustratingly nondescript bird. He flies up from an elevated perch (frequently the top of a low tree), spreads his tail feathers, holds his head up, stretches his legs downward, then floats down, as though parachuting, to alight on another perch — all the while singing.

He also has a nonaerial display: he elevates and spreads his tail, holds his head down, spreads his wings outward, and flutters his wings and tail.

NESTING. On the ground or low in a bush. Nest: an open cup of weeds, grass, and plant fibers. Both parents bring food to the nestlings. As common and widespread as this species is, much of its nesting behavior is essentially unknown.

VOICE. Call: a loud *tziip*. Song: a sweet, musical, insectlike buzzy trill preceded by two higher notes and followed by four lower ones. Actually, the song is a delightful melody. To some listeners it suggests the first notes of Beethoven's Fifth Symphony. It is heard repeatedly throughout the day and often at night. The primary song varies among individual males and can be used to identify individuals in the field (Schnase and Maxwell, 1989).

OTHER BEHAVIORS. Evidently Cassin's Sparrows do not require pools or other sources of drinking water, even in extreme hot and dry conditions. They tend to be secretive except when displaying. Their time and energy budgets have been the focus of an interesting modeling study (Schnase et al., 1991).

The flight-song display was studied in detail in the Southern High Plains (Anderson and Conway, 2000). It appears to have multiple functions, including territorial defense, mate attraction, and predator detection and avoidance.

Bachman's Sparrow, *Aimophila aestivalis.* Princ. distrib. Texas: Uncommon and local PR in e. Texas, in dry, open pine forests and adjacent overgrown fields. (A former name was "pinewoods sparrow.") Their niche is relatively narrow. In e. Texas, where Bachman's Sparrows occur in the same region as Field and Chipping Sparrows, the other two species tolerate a wider range of habitats.

FEEDING BEHAVIOR. They almost always forage on the ground, where they search for insects and seeds. They sometimes jump up for items in the lower branches of bushes.

Bachman's Sparrows move about slowly in a small feeding territory. In east Texas they eat seeds and insects that are larger than and of different kinds from those eaten by Field and Chipping Sparrows. Of the three, Field Sparrows are most capable of obtaining seeds from the heads of tall grasses (Allaire and Fisher, 1975).

VOICE. Call: a snakelike *tsissssp*, when alarmed. Song: a sweet trill followed by abrupt notes, suggesting a Field Sparrow. The song is exquisite and regarded as one of the finest of our sparrows' songs.

OTHER BEHAVIORS. They are secretive and difficult to spot or flush, apparently flying up to high perches only to sing.

Botteri's Sparrow, *Aimophila botterii*. Princ. distrib. Texas: Uncommon to locally common SR on the Lower Coast, usually in tall bunchgrass (Adams and Bryan, 1999). In Arizona, Botteri's Sparrows, which have relatively long tarsi, nest on the ground; in contrast, the closely related Cassin's Sparrows, which have shorter tarsi, nest in shrubs. (A longer tarsus—the lower part of the leg—facilitates running.)

FEEDING BEHAVIOR. They forage for insects and seeds by running or hopping on the ground.

COURTSHIP BEHAVIOR. Apparently unrecorded.

NESTING. Nest: a shallow open cup of grass on the ground. The breeding biology of this species is poorly known.

VOICE. Call: a colorless *tsip*. Song: a series of jumbled, canarylike notes that terminate in accelerating *chips*.

OTHER BEHAVIORS. They flush easily, sometimes flying to a post or other perch, but more often dropping immediately back down into the grass.

Rufous-crowned Sparrow, *Aimophila ruficeps*. Princ. distrib. Texas: Uncommon to locally common PR in w. half of Texas, in brushy canyons and hillsides, and almost always near rocks or boulders.

FEEDING BEHAVIOR. One or two birds usually feed together, slowly walking or hopping on the ground, sometimes flying up to low bushes. They typically restrict their activities to a relatively small feeding territory, where they search for insects and seeds.

COURTSHIP BEHAVIOR. Like Canyon Towhees, the pair perform a squeal duet when they reunite after being separated. Possibly this display is also used in the early stages of courtship.

NESTING. Usually well hidden in a grass clump or at the base of a shrub. Nest: from materials that include grass, weeds, twigs, and plant fibers, contained within a shallow depression so that the rim is level with the ground. Eggs: apparently incubated only by the female, who is a close sitter on the nest. When disturbed, she feigns injury with frenetic wing fluttering.

Both parents bring food to the nestlings. Many young birds remain with their parents for a month or more after leaving the nest. Such familial solidarity could be related to the pair's tendency to stay together on a permanent territory throughout the year.

VOICE. Call: a sharp *chirp chirp;* when alarmed, a harsh, nasal *deer deer deer.* Song: a soft, dry, wrenlike series of jumbled warbles and trills. This sprightly, staccato chittering is uttered almost too rapidly to follow; to some listeners it suggests the Lazuli Bunting's song.

OTHER BEHAVIORS. Rufous-crowned Sparrows usually stay hidden in dense cover; when flushed, they quickly find refuge in a clump of grass or behind rocks. On the other hand, they usually fly up to a conspicuous perch if an observer waits patiently, and they are inquisitive enough to respond to squeaking.

Their territories tend to be clumped, so that once one bird is located, others are likely to be found nearby.

American Tree Sparrow, *Spizella arborea.* Princ. distrib. Texas: Locally common WR in the Panhandle, in weedy fields, brushy areas, and along fencerows.

FEEDING BEHAVIOR. They forage for seeds (the principal item in their winter diet) by vigorously scratching on the ground, sometimes ascending to bushes and trees to feed. In winter they obtain water by eating snow.

VOICE. Call: a thin *tseep;* also, a musical *teedle witt,* used to maintain contact between individuals. Song: a metallic, canarylike *seet seet seetita sweet sweet.*

OTHER BEHAVIORS. In Texas, they occur in small flocks that apparently maintain a dominance hierarchy and defend a feeding territory.

Chipping Sparrow, *Spizella passerina.* Princ. distrib. Texas: Common to locally abundant PR in e. Texas, the Edwards Plateau, and Trans-Pecos Texas, in open woodlands, orchards, gardens, and parks. In the Trans-Pecos they inhabit montane pine-oak habitats. *See* Bachman's Sparrow.

FEEDING BEHAVIOR. These likable little birds generally feed on the ground, hopping about in search of insects and seeds, often oblivious to the presence of humans. Occasionally they fly up to capture insects in midair. They also feed in bushes and trees.

Chipping Sparrows have been shown to survive without water for up to 3 weeks, subsisting entirely on dry seeds.

COURTSHIP BEHAVIOR. Apparently, their courtship behavior has yet to be reported.

NESTING. In a tree, typically lower than 15 feet above ground (sometimes on the ground). Nest: an open cup of grass, rootlets, and weed stems, built by the female. The female incubates the eggs, and she is not easily flushed from the nest. Her mate brings her food at this time. Both parents feed the nestlings. About 5 percent of the males have more than one mate.

VOICE. Call: a simple *tzip*. Song (usually delivered from an elevated perch): a monotonous series of loud *chips* (on the same pitch) that varies in tempo, the slowest being in the range of human imitation, the fastest almost a cicada trill. Sometimes they sing at night.

Apparently the song functions primarily in courtship, as song output declines precipitously after pairing and resurges when a male loses his mate (Albrecht and Oring, 1995).

OTHER BEHAVIORS. These are among the tamest of our sparrows, sometimes learning to feed out of one's hand. After the breeding season they forage in flocks of up to 50 individuals.

Clay-colored Sparrow, *Spizella pallida.* Princ. distrib. Texas: Locally common M throughout Texas; locally common WR mostly in e. half and s. parts of the state; they frequent fencerows and weedy pastures (spring) and open shrubby areas (fall and winter). At the height of migration they also occur in urban parks and gardens.

FEEDING BEHAVIOR. These nondescript birds hop around on the ground searching for seeds and insects; now and then they fly up to low bushes to feed.

VOICE. Call: a soft *tzip*. Song: a flat, dry, monotonous, cicadalike buzz (*dzee dzee dzee dzee*), delivered from an elevated perch.

OTHER BEHAVIORS. Some flocks contain more than 50 individuals and include Brewer's and White-crowned Sparrows.

Brewer's Sparrow, *Spizella breweri.* Princ. distrib. Texas: Uncommon to common M and WR in w. third of Texas, in desert scrub and brushy areas.

FEEDING BEHAVIOR. They search for seeds and insects on the ground and in shrubs, sometimes surviving for long periods with no water.

VOICE. Call: a weak *tsiip*. Song (variable): a prolonged series of rapid, cicadalike buzzes, to some listeners suggesting a canary.

OTHER BEHAVIORS. When disturbed, they first try to escape on foot, then take flight along a zigzag path before quickly alighting on the ground or in a shrub. Wintering birds usually associate with other sparrows, including White-crowned, Black-throated, and Chipping Sparrows. Brewer's Sparrows are shy and wary on their northern nesting grounds, but wintering birds can become quite tame at feeders.

Field Sparrow, *Spizella pusilla.* Princ. distrib. Texas: Uncommon to common PR in the n.e. and c. part of Texas, less common to the w. They are commonly found in abandoned fields and pastures that are overgrown with trees and shrubs, as well as brushy clearings within woodlands. *See* Bachman's Sparrow. Uncommon to common M and WR in most parts of the state, chiefly in shrubby and grassy habitats.

FEEDING BEHAVIOR. They forage for seeds and insects on the ground and in low bushes, sometimes flying up to a grass stem and bending it to the ground to gather seeds.

COURTSHIP BEHAVIOR. Apparently unrecorded.

NESTING. The first birds to arrive on the nesting grounds are older males, which are followed by younger males, then females. Shrubs or low trees seem to be an essential element in their nesting habitat, although the nest is usually placed on the ground. Unlike Song and Chipping Sparrows, Field Sparrows rarely nest near human habitations. The female selects the site and builds the nest. Her mate brings nesting materials.

Nest: a woven cup of grasses. If nesting attempts fail because of predation or other disturbances, the pair may build another nest, this time in a higher location such as a bush or sapling. (One female attempted to nest 10 times in one season.) Eggs: incubated by the female, who feigns injury when the nest is threatened. The male generally feeds his mate while she incubates. Both parents bring food to the nestlings. After immature birds leave the nest they form small flocks that move about with impunity in the territories of neighboring males.

When birds are parasitized by cowbirds, the pair desert the nest and attempt to renest in another location.

VOICE. Call: a thin *tzeee;* also, an abrupt *chip,* given in alarm. Song: a series of sweet, plaintive notes that increase in tempo, like a bouncing ball, sometimes rendered as *chee chee chee chi chi chi chi chi chichichi,* the last notes becoming a trill that rises and falls in pitch until fading away. Birds sing persistently during the heat of the day, as well as at night.

When male Field Sparrows settle on a territory for the first time, they advertise the territory with two or more song types. With time they abandon most song types and retain the type most closely resembling that of their neighbors (Nelson, 1992).

OTHER BEHAVIORS. They show almost no aggression toward other birds. Some flocks in winter contain more than 100 individuals and include other sparrows.

Black-chinned Sparrow, *Spizella atrogularis.* Princ. distrib. Texas: Uncommon SR and locally common WR in mountains of w. Texas, usually on brushy mountain slopes but in winter occasionally descending to brushy canyons at lower elevations.

FEEDING BEHAVIOR. They forage mainly on the ground, moving about slowly and limiting their activities to a relatively small feeding territory. Occasionally they feed in low bushes. Their diet appears to consist exclusively of insects and seeds.

VOICE. Call: a weak *tseep,* and a louder, sharper *tchit.* Song (sung from a conspicuous perch): a series of plaintive, high-pitched, canarylike notes that suggest the Field Sparrow's song.

OTHER BEHAVIORS. Except during the breeding season, Black-chinned Sparrows associate in small flocks that include Chipping, Brewer's, and other sparrows. They are wary and tend to flee quickly when approached.

Vesper Sparrow, *Pooecetes gramineus.* Princ. distrib. Texas: Uncommon to common M and WR in most parts of the state, in fields, pastures, and roadsides.

FEEDING BEHAVIOR. Vesper Sparrows feed on insects and seeds that they locate as they hop or run across the ground.

VOICE. Call: a sharp *chirp.* Song: sweet and plaintive, two long notes and two high notes followed by descending trills and buzzes that suggest the song of a Song Sparrow (although to some listeners it appears more deliberate and more melodious). More than one observer has emphasized how striking the song sounds at the end of the day.

OTHER BEHAVIORS. They are not particularly shy, and they fly reluctantly when approached, preferring instead to run ahead of the intruder. Apparently they neither drink nor bathe in water. They obtain water from dew or food, and take frequent dust baths. At night they sometimes sleep in the ruts of roads. They occur in small flocks except during the nesting season.

Lark Sparrow, *Chondestes grammacus.* Princ. distrib. Texas: Uncommon to common M and SR throughout most of Texas, in open areas with scattered trees and bushes; locally uncommon WR except in w. parts and in the Panhandle.

FEEDING BEHAVIOR. Typically two to four forage together, even during the breeding season, in pastures, roadsides, or other open areas, walking or hopping around as they search for insects and seeds. (Relatively few birds both walk and hop.)

COURTSHIP BEHAVIOR. The male struts about with his head pointed up and his tail spread wide, suggesting a caricature of a gobbler. During his flight song, he spreads his tail and beats his wings rapidly.

NESTING. On the ground near a tall weed or clump of grass, or in a shrub or lower branches of a tree. Both males and females participate in nest building; the male brings twigs, weeds, grass, and other material to the site they have selected, and the female constructs the nest. Eggs: incubated by the female, who is provided food by her mate. Both parents feed the nestlings. Females disturbed at the nest feign injury by fluttering away with spread tail.

Lark Sparrows are only weakly territorial. They cease defending their territory when incubation begins.

VOICE. Call: a thin *tzipp*. Song: a rich, intense, and melodious gush of trills and buzzes that vary in volume and pitch; the male sings from the ground, from a bush or tree, or while in flight. He often sings at night.

OTHER BEHAVIORS. Their flight is quick and direct.

Black-throated Sparrow, *Amphispiza bilineata*. Princ. distrib. Texas: Common PR in w. half of Texas, in desert scrub, barren creosote flats, and other arid habitats. They are remarkably adapted to arid regions, apparently capable of tolerating the hottest, driest, and most inhospitable of habitats.

FEEDING BEHAVIOR. These sharp-looking birds, with their crisp black bib and prominent head stripes, run along the ground foraging for seeds and insects. They feed to a lesser extent on young shoots, berries, and fruits. Occasionally they fly up to capture passing insects.

Like many desert birds, they do not seem to need water, other than what they can extract from green plants and insects.

COURTSHIP BEHAVIOR. Apparently unrecorded.

NESTING. Rain apparently initiates the nesting season in some areas. This behavior needs to be documented if it occurs in Texas. Nest: usually well concealed in a shrub or on the ground; sturdy, bulky, and built of plant fibers, weeds, grass, and small twigs. Apparently both parents bring food to the nestlings.

VOICE. Call: an ascending, tinkling sound (*weeeet*). Song: a canarylike series of sweet trills and buzzes and often quite variable in structure. Some notes are metallic and bell-like. It is sung from the ground or a bush.

OTHER BEHAVIORS. They fly very close to the ground and through tunnellike openings in brushy areas. In winter, they associate with Brewer's, Sage, and White-crowned Sparrows.

Sage Sparrow, *Amphispiza belli.* Princ. distrib. Texas: Uncommon and local M and WR in Trans-Pecos Texas, in shrubby open areas, brushy gulches, and low mesquites.

FEEDING BEHAVIOR. Other than during the nesting season, they forage in small flocks that hop or walk on the ground in pursuit of insects and seeds. They sometimes scratch for food.

VOICE. Call: a *tsip,* given in alarm. Song: a high-pitched *tsit zoo TSEEE tsay.*

OTHER BEHAVIORS. One distinctive feature is that they cock their tail like a wren when they run; on a perch, they twitch their tail up and down like an Eastern Phoebe.

These are shy birds. They run swiftly for cover when flushed or fly to the top of a bush and then quickly descend to the ground. Winter flocks sometimes contain more than 25 individuals.

Lark Bunting, *Calamospiza melanocorys.* Princ. distrib. Texas: Uncommon to abundant M and WR in w. half of Texas and uncommon and local (and irregular) SR in the Panhandle, in pastures, short-grass prairies, and along roadsides.

FEEDING BEHAVIOR. Lark Buntings walk or run along the ground gleaning insects from the soil or from plants. They tenaciously pursue insects that they flush.

COURTSHIP BEHAVIOR. The male flies up to about 20 feet, extends his wings in an upward dihedral, and moves them in a jerky manner while descending to a perch. Sometimes several males sing together while flying with their crown and body feathers fluffed out.

NESTING. In a grassy area, often in a depression under overhanging grass or weeds. Nest: constructed of weeds, rootlets, grass, and plant down. Although males sometimes participate in incubating the eggs, this task is usually left to the female. Both parents feed the nestlings.

Females that arrive late in spring sometimes settle in a male's territory, along with his other mate or mates. These late arrivers generally lose weight because the male rarely assists them in raising their offspring.

VOICE. Call: a soft *hoo eeee*, sung in flight. Song: a series of rich, melodious slurs, clear piping notes, and trills.

OTHER BEHAVIORS. Lark Buntings migrate and winter in large flocks that may number more than 100 birds; they commonly fly across highways, through pastures, and in other open habitats, twisting and turning in unison while uttering their gentle flight call. Flocks flying across the prairie suggest the movement of a large wheel; birds at the rear of the flock flutter up to the front but then are immediately passed over by those they just overtook.

Breeding birds can be bold, but migrating and wintering flocks tend to be wary and easily flushed.

Savannah Sparrow, *Passerculus sandwichensis.* Princ. distrib. Texas: Uncommon to abundant M and WR in most parts of Texas, in prairies, meadows, and open fields.

FEEDING BEHAVIOR. They run or walk on the ground as they forage for insects and seeds, sometimes flying up to capture insects passing by. At times they also search the leaf litter by scratching with both feet, like a towhee.

VOICE. Call: a weak *tsiip.* Song: usually two to five short notes followed by several trills, some almost insectlike in quality, sometimes rendered as *tseet tseet tseet sweeeeeee yeear.* Birds sing from the top of a bush or small tree. They are infrequently heard in Texas.

OTHER BEHAVIORS. Wintering birds form loose flocks that spend most of the day on the ground. If intruders approach, birds flutter out of the grass, fly an erratic course for a short distance, then drop immediately back into the grass. They also flee by running like a mouse, which they do with remarkable speed.

Grasshopper Sparrow, *Ammodramus savannarum.* Princ. distrib. Texas: Locally common M, SR, and WR throughout most of Texas, in grasslands, pastures, and prairies.

FEEDING BEHAVIOR. These are possibly our most insectivorous sparrows (Pough, 1946). They usually feed alone on the ground, running or hopping as they search for prey, mainly insects, less often seeds. They show a clear preference for grasshoppers; moreover, some individuals select certain grasshopper species over others, regardless of their abundance (Joern, 1988).

COURTSHIP BEHAVIOR. The male displays with a low, fluttering flight.

Sometimes he does this silently and at other times while singing. In the latter case the female answers with a trill. The male also chases the female as part of the courtship.

NESTING. In a shallow depression at the base of a bush, weed, or clump of grass. Nest: a cup of dry grass lined with plant materials and animal hair, most likely built by the female. The back and sides form a slight dome that is woven into the overhanging grass or other plant. This dome forms an opening at the front while concealing the nest from above.

Eggs: incubated by the female. Both parents bring food to the nestlings. When disturbed at the nest, the female performs a captivating distraction display, fluttering a short distance across the ground, then feigning injury by spreading her wings and tail.

VOICE. Call: a thin *tillick.* Song: a thin *pitook zeeeeeeeeeeeeee,* remarkably like a grasshopper's song, and so high in pitch that it is inaudible to many persons. Birds flutter from one clump of grass to another while singing; they also deliver their song from a low perch. They sing persistently at night (especially moonlit nights) and all day long.

OTHER BEHAVIORS. Like the Savannah Sparrow, they flutter up when disturbed, zigzag low over the grass, then quickly drop down to hide. They do not form flocks in winter.

The male sings during his territorial display; he also lowers his head, crouches, and flutters his wings.

Henslow's Sparrow, *Ammodramus henslowii.* Princ. distrib. Texas: Locally uncommon M and WR in e. half of Texas (Shackelford, 2000, and Arnold and Garza, 1998).

FEEDING BEHAVIOR. They usually forage alone and on the ground, consuming insects and seeds.

VOICE. Song: an explosive buzz, rendered as *flee tsick,* remarkable for its short duration. It is possibly the shortest of our birdsongs.

OTHER BEHAVIORS. They escape danger by flying an erratic, undulating course, then suddenly dropping back into the grass. They also flee by running.

Le Conte's Sparrow, *Ammodramus leconteii.* Princ. distrib. Texas: Uncommon and local M and WR in e. half of Texas; their habitat and feeding behavior are essentially like that of Henslow's Sparrow.

VOICE. Call: a thin, high-pitched *zeeeeee.* Song: a short *tseek,* followed by an insectlike buzz and concluding with a *tchip.* It is rarely heard in Texas.

Nelson's Sharp-tailed Sparrow, *Ammodramus nelsoni.* Princ. distrib. Texas: Uncommon to locally common WR on the coast, chiefly in salt and brackish marshes.

FEEDING BEHAVIOR. They forage for insects and other invertebrates, less often for seeds, by walking or running through the marsh like a mouse, climbing on marsh plants, or walking on floating vegetation. They do not hop.

VOICE. Call: a short *chuck.* Song: a buzz or hiss, suggesting *tz tz tz sssss tzik,* usually sung while perched on a grass stalk or while in flight.

OTHER BEHAVIORS. They are generally secretive, but they respond to squeaking. They have an unusual mating behavior for a passerine: males rove about seeking receptive females and do not defend a territory (White, 1999).

Seaside Sparrow, *Ammodramus maritimus.* Princ. distrib. Texas: Uncommon to common PR on the coast, inhabiting only salt marshes. Their distribution is limited by the presence of rice rats (*Oryzomys*), which attack their nests (Post, 1981).

FEEDING BEHAVIOR. They walk or run by the water's edge, or beneath adjacent vegetation, following a zigzag course across the ground. They feed on insects and other invertebrates, as well as seeds, often probing the mud for food items, often in the wettest parts. It has been suggested that their relatively large feet enable them to run across muddy areas that are inaccessible to most small birds. Remarkably, they sometimes wade like shorebirds.

COURTSHIP BEHAVIOR. The male follows the female while raising his wings and singing; he also flutters up to about 25 feet and then descends while singing.

NESTING. In marsh vegetation, just a few inches above the water level at high tide. Nest: a cup of grass constructed by the female that she partially covers with a grass canopy. Eggs: incubated by the female. Both parents feed the nestlings; when threatened, they perform a distraction display by running and fluttering their wings.

VOICE. Call: an abrupt *chip,* as well as a thin *tzeep,* given when alarmed. Song: a buzzy *tup tup zee reeeeee.*

OTHER BEHAVIORS. They are quite secretive but can be lured by squeaking. When flushed, they hold the body erect and run like a rail through the vegetation; also, they flutter a short distance before dropping quickly back into hiding. When he encounters an intruding male in his territory (about 1–2 acres), the male displays by raising his wings.

Fox Sparrow, *Passerella iliaca.* Princ. distrib. Texas: Uncommon to locally common M and WR in e. half of Texas, in both urban and nonurban areas as long as there is dense undergrowth; they seem especially fond of wooded bottomlands.

FEEDING BEHAVIOR. These strikingly handsome sparrows forage on the ground like a towhee, jumping slightly forward before coming down quickly to scratch with both feet in the leaf litter, sometimes vigorously enough to dig a shallow hole in the ground. They eat primarily seeds and insects.

VOICE. Call: a prolonged *tsiipp,* or an abrupt *tcheck.* Song: a series of clear, rich, and melodious notes, generally regarded as one of the finest of our sparrows' songs. The flutelike notes rise in pitch, then fall to their conclusion.

The male has several songs. He sings one or more of them, then returns to sing the first one again. He usually sings from a concealed perch.

OTHER BEHAVIORS. They fly with quick, jerky movements, fanning their tail as they move from bush to bush.

Song Sparrow, *Melospiza melodia.* Princ. distrib. Texas: Common M and WR throughout most of Texas, in brushy habitats near streams, marshes, fencerows, and wet meadows.

FEEDING BEHAVIOR. They walk or run in a zigzag manner across the mud as they pick up insects and seeds. When foraging they sometimes scratch in the soil, walk out into very shallow waters, or hop up into bushes.

VOICE. Call: a nasal, slurred *chenk.* Song (infrequently heard in Texas): variable, usually beginning with three or four *cheet* notes and continuing with rapid trills. Their repertoire includes at least a dozen song types (Horning et al., 1993). They produce the song of the White-crowned Sparrow so faithfully that Whitecrowns cannot distinguish the imitation from the real song (Catchpole and Baptista, 1988).

Song Sparrows tend to sing simple songs throughout most of the day, their songs becoming more complex during territorial boundary disputes, as well as at dawn.

In fall, males utter the female's chitter call, possibly to communicate information to juveniles about the territory (Elekonich, 1998). It is of interest that females discriminate between song types from different males. They respond to their own mate's song by giving more solicitation displays and by displaying more intensely (O'Loghlen and Beecher, 1999).

OTHER BEHAVIORS. They fly with quick and jerky wingbeats, pumping their tail up and down as they go; however, when avoiding intruders they are more likely to sneak through the understory than take flight.

Lincoln's Sparrow, *Melospiza lincolnii.* Princ. distrib. Texas: Locally common M and WR in most parts of Texas, generally in dense thickets and overgrown fields, especially near water.

FEEDING BEHAVIOR. They usually hop on the ground as they forage for insects and seeds.

VOICE. Call: an abrupt *chep.* Song (rarely heard in Texas): sweet, gurgling notes that begin on a low pitch, rise, then abruptly drop, suggesting the House Wren's song.

OTHER BEHAVIORS. The behavior of this species on its Texas wintering grounds is essentially unknown. They are secretive and for the most part solitary in winter, although they respond to squeaking. Like several other ground sparrows, they often escape danger by running instead of taking flight.

Swamp Sparrow, *Melospiza georgiana.* Princ. distrib. Texas: Common M in e. half of Texas; uncommon WR in most of the state, more common in e. Texas and on the coast. They inhabit marsh edges and brushy wetlands, but in winter they sometimes stray to brushy habitats away from water.

FEEDING BEHAVIOR. Their principal food items are insects in summer and seeds in winter. They feed predominantly on the ground, but they are amazingly versatile in their foraging techniques, sometimes climbing in marsh vegetation (which they do with considerable skill) or wading like a sandpiper in shallow water.

In contrast to the Song Sparrow, which is a habitat generalist and a good colonizing species, the Swamp Sparrow is a specialist that is restricted to marsh habitats. When feeding preferences of captive birds are compared, Swamp Sparrows forage less when surrounded by novel objects than do Song Sparrows.

VOICE. Call: a sharp *chink.* Song: a rapid trill and a series of slurred *peets,* suggesting a Chipping Sparrow but louder and more musical. It is usually sung from an elevated perch.

OTHER BEHAVIORS. They are difficult to observe because they tend to stay on the ground under vegetation.

White-throated Sparrow, *Zonotrichia albicollis.* Princ. distrib. Texas: Common to abundant M and WR except in w. third of the state, in thickets, brushy roadsides, and second-growth woodlands, as well as parks and gardens.

FEEDING BEHAVIOR. White-throated Sparrows forage on the ground, usually near or under dense vegetation. They sometimes scratch for seeds and insects with both feet (like a towhee); at other times they climb up into shrubs and low trees. Proximity to cover seems to be essential when they select a feeding site. Apparently they eat more wild fruits than most sparrows.

VOICE. Call: a lisping *zzzt*, used to maintain contact when birds cannot see each other, and an abrupt *tzink*. Song: a series of melodious, melancholy whistles that are more or less on the same pitch. For generations, birders have rendered the song as *Old Sam Peabody Peabody Peabody*. The female also sings (*see* "Other behaviors").

Birds sing from the ground or a low perch, often at night. Wintering birds sing throughout the winter.

White-throated Sparrows apparently recognize the song of their own species by detecting the relative pitch interval of adjacent notes in the *Old Sam Peabody* song, just as we recognize the same melody if sung or played at a higher or lower pitch (Hurley et al., 1990).

OTHER BEHAVIORS. These are curious birds that respond to squeaking as well as to a whistled imitation of their song.

Wintering flocks have a stable dominance hierarchy, dominant birds generally having brighter and more contrasting plumage. This relationship supports the status-signaling hypothesis, which postulates that such badges signal rank in a flock and thus conserve energy by reducing the number or intensity of contests (Slotow et al., 1993).

Plumage differences also play a role in mate selection. White-throated Sparrows occur in two genetic variants, those with white crown stripes and those with tan crown stripes. White-striped birds are more aggressive and dominant, yet curiously, the white-striped females more often mate with the less dominant tan-striped males. This difference in mate selection could be due to the aggressive behavior white-striped males show toward white-striped females. White-striped females sing; tan-striped females generally do not. Moreover, brain parts associated with singing are larger in white-striped than in tan-striped birds (DeVoogd et al., 1995). Therefore, white-striped males may be driving off white-striped females that are singing, leaving these females to mate only with the less aggressive tan-striped males. This behavior is one of several in a complex mating system that evidently accounts for the maintenance of both white-striped and tan-striped birds in the population (Houtman and Falls, 1994).

Wintering birds often associate with other sparrows. Flight is quick and jerky, with some degree of tail-pumping.

Harris's Sparrow, *Zonotrichia querula.* Princ. distrib. Texas: Common WR in n. and c. Texas, in open woodlands with brushy areas.

FEEDING BEHAVIOR. They hop on the ground looking for seeds and insects, scratching vigorously in the leaf litter with both feet; they sometimes move up into bushes to feed on berries.

VOICE. Call: a loud, metallic *chink.* Song: in fall and winter, one or two plaintive, whistled notes in a minor key, sometimes followed by a note in a different key. In fall and winter birds sing in chorus before sunset, a pleasant experience for birders on quiet, cold evenings. In spring the song becomes more elaborate, although it retains its plaintive, musical quality.

OTHER BEHAVIORS. The amount of black on the crown and throat, which is under hormonal control, indicates social dominance. Individuals with more black are generally dominant. Signaling rank to subordinates is a method for reducing the number and intensity of contests for dominance.

Flocks show a strong site attachment to their feeding territories in winter. At this time they are usually unsuspicious and will readily come to feeders.

White-crowned Sparrow, *Zonotrichia leucophrys.* Princ. distrib. Texas: Abundant M and WR in w. half of Texas, less common to the e., in brushy hedgerows, overgrown fields, and thickets. They seem to favor mesquites when available.

FEEDING BEHAVIOR. They hop or run on the ground, foraging for seeds, buds, berries, and insects, scratching vigorously with both feet. Occasionally they fly out to capture insects in midair.

VOICE. Call: a soft *tzit.* Song: a rather melancholy, reflective whistle that terminates in a husky trill, sometimes rendered as *more wet wetter cheese eez.* They sing throughout the year.

OTHER BEHAVIORS. White-crowned Sparrows are relatively tame. In winter 10–15 individuals form flocks that move about quietly and deliberately. They establish a dominance hierarchy when they feed together: adult males dominate over adult females and adult females over immature males (which actually are larger). Immature females occupy the lowest rank (Keys and Rothstein, 1991). The size and location of their winter feeding ranges usually do not change each year. Territories tend to be smaller in dominant and older birds.

White-crowned Sparrows frequently raise their crown feathers to form a low crest that emphasizes the white on the head.

In California, females sing. Apparently, as in males, they do this to advertise territory or to attract a mate (Baptista et al., 1993).

The development of birdsong has been extensively studied in this species. Young males must hear other male White-crowned Sparrows singing before they can produce a full song. If they are presented with a tape containing both White-crowned and Song Sparrow songs, they ignore the Song Sparrow sounds and incorporate the White-crowned Sparrow sounds into their developing song.

Dark-eyed Junco, *Junco hyemalis.* Princ. distrib. Texas: Uncommon to locally abundant M and WR in most parts of the state except s. and coastal regions; common SR in Guadalupe Mountains. They frequent woodland edges, clearings, roadsides, and urban areas.

FEEDING BEHAVIOR. Juncos forage mainly on the ground, where they hop and run, occasionally stopping to scratch in the litter for seeds and insects. They are more likely to feed underneath birdfeeders than on top.

VOICE. Call: a soft clink, which has been likened to hitting two stones together, and a scold that is rendered as *tack tack tack.* Song: a musical trill that suggests a Chipping Sparrow's song; also, a rambling warble or twitter.

OTHER BEHAVIORS. These are likable, unwary little birds. They maintain winter flocks of 10–30 individuals, sometimes joining nuthatches and chickadees. Some wintering birds defend a feeding territory. They also maintain a stable dominance hierarchy in which adults normally dominate younger birds. However, immatures are dominant if they arrive on the wintering grounds and set up residency in an area before the adults arrive (Cristol et al., 1990).

McCown's Longspur, *Calcarius mccownii.* Princ. distrib. Texas: Irregular and locally common M and WR in n. parts of the state; locally abundant in the Panhandle. All longspurs show a strong preference for pastures, prairies, and other open areas.

FEEDING BEHAVIOR. Our four longspurs feed basically the same way: they run, walk, and occasionally hop as they take seeds and insects from the ground. McCown's and Chestnut-collared Longspurs, however, fly after insects they have flushed, a behavior apparently not recorded in the other two species.

VOICE. Call: a low, dry, rattling *chirrup chirrup.* Song: a loud, sweet warble, probably not heard in Texas.

OTHER BEHAVIORS. McCown's Longspurs form large wintering flocks. They tolerate more barren areas, such as dry lake beds and plowed fields, than do the other longspurs. They are almost always found in shorter grass and more open ground than Chestnut-collared Longspsurs, which prefer damper and more vegetated prairie. Wintering Lapland and Smith's Longspurs seem to prefer airports, stubble fields, and other open habitats.

Although they feed in pastures and prairies, all longspurs are attracted to bodies of water, where they can be observed on muddy banks. Wintering longspurs (especially McCown's and Lapland) frequently associate with Horned Larks.

The flight of longspurs is swift, strong, and undulating. When flushed, they fly off explosively and erratically, then drop suddenly back to the ground. In McCown's, at least, the male and female usually remain next to each other while foraging.

Lapland Longspur, *Calcarius lapponicus.* Princ. distrib. Texas: Locally common M and WR in e. half of Texas, locally abundant in the Panhandle. *See* McCown's Longspur.

FEEDING BEHAVIOR. *See* McCown's Longspur. They forage in flocks that may contain thousands of individuals.

VOICE. Call: a harsh rattle, rendered as *dikerik dikerik, psoo psoo,* as well as a softer *eeeelyu.* Song: a liquid, gurgling flight song similar to that of McCown's Longspur.

OTHER BEHAVIORS. *See* McCown's Longspur.

Smith's Longspur, *Calcarius pictus.* Princ. distrib. Texas: Irregular and locally common M and WR in c. and n.c. Texas, especially in areas containing very short grass. *See* McCown's Longspur.

FEEDING BEHAVIOR. *See* McCown's Longspur.

VOICE. Call (given in flight): a series of dry, rattling clicking notes. Song: resembles the call; given from the ground.

OTHER BEHAVIORS. *See* McCown's Longspur. They are not easily observed, as they flush readily and usually alight far from the observer. They generally associate in smaller flocks than the other longspurs, although enormous flocks are sometimes reported.

Chestnut-collared Longspur, *Calcarius ornatus*. Princ. distrib. Texas: Irregular and locally common M and WR throughout most of Texas. *See* McCown's Longspur.

FEEDING BEHAVIOR. *See* McCown's Longspur.

VOICE. Call: a musical twitter. Song: a short, weak, high-pitched twitter that has been compared to the Western Meadowlark's song. It is almost never heard in Texas.

OTHER BEHAVIORS. *See* McCown's Longspur.

CARDINALS, SALTATORS, AND ALLIES: FAMILY CARDINALIDAE

A thick bill enables them to crack seeds and nuts with considerable efficiency. When searching for food they hop among branches or on the ground.

During courtship, males assume an elaborate posture, or flutter their wings, often singing while doing this. They also engage in courtship flights and courtship feeding.

Northern Cardinal, *Cardinalis cardinalis*. Princ. distrib. Texas: Common to locally abundant PR in e. two-thirds of Texas, less common to the w.; they prefer dense brush, woodlands, and urban areas in the e. part of the state, and stream bottoms and tall bushes to the w. *See* Pyrrhuloxia.

FEEDING BEHAVIOR. They forage for seeds, berries, and insects, usually while hopping on the ground or moving slowly among trees and bushes.

COURTSHIP BEHAVIOR. Both males and females raise their head high and sway back and forth as they sing softly; the male also feeds the female.

Interestingly, females with brighter underwing plumage spend more time feeding their offspring than those having duller underwing plumage. This relationship supports the good-parent hypothesis, which proposes that color brightness signals to a potential mate how well she will care for her offspring (Linville et al., 1998).

NESTING. Well concealed in dense vines, shrubs, and low trees. Nest: an open cup shaped from weeds, grass, leaves, twigs, and other materials, built by the female. Eggs: incubated by the female.

The male brings food to his mate when she incubates the eggs and tends to the nestlings. Sometimes he does this without singing, but sometimes the female communicates her need for food by singing from the nest. If she sings the same song as he is singing, he is less likely to come to the nest than if she sings a different song (Halkin, 1997).

Both parents bring food to the nestlings; the male not only feeds them at higher rates than does the female but, curiously, he is the only one that increases feeding effort as nestlings grow older (Filliater and Breitwisch, 1997). Also, he continues feeding the young when his mate begins another nesting attempt.

VOICE. Call: a sharp *tchip* (softer in the female). Songs: rich, powerful whistles; two versions are rendered as *what cheer, what cheer* and *pretty pretty pretty.* More than 30 songs have been documented, and both sexes sing. Northern Cardinals typically deliver their songs from a high, conspicuous perch.

Songs differ geographically. In east Texas, populations in neighboring sapling, pole, and sawtimber stands use different syllable types and sing songs of different duration and complexity (Anderson and Conner, 1985).

The songs of the Pyrrhuloxia are generally sweeter, shorter, and musically more lyrical; the songs of the Carolina Wren, which at a distance suggest a Northern Cardinal, have a more rolling (less staccato) quality.

The female sings in spring after the male has established the territory, and almost always prior to nesting. She probably sings to strengthen the pair bond and to synchronize her reproductive cycle with his. Sometimes the two sing in duet.

OTHER BEHAVIORS. See Pyrrhuloxia. When not molested, Northern Cardinals become quite tame, even to the point of taking food from one's hand. In winter, they sometimes occur in flocks of more than 75 individuals. Occasionally they have been observed anting.

Pyrrhuloxia, *Cardinalis sinuatus.* Princ. distrib. Texas: Uncommon to locally common PR in w. and s. Texas, in mesquite scrub, desert washes, and streamside thickets. Where Pyrrhuloxias and Northern Cardinals overlap in range, this species tends to occupy drier and more open habitats, and the Northern Cardinal favors more mesic, less thorny vegetation, although both species often occupy the same bush. In s. Texas their territories overlap, apparently without aggressive interactions (Lemon and Herzog, 1969).

FEEDING BEHAVIOR. Like Northern Cardinals, they hop on the ground or in low vegetation while foraging for insects, seeds, and berries, but they spend less time on the ground than Northern Cardinals.

COURTSHIP BEHAVIOR. Courtship feeding appears to be the only courtship behavior that has been reported.

NESTING. Generally low in a tree or shrub. Nest: an open cup of weeds, thorny stems, grass, and other plant materials, constructed primarily by the female. The female incubates the eggs; at this time she is brought food by her mate. Both parents feed the nestlings, and both vigorously chase intruders from the nesting territory.

VOICE. Call: an abrupt, metallic *cheek,* much like the Northern Cardinal's but thinner and flatter. Song: also like the Northern Cardinal's, but usually less staccato, *queet queet queet queeyuuuu,* often sung from a conspicuous perch. The male advertises his territory by singing, most intensely in early morning. Apparently females sing very little. *See* Northern Cardinal.

OTHER BEHAVIORS. They are more gregarious than Northern Cardinals. In winter they forage in groups that leave from the nesting area. Some congregations in parts of the southwestern United States contain more than 1,000 individuals, but flocks in Texas are apparently much smaller. They seldom move very far from good cover; however, they are generally not difficult to locate, especially if one imitates their whistle. Their flight is jerky and is accompanied by noisy fluttering of the wings.

When establishing territories, males and females become very pugnacious and chase away other Pyrrhuloxias. The most aggressive males gain the best territories. The songs of Northern Cardinals and Pyrrhuloxias, which are crucial for the establishment and maintenance of territories, are probably less important for pair formation. Ritualistic duetting, which strengthens the pair bond, occurs only in the Northern Cardinal, as the female Pyrrhuloxia rarely sings.

The Pyrrhuloxia's ability to nest in more open situations may be the main (if not the only) factor that is important in reducing competition with Northern Cardinals (Gould, 1961).

Rose-breasted Grosbeak, *Pheucticus ludovicianus.* Princ. distrib. Texas: Uncommon to locally common M in e. half of Texas, in most wooded habitats, including gardens, second-growth woodlands, and streamsides; apparently migrants prefer mulberries in spring and pecans and oaks in fall.

FEEDING BEHAVIOR. These handsome birds forage in trees and bushes, where they take seeds, fruits, and berries. They occasionally come to the ground to feed. Observations elsewhere reveal that females tend to forage higher in trees than males, and they also hover more frequently; these behavioral differences should be looked for in Texas migrants.

VOICE. Call: a metallic *eenk.* Song (male): a long, liquid, robinlike carol, considered one of the finest songs within the family. The female's song is similar, but shorter and softer. Neither is heard often in Texas.

OTHER BEHAVIORS. Their deliberate, stiff, teetering mannerisms have been compared to movements in parrots.

Black-headed Grosbeak, *Pheucticus melanocephalus.* Princ. distrib. Texas: Common SR in mountains of w. Texas, in pine, juniper, and oak woodlands, sometimes descending to lower elevations to occupy trees and shrubs along streams and gullies; locally uncommon M in wooded habitats in w. half of the state.

FEEDING BEHAVIOR. They forage for seeds, berries, and insects, generally in trees and shrubs, less frequently on the ground. They crack pine seeds and other hard seeds with their powerful bill.

Black-headed Grosbeaks are among the very few birds that consume monarch butterflies, which they take in considerable quantities on the monarch's wintering grounds in Mexico. They commonly capture insects by hovering next to foliage or by flying out to capture them on the wing.

COURTSHIP BEHAVIOR. In spring, the male arrives about a week before the female. He sings almost continuously as he flies above the female with his wings and tail spread. He also performs this song flight when she incubates, evidently as a territorial display.

NESTING. In a tree or shrub, about 10 feet above ground. Nest: a bulky cup of twigs, pine needles, rootlets, and grasses, usually built by the female. Both parents incubate the eggs (but only the female at night); when incubating, both birds are difficult to flush from the nest. Both feed the nestlings and vigorously defend their nesting territories.

After leaving the nest, fledglings remain in nearby trees for about 2 weeks and are fed at this time by the parents.

VOICE. Call: a sharp *speek.* Song: rich, clear, whistled notes, with trills, suggesting a robin or Western Tanager, but richer in quality. It closely resembles the song of the Rose-breasted Grosbeak. The female's song is less complex but more variable; this sexual difference permits birds to recognize the sex of an individual by its song (Ritchison, 1983).

OTHER BEHAVIORS. Predation appears to be the major factor that limits breeding success in this species. In the Sandia Mountains of New Mexico, pairs that nest where egg-eating Western Scrub-Jays and Steller's Jays are uncommon are more than four times more likely to fledge their young than those in areas of high jay activity (Hill, 1988).

These birds become very tame, especially around campgrounds.

Blue Grosbeak, *Guiraca caerulea.* Princ. distrib. Texas: Locally common M and SR throughout Texas, in open woodlands, including dense mesquite scrub, streamside thickets, and second-growth woods. They are more likely to be seen in arid habitats than are Indigo Buntings.

FEEDING BEHAVIOR. They feed on many insects in summer, as well as seeds and grain (especially later in the season). They forage by hopping in low vegetation or on the ground, but they sometimes hover while picking insects from leaves and twigs or sally out to capture insects in midair.

COURTSHIP BEHAVIOR. Apparently unrecorded.

NESTING. Low in trees, vines, or shrubs. Nest: a compact cup built by the female from plant materials and embellished with rags, paper, or parts of a snake skin. Only the female incubates the eggs, and at this time she is fed by the male. The female generally feeds the nestlings, but if she begins a new nest before they are able to fly, the male feeds them. Both parents vigorously defend their nest.

VOICE. Call: an explosive, metallic *speenk,* and a colorless *tchuck.* Song: a jumbled series of warbles that lacks the two-note phrases of the Indigo Bunting's song; superficially it resembles the Painted Bunting's song, but it seems a little less sweet and delicate. The song has also been compared to the songs of the Orchard Oriole and the House Finch.

The male sometimes sings for long periods of time while sitting motionless in the top of a tall tree or bush.

OTHER BEHAVIORS. Generally these are quiet birds; when alarmed, they nervously flick and spread their tail. Prior to their fall migration they form flocks that feed in grain fields, rice fields, and grasslands. Their flight is swift and undulating and usually follows a low path between bushes.

Lazuli Bunting, *Passerina amoena.* Princ. distrib. Texas: Rare SR in the Panhandle; rare to locally common M elsewhere except in e. part of the state. *See* Indigo Bunting.

FEEDING BEHAVIOR. Essentially like the Indigo Bunting's. They sometimes bend grass stalks to the ground, then pick seeds from them.

VOICE. Call: similar to the Indigo Bunting's. Song: also resembles the Indigo Bunting's, but is "more rapid, less lively, and often with a few scratchy notes intermingled" (Oberholser and Kincaid, 1974).

Indigo Bunting, *Passerina cyanea.* Princ. distrib. Texas: Common to abundant PR and M in e. part of Texas, becoming less common to the w. Indigo Buntings favor river bottoms and canyons; Lazuli Buntings are more likely to be observed in dry, open hillsides with scattered shrubs. The two species are very closely related.

FEEDING BEHAVIOR. They favor dense thickets, where they forage for insects and seeds on the ground and in shrubs and trees.

COURTSHIP BEHAVIOR. The male performs a flight-song display, as well as courtship displays on the ground; during the terrestrial courtship displays he spreads his wings and dances around the female. He may have several mates at the same time.

NESTING. In a low tree or dense shrub. Nest: built by the female, an open cup of leaves, grass, bark strips, and weeds. The female incubates the eggs and is usually responsible for feeding the nestlings. If she begins a new nest before the young are able to fly, the male may participate in feeding them.

VOICE. Call: a metallic *tcheep,* and an abrupt, brittle *spit.* Song: a descending series of loud, clear *sweet sweet*s, the two-note phrases usually discernible even when the song varies. Males sing persistently throughout the day and deliver their song from a high perch as well as in flight. *See* Blue Grosbeak.

OTHER BEHAVIORS. They form foraging flocks in fall and winter. Migrating flocks on the coast sometimes number in the thousands. Their flight is quick and jerky; when perched, they regularly twitch their tail.

In east Texas, Indigo Buntings and Painted Buntings favor different tree species as singing perches (Kopachena and Crist, 2000).

Varied Bunting, *Passerina versicolor.* Princ. distrib. Texas: Rare and local SR in s. Texas, mainly along the Rio Grande; locally common on the s. Edwards Plateau west across the s. Trans-Pecos; they occur in dry riverbeds, arroyos, and on hillsides (Lockwood, 1995).

FEEDING BEHAVIOR. Probably similar to the Painted Bunting's.

COURTSHIP BEHAVIOR. Apparently unrecorded.

VOICE. Call: a sharp *tship.* Song: "a thin, crisp, energetic warbling similar to that of the Painted Bunting but more obviously phrased and less rambling" (Oberholser and Kincaid, 1974).

OTHER BEHAVIORS. Interestingly, some observers refer to this species as shy and secretive, others as not especially shy.

Painted Bunting, *Passerina ciris.* Princ. distrib. Texas: Uncommon M and SR throughout Texas, in open wooded areas, roadsides, and gardens.

FEEDING BEHAVIOR. These stunning birds forage for insects and seeds on the ground as well as in shrubs and low trees.

FIGURE 11. Courtship postures of the Painted Bunting. Courtship sequence (a–c), solicitation (d), and copulation (e). (S. M. Lanyon and C. F. Thompson, 1984, *Wilson Bulletin* 96: 396–407. With permission.)

COURTSHIP BEHAVIOR. The male spreads his wings and tail, fluffs his plumage, and performs stiff, jerky movements (Fig. 11).

NESTING. Less than 10 feet above ground, in low trees, dense shrubs, or vines. Nest: an open cup of weeds, grass, leaves, rootlets, and animal hair that is constructed by the female. Only she incubates the eggs and feeds the nestlings. After young birds leave the nest, the male may feed them if his mate begins nesting again. Some males have more than one mate.

VOICE. Call: a sharp *chirp chirp*. Song: a clear, sweet, rambling chant that slightly rises and falls in pitch.

OTHER BEHAVIORS. These are generally shy birds that normally hide in the underbrush except during the mating season. Males announce their territory by singing from a conspicuous perch, even at midday when many passerines are silent. In east Texas, Painted Buntings and Indigo

Buntings favor different tree species as singing perches (Kopachena and Crist, 2000).

Males also advertise their territory to intruding males by performing a fluttering flight display. Some encounters between rival males are so vicious that they result in the loss of an eye or even in death.

Their flight is quick and jerky.

Dickcissel, *Spiza americana.* Princ. distrib. Texas: Uncommon to common M and SR in e. two-thirds of Texas, sometimes locally abundant; they frequent meadows and prairies and show a special fondness for alfalfa fields.

FEEDING BEHAVIOR. They usually feed on the ground, less often in low vegetation. They eat insects, seeds, and cultivated grains.

COURTSHIP BEHAVIOR. Apparently unrecorded. The male may have more than one mate on his territory.

NESTING. Often a dozen or more pairs nest together in a loose colony. They place their nests on or very near the ground, or occasionally higher up in a shrub, and conceal it in weeds, alfalfa, clover, grass, or other plants. Nest: a bulky cup of leaves, weeds, rootlets, animal hair, and other materials that is built by the female. The female incubates the eggs and feeds the nestlings, which are dependent on parental care until they join premigratory flocks in late summer.

Sometimes birds move to a new locality and nest a second time. They show little fidelity to the previous year's nesting site and are erratic in changing nesting grounds; some years they are curiously absent in places where they were abundant the previous year.

VOICE. Call: an electric buzzer note given in flight. Song: a dry, emphatic *dick dick ciss ciss ciss,* usually delivered as the bird sits in a low tree or post and vigorously throws back his head.

OTHER BEHAVIORS. They generally occur in flocks (some very large) except when nesting.

BLACKBIRDS: FAMILY ICTERIDAE

This family comprises birds as diverse as bobolinks, meadowlarks, blackbirds, and orioles, yet one characteristic shared by all but a few species is that they form flocks at some time of the year; many species also nest in colonies. They were formerly referred to as troupials, derived from a word meaning "flock."

Most of our species feed on insects and seeds, but larger species like grackles have broader dietary tolerances that include small vertebrates and even garbage.

Courtship and territorial displays are accompanied by distinctive vocalizations. Some displays are extraordinary and involve bizarre postures and movements.

Bobolink, *Dolichonyx oryzivorus.* Princ. distrib. Texas: Locally uncommon M in e. half of Texas, as well as coastal and s. Texas; much more common in spring. They frequent marshes, rice fields, and grassy areas.

FEEDING BEHAVIOR. They usually forage in flocks, either on the ground or from a perch, such as a grass or weed stem. Apparently our migrants eat mainly grains and seeds, less frequently insects.

VOICE. Call: a metallic *peenk.* Song: a bubbling, tinkling series of notes that only remotely sounds like *bobolink.*

OTHER BEHAVIORS. Bobolinks are gregarious outside the breeding season and are often seen fluttering over meadows and cultivated fields.

Red-winged Blackbird, *Agelaius phoeniceus.* Princ. distrib. Texas: Uncommon to abundant PR throughout Texas, in freshwater and brackish marshes, wet marshy fields, roadside ditches, and other habitats that support cattails or other emergent plants.

FEEDING BEHAVIOR. These striking birds generally forage on the ground for seeds, grains, and, to a lesser extent, insects. They walk deliberately, occasionally hopping or running to keep up with a feeding flock; less often they perch in trees and shrubs. They frequently feed more than a mile from their wetland breeding grounds.

COURTSHIP BEHAVIOR. The male ascends into the air, hovers while fluttering his wings, then dives into the marsh vegetation. When given a choice, females select unmated males over mated males (Pribil and Picman, 1996), possibly because they detect that unmated males are able to provide more assistance in raising offspring.

NESTING. In marshy areas, wooded swamps, or virtually any habitat where dense vegetation adjoins a body of water. Sometimes males have more than one mate and defend a territory that contains several females. These females receive protection and some degree of assistance in raising their young. However, under certain circumstances, including a decrease in food supply, females pair with a second mate. Nest: built by the female, a bulky nest of grass, leaves, rootlets, and other plant materials that she ties firmly to standing vegetation. Eggs: incubated by the female. The male helps the female feed the nestlings. How much he assists her depends on the number of females in his harem.

VOICE. Call: a prolonged, high-pitched, nasal whistle (*peeanh*), which usually is an alarm note; also, a harsh *kacck,* uttered while flicking the tail. Song: a liquid, gurgling *konk a ree.*

Apparently the arrangement of notes in the female's call varies according to the stage of the breeding season, her breeding status, and the presence or absence of her mate in the territory (Armstrong, 1995).

OTHER BEHAVIORS. The male vigorously defends his territory against hawks, crows, and other large birds, sometimes riding on top of the intruder while violently pecking its back. His territorial display is intense. From an elevated and conspicuous perch he leans forward, droops his wings, spreads his tail, fluffs his body feathers, raises his red shoulder patches, and sings. During his song flight, he flies slowly and occasionally stalls in midair as he spreads his tail and lowers his head.

When the male's red shoulder patches are experimentally painted black, he no longer is capable of signaling his presence to neighboring males; within seconds, they invade and take over his territory.

When nesting, males change song types when they change perches. One interesting explanation for this behavior is the Beau Geste hypothesis: numerous songs coming from a territory suggest to potential intruders that several birds occupy the territory. (Beau Geste was a fictitious soldier who propped up dead men to deceive the enemy as to the number of soldiers present.) Alternatively, a large repertoire might signal to potential intruders that the holder of the territory is older and more able to defend his territory.

The enormity of postbreeding flocks (often containing cowbirds and several kinds of blackbirds and grackles) can be a nuisance around urban marshes and lakes. One mixed flock in Virginia contained an estimated 15 million birds. Males often flock apart from females and immature birds and sometimes roost together even during the breeding season.

Redwings sometimes fly miles to roost. Their flight is undulating, and flocks twist and turn with remarkable precision. The flock advances over a field by rolling like a wheel: birds feeding on the ground at the rear of the flock fly over those in front, who then fly over the birds that just passed over them.

There is some evidence that the brightness of the female's relatively pale shoulder patches signals status in female-female aggressive encounters.

Eastern Meadowlark, *Sturnella magna.* Princ. distrib. Texas: Uncommon to common PR in most parts of Texas, in prairies, meadows, open fields, and other grasslands, including grasslands at higher elevations in Trans-Pecos Texas. In some areas of the Midwest, where this and the Western Meadowlark overlap in range, the Western species appears to favor drier habitats with shorter grass. In the Northern High Plains, Western Meadowlarks nest on the relatively dry and short grassland in the uplands, whereas Easterns favor moist fields and floodplains, where vegetation is taller (Oberholser and Kincaid, 1974). The Eastern Meadowlarks that breed in the short-grass highland prairies of Trans-Pecos Texas present somewhat of an enigma.

FEEDING BEHAVIOR. Meadowlarks walk briskly and resolutely, feeding on insects and seeds that they pick up from the ground or from low vegetation. They sometimes probe the ground with their bill and occasionally eat animals killed by traffic.

COURTSHIP BEHAVIOR. While facing the female, the male puffs his breast feathers, spreads his tail, flicks his wings, and points his bill up to expose the black V on his chest. Apparently this mark is the critical releaser for the female's response. Sometimes he jumps up while in this posture. The female may respond with the same posture except that she remains on the ground.

NESTING. Nest: a small depression in dense grass or other low cover that the female lines with grass stems. She also builds a dome over the nest and makes runways or trails through the grass that lead to the nest. Eggs: incubated by the female. Both parents feed the nestlings, but because the male may have more than one mate, most of this responsibility falls on the female.

VOICE. Call: a harsh chatter, given in alarm, and a nasal *peeeent.* Song: a plaintive *tee you tee yeear,* sung from a tree or fence post. They also utter a bubbly flight song.

OTHER BEHAVIORS. They fly with quick wingbeats that alternate with stiff glides. In winter they generally occur in flocks. The male advertises his territory by singing.

Western Meadowlark, *Sturnella neglecta.* Princ. distrib. Texas: Common PR in w. half of Texas; locally abundant WR in w. Texas. *See* Eastern Meadowlark. The two species seem to differ very little in feeding, courtship, or nesting behavior.

VOICE. Call: an abrupt *chuck.* Song: a carol of rich, melodic, flutelike, liquid notes. Males appear relatively oblivious to neighboring birds. Unlike many species, Western Meadowlarks share no more song types with birds from adjacent territories than they do with distant birds (Horn and Falls, 1988).

Yellow-headed Blackbird, *Xanthocephalus xanthocephalus.* Princ. distrib. Texas: Common to locally abundant M in w. half of Texas; rare SR in the Panhandle. They occur in marshy areas, corrals, feedlots, pastures, and similar habitats, especially if water is nearby.

FEEDING BEHAVIOR. Flocks of these striking birds present an unforgettable picture as they feed on insects and seeds. They fly deliberately up to catch insects in midair and follow tractors to feed on insects that are flushed. They seem to be especially fond of muddy areas near water.

VOICE. Call: a low, hoarse *ka aaak.* Song: an unpleasant, harsh, and labored *klee klee klee kow cr r r r r,* along with buzzes and squeals. These sounds accompany the male's courtship display (Fig. 12), which can be seen in Texas in spring.

OTHER BEHAVIORS. They migrate and overwinter in flocks, some containing thousands of birds that roost noisily in marshes along with other blackbirds. Often adult males remain apart from the females and immature birds.

Rusty Blackbird, *Euphagus carolinus.* Princ. distrib. Texas: Locally uncommon M and WR in e. half of Texas, in wooded swamps and open fields, especially near water, as well as cattle feedlots. Yearly changes in the numbers of Texas birds may be due to fluctuations in rainfall.

FEEDING BEHAVIOR. They walk along on wet ground or wade in shallow water, actively searching for insects and (especially in winter) seeds and berries. Aquatic prey items include snails, crustaceans, and small fish. They consume more insects than any other of our blackbirds.

Large numbers of migrants and wintering birds often join other blackbirds to forage in dry fields, where they may follow tractors for insects that are flushed. Some flocks contain thousands of individuals.

VOICE. Call: a hoarse *kuuk.* Song: harsh and grating, like a rusty hinge (*koo ah leee*); they also produce high-pitched squeaks that alternate with gurgles and other sounds. Birds in a flock babble as they feed, and if flushed, they fly to a tree and continue their chorus.

OTHER BEHAVIORS. They fly more swiftly and in more compact groups than Red-winged Blackbirds.

FIGURE 12. Courtship displays in the Yellow-headed Blackbird. (R. W. Nero, 1963, *Wilson Bulletin* 75: 376–411. With permission.)

Brewer's Blackbird, *Euphagus cyanocephalus.* Princ. distrib. Texas: Common to locally abundant M and WR throughout most of Texas, becoming less common in e. Texas forests; they occur in arid and semiarid habitats such as deserts, dry fields, and prairies, less frequently around water. They adapt more readily than Rusty Blackbirds to human presence, and seem to benefit greatly from urban habitats such as parking lots.

FEEDING BEHAVIOR. They forage for seeds and insects on the ground and sometimes in shallow water. They follow tractors to take advantage of exposed insects, sometimes flying up to catch insects in midair. Flocks roll across fields like Red-winged Blackbirds.

VOICE. Call: a harsh *chek*, as well as a variety of whistles and squeaks. Song: a squeaky *kush eee*, sometimes compared to the sound of a rusty hinge.

OTHER BEHAVIORS. They form flocks with other blackbirds. Some large flocks stretch out a mile or more; others fly in tight formation. They fly more swiftly and with stronger wingbeats than Red-winged Blackbirds. When they walk, Brewer's Blackbirds jerk their head slightly forward.

Common Grackle, *Quiscalus quiscula.* Princ. distrib. Texas: Uncommon to locally common M, SR, and WR in e. half of Texas, becoming less common to the w. They occur in most open and semiopen habitats, such as farmland, swamps, marshes, suburban lawns, and parks, favoring more thickly wooded areas than the other two grackles.

FEEDING BEHAVIOR. Essentially like that of the Great-tailed Grackle. They have been observed preying on Ruby-crowned Kinglets, Ovenbirds, White-throated Sparrows, and House Sparrows.

COURTSHIP BEHAVIOR. The male partially spreads his wings and tail, fluffs out his body feathers, and utters a scraping song. Like the other grackles, he strikes a pose with his bill pointing straight up.

NESTING. They nest in colonies. Nest: concealed in a dense tree or shrub, usually near water, and typically no more than 20 feet above the ground. The female constructs a bulky cup of grass, weeds, and twigs, usually adding a little mud as cement. Eggs: incubated by the female. Both parents bring food (primarily insects) to the nestlings.

VOICE. Call: a loud *chaak*. Song: a squeaking *koguba leek*, likened to a rusty hinge. When vocalizing, they spread their tail, open their wings, and puff out their plumage.

OTHER BEHAVIORS. They are highly gregarious. Even during the nesting season individuals that are not incubating eggs roost together. Winter flocks are often enormous (up to several million) and include robins, Brown-headed Cowbirds, and various blackbirds.

Their flight is level and steady (though somewhat labored) rather than undulating, like the Red-winged Blackbird's. They walk in a stately manner, holding the head high and nodding it as they move. They have been observed anting by picking up a mothball, rubbing both wings with it, then dropping it.

Boat-tailed Grackle, *Quiscalus major.* Princ. distrib. Texas: Uncommon to abundant PR on the coast, except extreme s. parts, in marshes, flooded fields, mudflats, beaches, and other wet habitats. In Texas they are much less inclined than the Great-tailed Grackle to inhabit areas away from the coast. *See* Great-tailed Grackle.

According to Oberholser and Kincaid (1974), "A visitor to Galveston Island can readily view the differing habitats. The built-up areas and garbage dumps in and near the city of Galveston are inhabited by flocks of Great-tails; remote cattail ponds, *Spartina* marshes, and wet prairies are frequented by Boat-tails. The Boat-tail's avoidance of civilization is apparently confined to Texas." (In Florida they are common in urban areas.)

Great-tailed Grackles will nest in marshes in the absence of Boat-tailed Grackles, suggesting that the latter species is dominant, at least in this habitat.

FEEDING BEHAVIOR. Basically like the Great-tailed Grackle's.

COURTSHIP BEHAVIOR. It probably does not differ significantly from that of the Great-tailed Grackle, except that Boat-tailed Grackles hold their wings higher when fluttering, and group displaying by males is more common.

Female grackles more frequently solicit males that display to them in a colony, but they do not seem to favor high-ranking over low-ranking males or long-tailed over short-tailed males (Post, 1992, and Poston, 1997).

NESTING. In cattails, bushes, and other low plants, usually near water and less than 12 feet from the ground. Like the Great-tailed Grackle, they nest in colonies, and both sexes are sometimes promiscuous. Nest: built by the female, using twigs, grass, Spanish moss, and other materials to construct a large, bulky, open cup that is partially held together with mud. The female incubates the eggs and feeds the nestlings.

VOICE. The male makes a variety of guttural rattles, *chips, kwees, churrs,* and rasping clicks, as well as a loud wolf whistle. Some calls have been compared to those of the Red-winged Blackbird.

Almost certainly the different calls of the male Great-tailed and Boat-tailed Grackles prevent females from mating with the wrong species, which otherwise could occur where the ranges of the two species overlap (Houston, eastward).

OTHER BEHAVIORS. Winter flocks are smaller in the case of the Great-tailed Grackle, but other than different vocalizations and habitat preferences, the two species have almost identical behaviors.

Males exhibit an absolute dominance hierarchy that is stable for several years (Post, 1992). Older males dominate younger males, and a male's rank determines how closely he is allowed to approach a colony. Males of all ranks prevent lower-ranking individuals from approaching the colonies; thus these individuals are denied access to nesting females. Since a male may have to wait 6 or more years before he becomes an alpha male, most males die without acquiring a colony of females (Poston, 1997).

Great-tailed Grackle, *Quiscalus mexicanus.* Princ. distrib. Texas: Common to abundant PR throughout most of Texas, locally common in the Trans-Pecos. They are found in open wooded areas, pastures, and urban habitats, and invariably seem drawn to public parks. The availability of water nearby appears to be critical, and in desert areas they are common around rivers, irrigation ditches, and other waterways. *See* Boat-tailed Grackle.

Until a few decades ago, this and the Boat-tailed Grackle were considered conspecific and were called the Boat-tailed Grackle; however, vocal and behavioral differences between the two populations were shown to prevent interbreeding.

FEEDING BEHAVIOR. They usually feed in flocks. They walk on lawns and other grassy areas, wade in shallow water, or move up into trees and shrubs (where they hunt for bird eggs). Their extraordinarily varied diet consists of insects, snails, crayfish, and other invertebrates; it sometimes includes lizards, young of other species, seeds, berries, and fruits. On garbage dumps they eat virtually any food item, including meat, potatoes, and vegetables. They often soak dry breadcrumbs in water before eating them.

COURTSHIP BEHAVIOR. The male utters *chee chee chee* notes while chasing the female. He advertises his territory with several displays. In one, he fluffs his body feathers, spreads his wings and tail, vibrates his wings to make a crashing sound, and calls loudly. He concludes by posing stiffly with his bill pointed to the sky, which may be a ritualized drinking movement. This posture also functions as a threat display. Often two males posture like this for several minutes.

NESTING. Great-tailed Grackles nest in colonies, some containing several thousand birds. Both males and females may have more than one mate. Males are rarely vigorous in disputing nesting borders, but females regularly squabble over nesting sites. Nest: built by the female, usually in dense trees or shrubs near water, using grass, cattails, twigs, and other materials to construct a bulky, open cup. Some of these materials are stolen from nearby females. The female incubates the eggs and feeds the nestlings. Male nestlings require more food than their female counterparts.

VOICE. The male's calls are loud, harsh, and discordant. They include squeaks, whistles, chatters, and a variety of sounds that apparently only this species can produce. The female typically utters a quiet *cluck*.

OTHER BEHAVIORS. They are highly gregarious. Flocks returning at sunset to roost often string out in long lines. "On clear afternoons, groups fly one hundred or more feet above tree and house tops, in contrast to morning flights of birds which usually 'hedge-hop'" (Oberholser and Kincaid, 1974).

Bronzed Cowbird, *Molothrus aeneus.* Princ. distrib. Texas: Common SR in s. part of the state, in farmlands, parks, feedlots, and other open areas.

FEEDING BEHAVIOR. Essentially like the Brown-headed Cowbird. They have been observed foraging on the backs of cattle.

COURTSHIP BEHAVIOR. The male's courtship display is singular. He begins slowly by hunching over, puffing out his body feathers (especially his neck feathers), and walking stiffly around the female. As the display intensifies, he begins to vibrate his wings. Soon the vibrations turn into rapid, powerful wingbeats that lift him a foot or two into the air, where he briefly suspends himself like a hovering helicopter. He concludes the display by gracefully dropping to the ground.

On the ground he spreads and lowers his tail, arches his wings, bends his head downward, and bounces up and down.

The female's reaction to these amazing antics is basically no reaction at all. She appears inattentive and unmoved and does not even momentarily deviate from what she is doing—walking around looking for seeds and insects in the grass. Soon after, both birds fly away and usually repeat the process within the hour.

NESTING. Like the Brown-headed Cowbird, the female must lay her eggs in the nests of other birds; however, in this species she sometimes punctures the host's eggs before depositing her own. Bronzed Cowbirds generally victimize smaller birds like vireos, but they also parasitize larger species such as Yellow-billed Cuckoos and Couch's Kingbirds (Clotfelter and Brush, 1995). Eggs: one egg per day, up to several weeks (in different nests). Host birds, which include thrashers, vireos, and orioles, incubate the eggs and raise the young cowbirds. Cowbird nestlings develop rapidly, giving them a competitive edge on the host species. They leave the nest in less than 2 weeks.

VOICE. Call: a guttural *chuck*. Their song is lower and wheezier than the Brown-headed Cowbird's. Unlike the Brown-headed Cowbird, they are generally silent, although flocks sometimes utter squeaky and creaky notes.

OTHER BEHAVIORS. They are gregarious except when breeding, usually forming flocks with fewer than 50 birds (although some flocks contain more than 500 individuals).

Brown-headed Cowbird, *Molothrus ater.* Princ. distrib. Texas: Common PR in all parts of Texas, more numerous in winter, in fields, prairies, farms, and other semiopen habitats. They generally avoid dense forests.

FEEDING BEHAVIOR. Cowbirds forage on the ground for insects and seeds, commonly walking next to cattle and other livestock to capture insects flushed by these animals.

COURTSHIP BEHAVIOR. The male fluffs up his body feathers, spreads his wings and tail, and bows. He sings as he does this. Sometimes several males display and sing together.

Among the male's several displays is an aerial display: he ruffles his feathers, bends his head, arches his wings, and sings. Both sexes are promiscuous.

NESTING. This species and the Bronzed Cowbird lay their eggs in other birds' nests, forcing the foster parents to raise their offspring. The female Brown-headed Cowbird sometimes removes one of the host's eggs, insuring that her own young receive the host's undivided attention. She also may pierce the host's eggs as well as any previously laid by other cowbirds. Her own eggs have an unusually strong shell, which no doubt reduces the chances that the host birds will puncture them.

Cowbird eggs have a short incubation period. By hatching early and also by being more vocal, cowbird nestlings dominate the intrabrood hierarchies and compete more successfully for food. Furthermore, they sometimes kill the host nestlings. A 25-second video segment of a cowbird nestling ejecting an Indigo Bunting nestling can be viewed at www.facstaff. bucknell.edu/ddearbor/cowbird.html (Dearborn, 1996).

A single female has been known to lay more than 70 eggs in a season, and more than 140 North American bird species have successfully raised cowbirds.

VOICE. Call: a harsh, dry rattle, and a high whistle (*so so sweet*), given in flight. Song: a liquid *glug glug gleee*.

Because Brown-headed Cowbirds do not raise their own young, males cannot directly teach their offspring how to sing. Consequently, cowbirds raised in isolation have basically the same song as those raised out-of-doors. Nonetheless, their songs can be modified. They are shaped by male-male dominance interactions as well as the responses of females to their songs. Interestingly, females respond more intensely to songs from males raised in isolation than to songs from males raised in cages that contain dominant males. It appears that subdominant males learn to suppress their songs to avoid being attacked by dominant males, making their songs less potent than those of birds raised in isolation.

OTHER BEHAVIORS. They form enormous flocks after the breeding season. These flocks are astonishing in their ability to maneuver together.

In one study, individual cowbirds caught in the same trap were more closely related (based on DNA fingerprinting) than would be expected at random; thus, although females leave the nest after depositing eggs, the nestlings apparently identify their mothers in some manner, as they associate more closely with them than with other females. Possibly females monitor the development of their offspring in the foster parent's nest (Hahn and Fleischer, 1995).

Females often return to a nest to lay another egg, indicating that they memorize the location of the nest. Males do not participate in locating nests. Accordingly, the size of the hippocampal complex (a part of the brain involved in memory) is relatively larger in females than in males (Sherry et al., 1993).

Orchard Oriole, *Icterus spurius.* Princ. distrib. Texas: Uncommon to common M and SR throughout much of the state, more common in n., e., and c. Texas. They frequent orchards, parks, yards, forest edges, and other open areas where there are scattered trees.

FEEDING BEHAVIOR. Orchard Orioles restlessly glean the foliage of trees and bushes, primarily searching for insects. Later in the summer they forage for berries and seeds, and they sometimes visit flowers for insects and nectar.

COURTSHIP BEHAVIOR. Apparently the male's courtship behavior is limited to singing. He sings high in the treetops, but also while hopping from branch to branch at lower levels.

The black-throated subadult males mate, but less frequently than adult males. When they do so, it is usually after most adult males have already mated (Enstrom, 1993).

NESTING. Orchard Orioles are not particularly territorial, with more than one pair sometimes nesting in the same tree. They generally place the nest in the fork of a horizontal branch of a tree. Nest: a hanging basket woven with grass and plant fibers, often with plant down lining the bottom, built by the female with help from her mate. Apparently only the female incubates the eggs, and presumably she is fed by her mate at this time. Both parents feed the nestlings, and sometimes each parent cares for different nestlings.

VOICE. Call: a guttural clucking and a squeaky chatter. Song: a series of loud, clear, slurred notes (especially toward the end) that suggest an American Robin or, in central Texas, a House Finch.

OTHER BEHAVIORS. Although they often bury themselves in dense foliage, they are not particularly shy and will perch conspicuously on an exposed limb, especially when singing. They rarely descend to the ground.

Hooded Oriole, *Icterus cucullatus.* Princ. distrib. Texas: Locally common SR along the Rio Grande and also locally in Trans-Pecos Texas, where they favor trees along rivers and streams. In s. Texas they seem to prefer palm trees in urban areas.

FEEDING BEHAVIOR. They slowly and deliberately glean trees and shrubs for insects, less often for berries and fruits; in towns they frequently visit hummingbird feeders as well as flowers. Sometimes they pierce the base of the flower for nectar. When picking caterpillars from beneath leaves, they may hang upside down like chickadees. Although they do much feeding in low growth, they rarely come to the ground.

COURTSHIP BEHAVIOR. The male assumes exaggerated postures and engages in bowing movements, including pointing his bill straight up. He also sings softly as he moves around or chases the female. Sometimes the female responds with similar displays.

NESTING. Typically in a palm or yucca. Nest: woven by the female, a hanging pouch of grass, plant fibers, hair, and feathers that are brought to her by her mate. She sews the nest to the underside of a large leaf or binds it to a clump of Spanish moss, mistletoe, or other plant. Nests suspended from a palm tree have their entrance on the side. Eggs: incubated by the female. The parents feed regurgitant to the nestlings for 4–5 days.

VOICE. Call: a rough chatter, similar to that of Bullock's Oriole. Song: a jumbled series of throaty whistles, chatters, and sometimes calls of other birds.

OTHER BEHAVIORS. They rarely form groups larger than the family. These are shy, restless birds that generally stay hidden in the foliage. They fly with powerful wingbeats and usually follow a direct course.

Altamira Oriole, *Icterus gularis.* Princ. distrib. Texas: Uncommon PR in extreme s. Texas, in native woodlands near the Rio Grande.

FEEDING BEHAVIOR. These colorful birds normally forage high in trees and shrubs, less frequently at lower levels. They move about slowly and deliberately, gleaning for insects, berries, and small fruits, and sometimes visiting feeders for sugar water. They have been observed investigating automobile radiators for dead insects.

COURTSHIP BEHAVIOR. Apparently unrecorded.

NESTING. Nest: a pendant bag 1–2 feet long conspicuously suspended at the end of a branch, sometimes as high as 80 feet above the ground. It is woven by the female from grass, bark, Spanish moss, and other materials; she may take 3 weeks to complete it. The entrance is at the top. Most likely both parents incubate the eggs. Both bring food to the nestlings, but the male alone feeds the nestlings if the female begins a new nest.

VOICE. Call: a harsh, fussing *ike ike ike,* often the best indicator of their presence, as they typically stay hidden in foliage. Song: a disjointed series of whistles and other sounds, some flutelike in quality. It has been compared to the Baltimore Oriole's song. Both males and females are very vocal, and their loud song carries a considerable distance.

OTHER BEHAVIORS. Their flight is generally quick and jerky and rarely prolonged.

Audubon's Oriole, *Icterus graduacauda.* Princ. distrib. Texas: Locally uncommon PR in s. Texas, in native woodlands (Brush, 2000).

FEEDING BEHAVIOR. Essentially like that of the Altamira Oriole, although it probably spends more time foraging at lower levels and sometimes drops to the ground to feed.

COURTSHIP BEHAVIOR. Apparently unrecorded in detail. The male and female generally stay together throughout the year, but most birds pair anew each year.

NESTING. Typically 5–15 feet above the ground, well concealed in the outer branches of a tree, and firmly attached to adjacent twigs. Nest: a hanging pouch constructed from long grass stems that are woven while still green. It is relatively shallow, as oriole nests go, and is small compared with the body size of the female. It also has a small entrance. Only the female incubates the eggs, but both parents feed the nestlings.

VOICE. Call: a harsh *ike ike ike,* similar to the Altamira Oriole's call note, but higher pitched. Song: several slow, sweet, melancholy whistles, each on a different pitch, like a young boy idly whistling. It is sometimes rendered as *peut pou it.* This species is much less vocal than the Altamira Oriole.

OTHER BEHAVIORS. Audubon's Orioles are quiet, secretive, and slow moving.

Baltimore Oriole, *Icterus galbula.* Princ. distrib. Texas: Locally uncommon SR in n. parts of the state, and common M in e. half. *See* Bullock's Oriole. Except for a few differences in their vocalizations, the two species (formerly considered a single species, the Northern Oriole) have almost identical behaviors.

Apparently their reproductive behavior is like that of Bullock's Oriole, although their nests are slightly deeper.

Bullock's Oriole, *Icterus bullockii.* Princ. distrib. Texas: Locally common M and SR in w. half of the state, in open and riparian woodlands, as well as areas around human habitations that are open and support scattered trees.

FEEDING BEHAVIOR. They glean for insects by moving about in trees and shrubs, sometimes sallying out for passing insects. They also visit flowers for nectar and birdfeeders for sugar water and fruit.

COURTSHIP BEHAVIOR. The male faces the female, stretches upright, partially spreads his wings and tail, and bows deeply. Males also noisily chase females and other males.

VOICE. Call: a dry, rolling chatter, given frequently throughout the day, usually in response to alarm; also, a clear piping whistle. Song: a melodic series of clear, flutelike whistles. The vocalizations of the Baltimore and Bullock's Orioles differ in structure and sound quality, the song of Bullock's usually being regarded as less musical.

OTHER BEHAVIORS. The male advertises his territory by singing. They occur alone or in pairs, except in fall and spring when they migrate in flocks. Their flight is swift and uneven.

Scott's Oriole, *Icterus parisorum.* Princ. distrib. Texas: Uncommon to common SR in Trans-Pecos Texas, becoming less common toward the Edwards Plateau; they occur in arid woodlands and thickets, generally avoiding barren desert habitats. In Big Bend National Park they occur from the Rio Grande (1,700 feet above sea level) to about 7,800 feet. They seem to avoid lower elevations, and in Texas, they rarely occur below 1,000 feet.

FEEDING BEHAVIOR. They search for insects, seeds, and berries as they quietly and deliberately move through treetops. They also visit flowers and hummingbird feeders. In fall as many as a dozen individuals may feed together.

COURTSHIP BEHAVIOR. Apparently unrecorded.

NESTING. In a tree, sometimes concealed in mistletoe or other clumped vegetation. Nest: a hanging basket woven from grass, yucca, and other plant fibers, usually built by the female. The cup is not as deep as that of many orioles. The female incubates the eggs, and both parents feed regurgitant to the nestlings.

VOICE. Call: a rough *tschak*. Song: a short series of extraordinarily rich, whistled notes, suggesting the song of the Western Meadowlark; it is sung repeatedly throughout the day. Females and first-year males also sing.

OTHER BEHAVIORS. Their flight is low and undulating.

FRINGILLINE AND CARDUELINE FINCHES AND ALLIES: FAMILY FRINGILLIDAE

Fringillids are noted for their stout bill, which gives them a decided advantage when feeding on seeds and nuts. As a group they are probably more strictly vegetarian than most seed-eating birds (which readily consume insects in summer).

Many forage in groups that clamber about on limbs, bushes, and weed stalks. This behavior contrasts markedly with that of buntings and sparrows, most of which forage almost exclusively on the ground.

The male's courtship displays include song flights that are performed as he circles near the female; he also feeds the female (courtship feeding). Both males and females assume specific postures during courtship.

Purple Finch, *Carpodacus purpureus.* Princ. distrib. Texas: Locally uncommon in winter in e. half of Texas, probably too irregular to be classified as a WR. They favor woodlands, streamside thickets, and heavily vegetated canyons, generally avoiding open prairies with scattered trees (a habitat frequented by House Finches).

FEEDING BEHAVIOR. They forage by climbing in trees and hopping on the ground. Wintering birds feed mainly on seeds, buds, and berries. They are regular visitors to feeding stations.

VOICE. Call: a metallic, distinctive *kriik,* often given in flight. Song: a rich, liquid warble similar to the House Finch's, but lower in pitch, more run together, and less sustained. Males in a flock sometimes sing in chorus.

OTHER BEHAVIORS. After breeding, males and females may form separate flocks of 20–30 birds; these often associate with Pine Siskins and other birds. Their flight is undulating. *See* House Finch.

Cassin's Finch, *Carpodacus cassinii.* Princ. distrib. Texas: Locally uncommon and irregular winter visitor in w. third of Texas, chiefly in wooded areas at elevations above 6,000 feet. In winter they descend to lower elevations.

FEEDING BEHAVIOR. Similar to the House Finch's. They feed either in treetops or on the ground, but rarely in bushes. Flocks flit nervously from tree to tree while feeding.

VOICE. Call: a clear *djeep*, repeated two or three times; also, a *dee jee yip*, usually given in flight. Song: lively, complex, and often containing sounds appropriated from other birds. It sounds like a hybrid of the songs of the Purple Finch and House Finch, as it is intermediate in speed and pattern, yet more varied than either. It is sung from a perch or while in flight.

OTHER BEHAVIORS. They are gregarious except during the breeding season. *See* House Finch.

House Finch, *Carpodacus mexicanus*. Princ. distrib. Texas: Common PR in w. half of Texas, becoming less common to the e. They occur in semiopen habitats, including urban parks and gardens, and generally in the vicinity of water. They usually avoid deep forests as well as treeless prairies.

FEEDING BEHAVIOR. These charming birds forage in trees and shrubs as well as on the ground. They eat plant materials, commonly seeds, berries, and buds; they readily come to feeding stations and sometimes take sugar water from hummingbird feeders.

COURTSHIP BEHAVIOR. The male raises his tail, droops his wings, and erects his head and crest feathers, all the while singing (Fig. 13). He also regurgitates seeds for his mate after she solicits with a begging posture (Hooge, 1990), and he follows her while singing.

NESTING. House Finches take advantage of a variety of potential nesting sites, including eaves, ivy on buildings, cactus, hanging planters, and deserted nests of other birds. Nest: normally built by the female, consisting of weeds, grass, twigs, leaves, and bits of debris such as string and paper. Eggs: incubated by the female. Both parents feed regurgitated seeds to the nestlings.

VOICE. Call: a musical *queenk*, frequently given in flight, as well as a harsh chattering. Song: a cheery cascade of warbled notes, also given in flight, that rise and fall in pitch. Males can be heard on warm days, even in winter; sometimes female House Finches sing.

In California, song types of neighboring males more closely resemble each other than those from nonneighbors (Bitterbaum and Baptista, 1979). This relationship should be looked for in Texas birds. Also, California birds sing more complex songs than those farther east.

House Finches that extend their range often appropriate elements from

FIGURE 13. House Finch postures. Typical song posture of male (a), courtship display (b), combat behavior (c). (W. L. Thompson, 1960, *Condor* 62: 245-271. With permission.)

other species' songs, including the Common Canary (Payne et al., 1988). Moreover, they sometimes teach this modified song to their progeny (thereby defining the behavior as cultural), which probably explains the inclusion of Orchard Oriole song elements in the repertoire of Texas House Finches.

OTHER BEHAVIORS. House Finches have a matriarchal society. Females are smaller, but they have a higher status than males and dominate them. Not surprisingly, male House Finches (as well as Purple and Cassin's Finches) do not defend territories; the female chooses the nest site and defends it without assistance from her mate. Since reproductive success appears to depend on the female's ability to secure and defend high-quality nest sites, natural selection may have favored genes for aggression in females (Belthoff and Gauthreaux, 1991).

Male plumage varies from pale yellow to bright red (depending on diet). Females generally select the most colorful males when given a choice, possibly because the brighter plumage signals that these males are better competitors for food resources (some of which contain the carotenoid pigment responsible for plumage color). Therefore, brightness may be an indicator (an honest advertisement) to the female of his potential to provide good genes as well as good parental care to her offspring (Hill, 1990, 1993; however, see McGraw and Hill, 2000, and Zahn and Rothstein, 2001).

Males also select the more brightly colored females, but the reason is not clear, as brightly colored females do not demonstrate superior overwinter survival, reproductive success, or general physical condition (Hill, 1993). Postbreeding flocks are common.

Red Crossbill, *Loxia curvirostra.* Princ. distrib. Texas: Irregular, rare SR in Guadalupe and Davis mountains, rarely far from conifers; irregular winter visitor in many parts of Texas, in some years appearing in large numbers.

FEEDING BEHAVIOR. Flocks fly restlessly from tree to tree, then quickly settle down in a conifer. They use their feet and bill, like a parrot, to crawl over limbs and feed on cones.

Their unusual bill helps them pry open evergreen cones and extract seeds while the cone is still attached to the tree. They insert the crossed bill tips between the scales of the cone and spread the bill tips to pry the scales apart. The seeds are then removed with the tongue. Birds are either right-handed or left-handed, depending on the way the mandibles cross. The mandibles do not begin to cross until a few weeks after hatching.

Bill size varies among individuals, and there is a strong tendency for crossbills to select conifers that accommodate their bill size. Birds with larger bills tend to select trees with larger cones (most pines), whereas birds with smaller bills prefer smaller cones (spruce). They have a curious appetite for calcium and will pick mortar from brick walls. Also, many have been killed by automobiles while eating salt placed on highways in winter.

VOICE. Call: a sharp, chicklike *kip kip,* given in flight, as well as twittering notes as they feed. Song: short whistles and warbles, heard as early as January.

OTHER BEHAVIORS. These are not particularly shy birds. They can be approached by a careful observer, and some have been trained to eat out of one's hand. Their undulating flight is swift, prolonged, and often high above the trees.

This variable group of birds (now regarded as a single species, the Red Crossbill) actually may represent eight different species that recognize each other on the basis of subtle characters that include slight differences in call notes.

Pine Siskin, *Carduelis pinus.* Princ. distrib. Texas: Common to abundant M and WR throughout most of Texas, occurring irregularly except in the Panhandle. They are found in open woodlands, river bottoms, gardens, and parks.

FEEDING BEHAVIOR. These lively birds forage in trees, shrubs, and weedy fields, where they search for seeds, insects, buds, and shoots. They cling upside down from a twig, like a chickadee, or hop on the ground, brushing aside leaves with their bill. In trees, close-knit flocks tend to progress downward from the top, then fly to the next tree to begin another descent. Their fondness for salt often leads to their death on highways in winter.

VOICE. Call: a distinctive, nasal *ze e e e e e e,* that rises in pitch and intensity, and a sharp *djit a dit,* given in flight. They call when foraging in dense foliage, but not in open fields. Song: a jumbled series of canarylike buzzes and chatterings.

OTHER BEHAVIORS. They forage in flocks (some as large as 200 or more), even during the nesting season. In winter they often associate with goldfinches. These trusting birds readily come to birdfeeders and can be taught to take food from one's hand.

Compact flocks fly high above the treetops. Their flight is quick and undulating.

Lesser Goldfinch, *Carduelis psaltria.* Princ. distrib. Texas: Uncommon to locally common SR in w. half of Texas, in open woodlands, gardens, river bottoms, and comparable habitats, as long as there is water in the vicinity.

FEEDING BEHAVIOR. *See* American Goldfinch.

COURTSHIP BEHAVIOR. In his song-flight display, the male spreads his tail and rapidly flaps his wings. He also feeds the female.

NESTING. In a vertical fork of a tree or shrub. Nest: a compact woven cup of grass and plant fibers that is built mostly (possibly entirely) by the female. Only the female incubates the eggs, and at this time she is fed re-gurgitant by her mate. Both parents feed the nestlings.

VOICE. Call: a plaintive *tee yee,* in tonal quality suggesting the Eastern Wood-Pewee's song. Song: softer and less run together than the American Goldfinch's (Oberholser and Kincaid, 1974).

OTHER BEHAVIORS. *See* American Goldfinch.

American Goldfinch, *Carduelis tristis.* Princ. distrib. Texas: Locally un-common SR in extreme n.e. Texas; uncommon to abundant M and WR

FIGURE 14. Hostile encounter in American
Goldfinches. (E. L. Coutlee, 1967, *Wilson Bulletin* 79:
89–107. With permission.)

in most parts of the state. They frequent roadsides, patches of woods, gardens, and other open areas.

FEEDING BEHAVIOR. Both of our goldfinches actively forage in trees, shrubs, and weeds, but Lessers seldom feed in the treetops. Both species eat seeds, insects, buds, and shoots. They seem especially fond of thistles and other composites, often clinging to dried flower heads while feeding. They also eat salt.

VOICE. Call: a sweet *zwee zweee e e e e*, and, in flight, *per chic o reee;* also, a plaintive, inquisitive *dear me, see me*. Song: a prolonged series of canarylike trills, twitters, and ascending *sweees*.

OTHER BEHAVIORS. Hostile encounters involve characteristic postures (Fig. 14). Winter flocks sometimes contain more than 400 individuals, and frequently include Pine Siskins. Mated pairs often stay together in these flocks. Their flight is swift and markedly undulating.

Evening Grosbeak, *Coccothraustes vespertinus*. Princ. distrib. Texas: Irregular winter visitor throughout most of Texas; they are absent most years, but winter irruptions bring thousands of individuals to the state. They are observed in open woodlands and other semiopen habitats.

FEEDING BEHAVIOR. Winter flocks forage in trees, shrubs, and occasionally on the ground, eating mainly seeds, which they crack with their enormous bill. They are especially fond of sunflower seeds. They also eat berries and fruits. They consume dirt and gravel, as well, evidently for salt and minerals.

VOICE. Call: a shrill *kleeee*, given in flight. Song: suggests the House Finch's; it is almost never heard in Texas.

OTHER BEHAVIORS. These bold, highly gregarious birds readily come to feeders. Their flight is powerful and undulating. The species is inappropriately named; at one time it was mistakenly believed that these birds sing at night.

OLD WORLD SPARROWS: FAMILY PASSERIDAE

House Sparrow, *Passer domesticus.* Princ. distrib. Texas: Locally abundant PR in all parts of the state, in cities, farms, feedlots, and virtually any habitat where human structures are present. This species was introduced from Europe during the last century. Their decline in numbers this century has been attributed to decreased horse travel and the resulting decrease in the amount of horse manure on the streets, which provided undigested grain.

FEEDING BEHAVIOR. They forage on the ground, where they hop around eating seeds and grain. In urban areas they scavenge discarded human food, dead insects, and virtually any edible food item they can find.

COURTSHIP BEHAVIOR. The male raises his tail, droops his wings, puffs out his chest, and hops near the female; while doing this he bows and chirps incessantly.

NESTING. Often in small colonies, where the male defends a small territory around the nest. Nest: in virtually any enclosed space, including rain gutters, birdhouses, eaves, traffic sign supports, Cliff Swallow nests, and holes in buildings. Away from urban areas or when nesting sites are scarce, they nest in trees. The nest generally conforms to the cavity, but when not restricted, it tends to become a globular mass of grass, twigs, and debris, with an entrance on one side. Both parents build the nest, and both incubate the eggs and feed the fledglings.

VOICE. Call: monotonous cheeps, twitters, and chirps. Song: a prolonged series of chirps. Captive birds have been taught to sing canaries' songs.

OTHER BEHAVIORS. The black throat patch is a badge of status: it is larger in older individuals and in birds that are in good physical condition regardless of age. Thus, badge size generally (though not always) functions as an honest signal for competing males; moreover, males with enlarged badges acquire more nest sites (Veiga, 1995; however, see Whitekiller et al., 2000).

House Sparrows are formidable competitors with our native species. Being bold, boisterous, intelligent, and aggressive, they can easily displace bluebirds, swallows, and other species from their nests and kill their young.

Under certain circumstances they kill their own young by pecking them on the head. Males that have recently lost a mate are more likely to engage in this infanticide (Veiga, 1990).

In spite of this darker side of their temperament, they apparently play at times. Birds have been observed dropping pebbles from a rooftop, then pausing to listen for the pebbles to hit the pavement.

Outside the nesting season thousands of these highly gregarious birds frequently roost together in trees and shrubs.

APPENDIX
SPECIES NOT TREATED

The following list includes species whose distributional status in Texas has been reviewed by the Texas Bird Records Committee of the Texas Ornithological Society (Lockwood, 2000). Species on List A (rarities) are indicated with an asterisk; those on List B, which are under study by the committee, are followed by (B). Species not on those two lists but that are indicated as rare in Texas by the TOS Checklist (Texas Ornithological Society, 1995) are followed by (TOS).

Most species on List A and List B, as well as most species listed as rare in the TOS Checklist, were not included in the species accounts in this book. Exceptions were made with regard to the following rare species, which one or more reviewers recommended. Clark's Grebe (B), Audubon's Shearwater (B), Band-rumped Storm-Petrel (B), Glossy Ibis (TOS), Muscovy Duck (B), *Masked Duck, *Northern Jacana, Common Black-Hawk (B), Aplomado Falcon (B), *Glaucous Gull, Bridled Tern (B), Eurasian Collared-Dove (B), Spotted Owl (B), Tropical Kingbird (B), and Northern Shrike (B).

*Red-throated Loon, *Gavia stellata*
  Pacific Loon, *Gavia pacifica* (B)
*Yellow-billed Loon, *Gavia adamsii*
*Red-necked Grebe, *Podiceps grisegena*
*Yellow-nosed Albatross, *Thalassarche chlororhynchos* (not included on main AOU checklist)
*White-chinned Petrel, *Procellaria aequinoctialis* (not included on main AOU checklist)
*Black-capped Petrel, *Pterodroma hasitata*
*Stejneger's Petrel, *Pterodroma longirostris*
  Cory's Shearwater, *Calonectris diomedea* (B)
*Greater Shearwater, *Puffinus gravis*
*Sooty Shearwater, *Puffinus griseus*
*Manx Shearwater, *Puffinus puffinus*
*Wilson's Storm-Petrel, *Oceanites oceanicus*
*Leach's Storm-Petrel, *Oceanodroma leucorhoa*
*Red-billed Tropicbird, *Phaethon aethereus*

*Blue-footed Booby, *Sula nebouxii*
*Brown Booby, *Sula leucogaster*
*Red-footed Booby, *Sula sula*
*Jabiru, *Jabiru mycteria* (See Brush, 1999)
*Greater Flamingo, *Phoenicopterus ruber*
*Brant, *Branta bernicla*
*Trumpeter Swan, *Cygnus buccinator*
*Eurasian Wigeon, *Anas penelope*
*American Black Duck, *Anas rubripes*
*White-cheeked Pintail, *Anas bahamensis*
*Garganey, *Anas querquedula*
*King Eider, *Somateria spectabilis*
*Harlequin Duck, *Histrionicus histrionicus*
*Barrow's Goldeneye, *Bucephala islandica*
*Snail Kite, *Rostrhamus sociabilis*
*Northern Goshawk, *Accipiter gentilis*
*Crane Hawk, *Geranospiza caerulescens*
*Roadside Hawk, *Buteo magnirostris*
*Short-tailed Hawk, *Buteo brachyurus*
*Collared Forest-Falcon, *Micrastur semitorquatus*
*Paint-billed Crake, *Neocrex erythrops*
*Spotted Rail, *Pardirallus maculatus*
*Double-striped Thick-knee, *Burhinus bistriatus*
*Collared Plover, *Charadrius collaris*
*Wandering Tattler, *Heteroscelus incanus*
*Eskimo Curlew, *Numenius borealis*
*Surfbird, *Aphriza virgata*
*Red-necked Stint, *Calidris ruficollis*
*Sharp-tailed Sandpiper, *Calidris acuminata*
*Purple Sandpiper, *Calidris maritima*
*Curlew Sandpiper, *Calidris ferruginea*
*Ruff, *Philomachus pugnax*
*Red Phalarope, *Phalaropus fulicaria*
 Pomarine Jaeger, *Stercorarius pomarinus* (B)
 Parasitic Jaeger, *Stercorarius parasiticus* (B)
*Long-tailed Jaeger, *Stercorarius longicaudus*
 Little Gull, *Larus minutus* rare WR (TOS)
*Black-headed Gull, *Larus ridibundus* (See White, 1999)
*Heermann's Gull, *Larus heermanni*

*Black-tailed Gull, *Larus crassirostris* (AOU checklist does not include Texas)
*Mew Gull, *Larus canus*
California Gull, *Larus californicus* (B)
*Thayer's Gull, *Larus thayeri*
*Iceland Gull, *Larus glaucoides*
Lesser Black-backed Gull, *Larus fuscus* (B)
*Slaty-backed Gull, *Larus schistisagus*
*Yellow-footed Gull, *Larus livens* (AOU checklist does not include Texas)
*Western Gull, *Larus occidentalis*
*Great Black-backed Gull, *Larus marinus*
*Kelp Gull, *Larus dominicanus* (not included for North America on the main AOU checklist)
Sabine's Gull, *Xema sabini* (B)
Black-legged Kittiwake, *Rissa tridactyla* (B)
*Elegant Tern, *Sterna elegans*
*Roseate Tern, *Sterna dougallii*
*Arctic Tern, *Sterna paradisaea*
*Brown Noddy, *Anous stolidus*
*Black Noddy, *Anous minutus*
White-crowned Pigeon, *Columba leucocephala* presumptive species (TOS)
*Ruddy Ground-Dove, *Columbina talpacoti*
*Ruddy Quail-Dove, *Geotrygon montana*
*Mangrove Cuckoo, *Coccyzus minor*
*Dark-billed Cuckoo, *Coccyzus melacoryphus*
*Snowy Owl, *Nyctea scandiaca*
*Northern Pygmy-Owl, *Glaucidium gnoma*
*Mottled Owl, *Ciccaba virgata*
*Stygian Owl, *Asio stygius*
*Northern Saw-whet Owl, *Aegolius acadicus*
*White-collared Swift, *Streptoprocne zonaris*
*Green Violet-ear, *Colibri thalassinus*
*Green-breasted Mango, *Anthracothorax prevostii*
*Broad-billed Hummingbird, *Cynanthus latirostris*
*White-eared Hummingbird, *Hylocharis leucotis*
*Berylline Hummingbird, *Amazilia beryllina*
*Violet-crowned Hummingbird, *Amazilia violiceps*
*Costa's Hummingbird, *Calypte costae*

*Allen's Hummingbird, *Selasphorus sasin*
*Elegant Trogon, *Trogon elegans*
*Lewis's Woodpecker, *Melanerpes lewis*
 Williamson's Sapsucker, *Sphyrapicus thyroideus* (B)
*Red-breasted Sapsucker, *Sphyrapicus ruber*
*Ivory-billed Woodpecker, *Campephilus principalis*
*Greenish Elaenia, *Myiopagis viridicata*
*Tufted Flycatcher, *Mitrephanes phaeocercus*
*Greater Pewee, *Contopus pertinax*
*Dusky-capped Flycatcher, *Myiarchus tuberculifer*
*Sulphur-bellied Flycatcher, *Myiodynastes luteiventris*
*Piratic Flycatcher, *Legatus leucophaius*
*Thick-billed Kingbird, *Tyrannus crassirostris*
*Gray Kingbird, *Tyrannus dominicensis*
*Fork-tailed Flycatcher, *Tyrannus savana*
*Rose-throated Becard, *Pachyramphus aglaiae*
*Masked Tityra, *Tityra semifasciata*
*Yellow-green Vireo, *Vireo flavoviridis*
*Black-whiskered Vireo, *Vireo altiloquus*
*Yucatan Vireo, *Vireo magister*
 Pinyon Jay, *Gymnorhinus cyanocephalus* rare winter visitor (TOS)
*Clark's Nutcracker, *Nucifraga columbiana*
*Black-billed Magpie, *Pica pica*
*Gray-breasted Martin, *Progne chalybea*
*Black-capped Chickadee, *Poecile atricapillus*
*American Dipper, *Cinclus mexicanus*
*Northern Wheatear, *Oenanthe oenanthe*
*Orange-billed Nightingale-Thrush, *Catharus aurantiirostris*
 Clay-colored Robin, *Turdus grayi* (B)
*White-throated Robin, *Turdus assimilis*
*Rufous-backed Robin, *Turdus rufopalliatus*
*Varied Thrush, *Ixoreus naevius*
*Aztec Thrush, *Ridgwayia pinicola*
*Black Catbird, *Melanoptila glabrirostris*
*Bohemian Waxwing, *Bombycilla garrulus*
*Gray Silky-flycatcher, *Ptilogonys cinereus*
*Olive Warbler, *Peucedramus taeniatus*
 Hermit Warbler, *Dendroica occidentalis* rare M (TOS)
*Connecticut Warbler, *Oporornis agilis*
*Gray-crowned Yellowthroat, *Geothlypis poliocephala*

*Red-faced Warbler, *Cardellina rubrifrons*
*Slate-throated Redstart, *Myioborus miniatus*
*Golden-crowned Warbler, *Basileuterus culicivorus*
*Rufous-capped Warbler, *Basileuterus rufifrons*
*Flame-colored Tanager, *Piranga bidentata*
*Yellow-faced Grassquit, *Tiaris olivacea*
*Baird's Sparrow, *Ammodramus bairdii* (See Jones et al., 1997)
*Golden-crowned Sparrow, *Zonotrichia atricapilla*
*Yellow-eyed Junco, *Junco phaeonotus*
*Snow Bunting, *Plectrophenax nivalis* (See White, 2001)
*Crimson-collared Grosbeak, *Rhodothraupis celaeno*
*Blue Bunting, *Cyanocompsa parellina*
*Shiny Cowbird, *Molothrus bonariensis*
*Black-vented Oriole, *Icterus wagleri*
*Pine Grosbeak, *Pinicola enucleator*
*White-winged Crossbill, *Loxia leucoptera*
*Common Redpoll, *Carduelis flammea*
*Lawrence's Goldfinch, *Carduelis lawrencei*

REFERENCES

General References

Alcock, J. 1998. *Animal Behavior: An Evolutionary Approach.* 6th ed. Sunderland, Mass.: Sinauer Assoc.

American Ornithologists' Union. 1998. *Check-list of North American Birds: The Species of Birds of North America from the Arctic through Panama, including the West Indies and Hawaiian Islands.* Washington, D.C.: American Ornithologists' Union.

Bejder, L., D. Fletchert, and S. Brager. 1998. A method for testing association patterns of social animals. *Anim. Behav.* 56: 719–725.

Bent, A. C. 1919–1968. *Life Histories of North American Birds.* Washington, D.C.: Smithsonian. (Reprinted by Dover, New York.)

Berthelsen, P. S., and L. M. Smith. 1995. Nongame bird nesting on CRP lands in the Texas Southern High Plains. *J. Soil Water Conserv.* 50: 672–675.

Birds of North America. 1992 and subsequent years. Nos. 1–560. Philadelphia: American Ornithologists' Union and Academy of Natural Sciences.

Butler, A. B., and W. Hodos. 1996. *Comparative Vertebrate Neuroanatomy.* New York: J. Wiley.

Casto, S. D. 1996. Use of dog food by birds in southern Texas. *Bull. Texas Ornith. Soc.* 29: 46–47.

Dixon, K. L. 1959. Ecological and distributional relations of desert scrub birds of western Texas. *Condor* 61: 397–409.

Drickamer, L. C., S. H. Vessey, and D. Meikle. 1996. *Animal Behavior.* 4th ed. Dubuque, Iowa: Wm. C. Brown Publishers.

Ehrlich, P. R., D. S. Dobkin, and D. Wheye. 1988. *The Birder's Handbook.* New York: Simon & Schuster.

Eibl-Eibesfeldt, I. 1970. *Ethology: The Biology of Behavior.* New York: Holt, Rinehart & Winston.

Ficken, R. W., and M. S. Ficken. 1966. A review of some aspects of avian field ethology. *Auk* 83: 637–661.

Gill, F. B. 1994. *Ornithology.* 2nd ed. New York: W. H. Freeman & Co.

Goodenough, J., B. McGuire, and R. A. Wallace. 1993. *Perspectives on Animal Behavior.* New York: John Wiley & Sons.

Grier, J. W., and T. Burk. 1992. *Biology of Animal Behavior.* 2nd ed. St. Louis, Mo.: Mosby Year Book.

Hauser, D. C. 1957. Some observations on sun-bathing in birds. *Wilson Bull.* 69: 78–90.

Howell, S. N. G., and S. Webb. 1995. *A Guide to the Birds of Mexico and Northern Central America.* New York: Oxford University Press.

Kandel, E. R., J. H. Schwartz, and T. M. Jessel. 1995. *Essentials of Neural Science and Behavior.* Norwalk, Conn.: Appleton & Lange.

Kaufmann, K. 1996. *Lives of North American Birds.* Boston: Houghton Mifflin.

Kozma, J. M., and N. E. Mathews. 1997. Breeding bird communities and nest plant selection in Chihuahuan Desert habitats in south-central New Mexico. *Wilson Bull.* 109: 424–436.

Krebs, J. R., and N. B. Davies, eds. 1997. *Behavioural Ecology: An Evolutionary Approach.* 4th ed. Oxford: Blackwell Science.

Lockwood, M. 2000. Texas Bird Records Committee Report for 1999. *Bull. Texas Ornith. Soc.* 33: 13–22.

———. 2001. *Birds of the Texas Hill Country.* Austin: University of Texas Press.

Manning, A., and M. S. Dawkins. 1998. *An Introduction to Animal Behavior.* 5th ed. New York: Cambridge University Press.

McFarland, D., ed. 1987. *The Oxford Companion to Animal Behavior.* New York: Oxford University Press.

Mock, D. W., and G. A. Parker. 1998. Siblicide, family conflict and the evolutionary limits of selfishness. *Anim. Behav.* 56: 1–10.

Moynihan, M. 1955. Types of hostile display. *Auk* 72: 247–249.

National Geographic Society. 1999. *Field Guide to the Birds of North America.* 3rd ed. Washington, D.C.: National Geographic Society.

Oberholser, H. C., and E. B. Kincaid Jr. 1974. *The Bird Life of Texas.* 2 vols. Austin: University of Texas Press.

Palmer, R. S., ed. 1962–1988. *The Handbook of North American Birds,* vols. 1–5. New Haven: Yale.

Pough, R. H. 1946, 1951. *Audubon Guides: All the Birds of Eastern and Central North America.* New York: Doubleday & Co.

Seyffert, K. D. 1985. The breeding birds of the Texas Panhandle. *Bull. Texas Ornith. Soc.* 18: 7–19.

Seyffert, K. D. 2001. *Birds of the Texas Panhandle.* College Station: Texas A&M Press.

Shackelford, C. E., and R. N. Conner. 1996. Woodland birds in three different forest types in eastern Texas. *Bull. Texas Ornithol. Soc.* 29: 11–17.

Terres, J. K. 1980. *The Audubon Society Encyclopedia of North American Birds.* New York: Alfred A. Knopf.

Texas Ornithological Society. 1995. *Checklist of the Birds of Texas.* 3rd ed. Austin: Texas Ornithological Society.

Vega, J. H., and J. H. Rappole. 1994. Composition and phenology of an avian community in the Rio Grande Plain of Texas. *Wilson Bull.* 106: 366–380.

Wauer, R. H. 1985. *A Field Guide to Birds of the Big Bend.* Austin: Texas Monthly Press.

Weller, M. W., E. H. Smith, and R. M. Taylor. 1996. Waterbird utilization of a freshwater impoundment on a coastal Texas wildlife refuge. *Texas J. Sci.* **48:** 319–328.

Welty, J. C. 1982. *The Life of Birds.* 3rd ed. Philadelphia: CBS College Publishing.

Wilson, E. O. 1975. *Sociobiology: The New Synthesis.* Cambridge, Mass.: Harvard University.

Grebes: Family Podicipedidae

Forbes, M. R. L., and C. D. Ankney. 1987. Hatching asynchrony and food allocation within broods of Pied-Billed Grebes, *Podilymbus podiceps. Canad. J. Zool.* **65:** 2872–2877.

Lockwood, M. W. 1992. First breeding record of *Aechmophorus* grebes in Texas. *Bull. Texas Ornith. Soc.* **25:** 64–66.

McAllister, N. M. 1958. Courtship, hostile behavior, nest-establishment, and egg laying in the Eared Grebe (*Podiceps caspicus*). *Auk* **75:** 290–311.

Nuechterlein, G. L., and D. Buitron. 1992. Vocal advertising and sex recognition in Eared Grebes. *Condor* **94:** 937–943.

Ryan, M. R., and P. A. Heagy. 1980. Sunbathing behavior of the Pied-Billed Grebe. *Wilson Bull.* **92:** 409–412.

Rylander, M. K. 1991. Pied-billed Grebe mistakes barn roof for surface water. *Bull. Texas Ornith. Soc.* **23:** 30.

Storer, R. W. 1969. The behavior of the Horned Grebe in spring. *Condor* **71:** 180–205.

———. 1976. The behavior and relationships of the Least Grebe. *Trans. San Diego Soc. Nat. Hist.* **18:** 113–126.

Pelicans: Family Pelecanidae

Abraham, C. L., and R. M. Evans. 1999. Metabolic costs of heat solicitation calls in relation to thermal need in embryos of American White Pelicans. *Anim. Behav.* **57:** 967–975.

Arnqvist, G. 1992. Brown Pelican foraging success related to age and height of dive. *Condor* **94:** 521–522.

Evans, R. M. 1992. Embryonic and neonatal vocal elicitation of parental brooding and feeding responses in American White Pelicans. *Anim. Behav.* **44:** 667–675.

McMahon, B. F., and R. M. Evans. 1992. Nocturnal foraging in the American White Pelican. *Condor* **94:** 101–109.

Pinson, D., and H. Drummond. 1993. Brown Pelican siblicide and the prey-size hypothesis. *Behav. Ecol. Sociobiol.* **32:** 111–118.

Schaller, G. B. 1964. Breeding behavior of the White Pelican at Yellowstone Lake, Wyoming. *Condor* **66:** 3–23.

Cormorants: Family Phalacrocoracidae
Darters: Family Anhingidae
Frigatebirds: Family Fregatidae

Allen, T. T. 1961. Notes on the breeding behavior of the Anhinga. *Wilson Bull.* 73: 115–125.

Francis, A. M. 1981. Wing- and tail-flapping in Anhingas: A possible method for drying in the absence of sun. *Auk* 98: 834.

Hennemann, W. W. III. 1982. Energetics and spread-winged behavior of Anhingas in Florida. *Condor* 84: 91–96.

———. 1988. Energetics and spread-winged behavior in Anhingas and Double-crested Cormorants: The risks of generalization. *Amer. Zool.* 28: 845–851.

Telfair, R. C. II. 1995. Neotropic Cormorant (*Phalacrocorax brasilianus*) population trends and dynamics in Texas. *Bull. Texas Ornithol. Soc.* 28: 7–16.

Telfair, R. C. II., C. D. Frentress, and B. G. Davis. 1982. Food of fledgling Olivaceous Cormorants (*Phalacrocorax olivaceus*) in East Central Texas. *Bull. Texas Ornith. Soc.* 15: 16.

Winkler, H. 1983. The wing-spreading behavior of the Little Cormorant, *Phalacrocorax niger. J. für Ornithologie* 124: 177–186.

Herons, Ibises, Storks, American Vultures, and Allies:
Order Ciconiiformes

Chavez-Ramirez, F., and R. D. Slack. 1995. Differential use of coastal marsh habitats by non-breeding wading birds. *Colonial Waterbirds* 18: 166–171.

Smith, J. P. 1995. Foraging flights and habitat use of nesting wading birds (*Ciconiiformes*) at Lake Okeechobee, Florida. *Colonial Waterbirds* 18: 139–158.

Herons, Bitterns, and Allies: Family Ardeidae

DuBowy, P. J. 1996. Effects of water levels and weather on wintering herons and egrets. *Southwest. Nat.* 41: 341–347.

Erwin, R. M., J. S. Hatfield, and W. A. Link. 1991. Social foraging and feeding environment of the Black-crowned Night Heron in an industrialized estuary. *Bird Behav.* 9: 94–102.

Gibbs, J. P., and L. K. Kinkel. 1997. Determinants of the size and location of Great Blue Heron colonies. *Colonial Waterbirds* 20: 1–7.

Katzir, G., and G. R. Martin. 1998. Visual fields in the Black-crowned Night Heron *Nycticorax nycticorax:* Nocturnality does not result in owl-like features. *Ibis* 140: 157–162.

Katzir, G., T. Strod, E. Schechtman, S. Harelis, and Z. Arad. 1999. Cattle Egrets are less able to cope with light refraction than are other herons. *Anim. Behav.* 57: 687–694.

King, T., and D. LeBlanc. 1995. Foraging behaviors of Snowy Egrets (*Egretta thula*) and Yellow-Crowned Night-Herons (*Nyctanassa violacea*) in South Louisiana. *Colonial Waterbirds* 18: 224–225.

Lotem, A., E. Schechtman, and G. Katzir. 1991. Capture of submerged prey by Little Egrets, *Egretta garzetta garzetta*: Strike depth, strike angle, and the problem of light refraction. *Anim. Behav.* 42: 341–346.

McCrimmon, D. A. 1974. Stretch and snap displays in the Great Egret. *Wilson Bull.* 86: 165–167.

McKilligan, N. G. 1990. Promiscuity in the Cattle Egret (*Bubulcus ibis*). *Auk* 107: 334–341.

Mock, D. W. 1976. Pair-formation displays of the Great Blue Heron. *Wilson Bull.* 88: 185–376.

———. 1978. Pair-formation displays of the Great Egret. *Condor* 80: 159–172.

Morrison, M. L., and E. Shanley Jr. 1978. Breeding success of Great Egrets on a dredged material island in Texas. *Bull. Texas Ornith. Soc.* 11: 17–18.

Palmer, P. C. 1989. Great Blue Heron kills and carries off an eastern cottontail rabbit. *Bull. Texas Ornith. Soc.* 22: 17–18.

Riegner, M. F. 1982. Prey handling in Yellow-crowned Night Herons. *Auk* 99: 380–381.

Rodgers, J. A. Jr. 1977. Breeding displays of the Louisiana Heron. *Wilson Bull.* 89: 266–285.

Rustay, C. 1998. Great Blue Heron predation on a Solitary Sandpiper. *New Mexico Ornithol. Soc. Bull.* 26: 98.

Shanley, E. Jr. 1984. Snowy Egrets and Great Egrets using foot-dragging behavior in aerial feeding. *Bull. Texas Ornith. Soc.* 17: 17–18.

Watts, B. D. 1995. Incidence of breeding by Yellow-Crowned Night Herons in juvenal plumage. *Colonial Waterbirds* 18: 222–223.

Wiese, J. H. 1976. Courtship and pair formation in the Great Egret. *Auk* 93: 709–724.

Wiggins, D. A. 1991. Foraging success and aggression in solitary and group-feeding Great Egrets (*Casmerodius albus*). *Colonial Waterbirds* 14: 176–179.

Ibises and Spoonbills: Family Threskiornithidae
Storks: Family Ciconiidae

Amat, J. A., and F. D. Rilla. 1994. Foraging behavior of White-faced Ibises (*Plegadis chihi*) in relation to habitat, group size, and sex. *Colonial Waterbirds* 17: 42–49.

Johnston, J. W., and K. L. Bildstein. 1990. Dietary salt as a physiological constraint in White Ibis breeding in an estuary. *Physiol. Zool.* 63: 190–207.

Kushlan, J. A. 1977. Foraging behavior of the White Ibis. *Wilson Bull.* 89: 342–345.

Rojas, L. M., R. McNeil, T. Cabana, and P. Lachapelle. 1997. Diurnal and nocturnal visual function in two tactile foraging water birds: The American White Ibis and the Black Skimmer. *Condor* 99: 191–200.

White, M. 1999. Jabiru: Texas Review Species. *Texas Birds* 1: 34–37.

New World Vultures: Family Cathartidae

Buckley, N. J. 1997. Experimental tests of the information-center hypothesis with Black Vultures (*Coragyps atratus*) and Turkey Vultures (*Cathartes aura*). *Behav. Ecol. Sociobiol.* **41**: 267–279.

Buckley, N. J. 1998. Interspecific competition between vultures for preferred roost positions. *Wilson Bull.* **110**: 122–125.

Prior, K. A., and P. J. Weatherhead. 1991. Turkey Vultures foraging at experimental food patches: A test of information transfer at communal roosts. *Behav. Ecol. Sociobiol.* **28**: 385–390.

Ducks, Geese, and Swans: Family Anatidae

Adair, S.E., J. L. Moore, and W. H. Kiel Jr. 1996. Wintering diving duck use of coastal ponds: An analysis of alternative hypotheses. *J. Wildl. Manage.* **60**: 83–93.

Anderson, J. T., and T. C. Tacha. 1999. Habitat use by Masked Ducks along the Gulf Coast of Texas. *Wilson Bull.* **111**: 119–121.

Anderson, J. T., L. M. Smith, and D. A. Haukos. 2000. Food selection and feather molt by nonbreeding American Green-winged Teal in Texas playas. *J. Wildl. Manage.* **64**: 222–230.

Bolen, E. G., and M. K. Rylander. 1973. Copulatory behavior in *Dendrocygna*. *Southwest. Nat.* **83**: 348–350.

Custer, C. M., T. W. Custer, and P. J. Zwank. 1997. Migration chronology and distribution of Redheads on the Lower Laguna Madre, Texas. *Southwest. Nat.* **42**: 40–51.

Davis, E. S. 1997. The Down-up display of the Mallard: One display, two orientations. *Anim. Behav.* **53**: 1025–1034.

Delnicki, D. 1983. Mate changes by Black-bellied Whistling Ducks. *Auk* **100**: 728–729.

Fedynich, A. M, R. D. Godfrey Jr., and E. G. Bolen. 1989. Homing of anatids during the nonbreeding season to the Southern High Plains. *J. Wildl. Manage.* **53**: 1104–1110.

Gawlick, D. E., and R. D. Slack. 1996. Comparative foraging behavior of sympatric Snow Geese, Greater White-fronted Geese, and Canada Geese during the non-breeding season. *Wilson Bull.* **108**: 154–159.

Grand, J. B. 1989. Habitat selection and social structure of Mottled Ducks in a Texas coastal marsh. Abstract in *Dissertation Abstracts International* (Sci. & Eng.) **50**: 88B.

Hohman, W. L., T. M. Stark, and J. L. Moore. 1996. Food availability and feeding preferences of breeding Fulvous Whistling-ducks in Louisiana ricefields. *Wilson Bull.* **108**: 137–150.

Johnsgard, P. A., and M. Carbonell. 1996. *Ruddy Ducks and Other Stifftails: Their Behavior and Biology*. Norman: University of Oklahoma Press.

Johnson, W. P., and F. C. Rohwer. 1998. Pairing chronology and agonistic behaviors of wintering Green-winged Teal and Mallards. *Wilson Bull.* 110: 311–315.

Johnson, W. P., F. C. Rohwer, and M. Carloss. 1996. Evidence of nest parasitism in Mottled Ducks. *Wilson Bull.* 108: 187–189.

Littlefield, C. D. 1996. Waterfowl mortality from castor bean poisoning in northwestern Texas. *Southwest. Nat.* 41: 445–446.

McKinney, F. 1965. The displays of the American Green-winged Teal. *Wilson Bull.* 77: 112–121.

Miller, D. L., F. E. Smeins, and J. W. Webb. 1996. Mid-Texas coastal marsh change (1939–1991) as influenced by Lesser Snow Goose herbivory. *J. Coastal Res.* 12: 462–476.

Mitchell, C. A., T. W. Custer, and P. J. Zwank. 1994. Herbivory on shoalgrass by wintering Redheads in Texas. *J. Wildl. Manage.* 58: 131–141.

O'Neill, E. J. 1947. Waterfowl grounded at the Muleshoe National Wildlife Refuge. *Auk* 64: 457.

Raikow, R. J. 1973. Locomotor mechanisms in North American ducks. *Wilson Bull.* 85: 295–307.

Ray, J. D., and H. W. Miller. 1997. A concentration of small Canada Geese in an urban setting at Lubbock, Texas. *Southwest. Nat.* 42: 68–73.

Ripley, S. D. 1963. Courtship in the Ring-necked Duck. *Wilson Bull.* 75: 373–375.

Robertson, D. G., and R. D. Slack. 1995. Landscape change and its effects on the wintering range of a Lesser Snow Goose, *Chen caerulescens caerulescens*, population: A review. *Biol. Conserv.* 71: 179–185.

Rupert, J. R., and T. Brush. 1996. Red-breasted Merganser, *Mergus serrator*, nesting in southern Texas. *Southwest. Nat.* 41: 199–200.

Rylander, M. K., and E. G. Bolen. 1970. Ecological and anatomical adaptations of North American tree ducks. *Auk* 87: 72–90.

Sorenson, M. D. 1991. The functional significance of parasitic egg laying and typical nesting in Redhead Ducks: An analysis of individual behaviour. *Anim. Behav.* 42: 771–796.

Tietje, W. D., and J. G. Teer. 1996. Winter feeding ecology of Northern Shovelers on freshwater and saline wetlands in South Texas. *J. Wildl. Manage.* 60: 843–855.

Weller, M. W., and D. L. Weller. 1995. Observations on feeding of wintering Lesser Scaup in relation to physical structures in Corpus Christi Bay, Texas. *Bull. Texas Ornith. Soc.* 28: 56–59.

Wilson, B. C., and P. D. Levesque. 1996. Egg carrying by a Mottled Duck. *Southwest. Nat.* 41: 444–445.

Hawks, Kites, Eagles, and Allies: Family Accipitridae

Allison, P. S., A. W. Leary, and M. J. Bechard. 1995. Observations of wintering Ferruginous Hawks (*Buteo regalis*) feeding on prairie dogs (*Cynomys ludovicianus*) in the Texas Panhandle. *Texas J. Sci.* 47: 235–237.

Benson, R. H., and C. E. Davis. 1985. Vegetation at 15 Bald Eagle nests in Texas. *Bull. Texas Ornith. Soc.* 18: 22–26.

Boal, C. W., and R. L. Spaulding. 2000. Helping at a Cooper's Hawk nest. *Wilson Bull.* 112: 275–277.

Bolen, E. G., and D. Flores. 1993. *The Mississippi Kite: Portrait of a Southern Hawk.* Austin: University of Texas Press.

Brown, R. E., J. H. Williamson, and D. B. Boone. 1997. Swallow-tailed Kite nesting in Texas: Past and present. *Southwest. Nat.* 42: 103–105.

Brush, T. 1999. The Hook-billed Kite. *Texas Birds* 1: 26–32.

Bush, M. E., and F. R. Gehlbach. 1978. Broad-winged Hawk nest in central Texas: Geographic record and novel aspects of reproduction. *Bull. Texas Ornith. Soc.* 11: 41–43.

Dawson, J. W., and R. W. Mannan. 1991. Dominance hierarchies and helper contributions in Harris' Hawks. *Auk* 108: 649–660.

———. 1991. The role of territoriality in the social organization of Harris' Hawks. *Auk* 108: 661–672.

Dickinson, V. M. 1995. Red imported fire ant predation on Crested Caracara nestlings in South Texas. *Wilson Bull.* 107: 761–762.

Farquhar, C. C. 1993. Individual and intersexual variation in alarm calls of the White-tailed Hawk. *Condor* 95: 234–239.

Gerstell, A. T., and J. C. Bednarz. 1999. Competition and patterns of resource use by two sympatric raptors. *Condor* 101: 557–565.

Grubb, T. C. 1977. Why Ospreys hover. *Wilson Bull.* 89: 149–150.

Kerlinger, P., and S. A. Gauthreaux Jr. 1985. Seasonal timing, geographic distribution, and flight behavior of Broad-winged Hawks during spring migration in South Texas: A radar and visual study. *Auk* 102: 735–743.

Morrison, M. L. 1978. Breeding characteristics, eggshell thinning, and population trends of White-tailed Hawks in Texas. *Bull. Texas Ornith. Soc.* 11: 35–40.

Mueller, H. C., N. S. Mueller, D. D. Berger, G. Allez, W. G. Robichaud, and J. L. Kaspar. 2000. Age and sex differences in the size of prey of the Sharp-shinned Hawk. *J. Field Ornithol.* 71: 399–408.

Palmer, P. C. 1999. Talon-grappling flight displays of White-tailed Hawks. *Texas Birds* 1: 9.

Rice, W. R. 1982. Acoustical location of prey by the Marsh Hawk: Adaptation to concealed prey. *Auk* 99: 403–413.

Rosenfield, R. N., and J. Bielefeldt. 1991. Undescribed bowing display in the Cooper's Hawk. *Condor* 93: 191–193.

Shackelford, C. E., D. Saenz, and R. R. Schaefer. 1996. Sharp-shinned Hawks nesting in the pineywoods of eastern Texas and western Louisiana. *Bull. Texas Ornith. Soc.* 29: 23–25.

Temeles, E. J. 1990. Interspecific territoriality of Northern Harriers: The role of kleptoparasitism. *Anim. Behav.* 40: 361–366.

Wauer, R. H. 1999. White-tailed Hawk: A Texas specialty. *Texas Birds* 1: 5–8.

Willis, E. O. 1963. Is the Zone-tailed Hawk a mimic of the Turkey Vulture? *Condor* **65**: 313–317.

Caracaras and Falcons: Family Falconidae

Ardia, D. R., and K. L. Bildstein. 1997. Sex-related differences in habitat selection in wintering American Kestrels, *Falco sparverius*. *Anim. Behav.* **53**: 1305–1311.

Dickinson, V. M., and K. A. Arnold. 1996. Breeding biology of the Crested Caracara in South Texas. *Wilson Bull.* **108**: 516–523.

Lasley, G. W. 1982. High numbers of Caracara found at temporary roost. *Bull. Texas Ornith. Soc.* **15**: 18–20.

Mueller, H. C. 1971. Displays and vocalizations of the Sparrow Hawk. *Wilson Bull.* **83**: 249–254.

Perez, C. J., P. J. Swank, and D. W. Smith. 1996. Survival, movements, and habitat use of Aplomado Falcons released in southern Texas. *J. Raptor Res.* **30**: 175–182.

Smallwood, J. A. 1989. Prey preferences of free-ranging American Kestrels, *Falco sparverius*. *Anim. Behav.* **38**: 712–728.

Villarroel, M., D. M. Bird, and U. Kuhnlein. 1998. Copulatory behaviour and paternity in the American Kestrel: The adaptive significance of frequent copulations. *Anim. Behav.* **56**: 289–299.

Gallinaceous Birds: Order Galliformes

Drovetski, S. V. 1996. Influence of the trailing-edge notch on flight performance of Galliforms. *Auk* **113**: 802–810.

Leonard, M. L., and L. Zanette. 1998. Female mate choice and male behaviour in domestic fowl. *Anim. Behav.* **56**: 1099–1105.

Curassows and Guans: Family Cracidae

Christensen, Z. D., D. P. Pence, and G. Scott. 1978. Notes on food habits of the Plain Chachalaca from the Lower Rio Grande Valley. *Wilson Bull.* **90**: 647–648.

Marion, W. R., and R. J. Fleetwood. 1978. Nesting ecology of the Plain Chachalaca in South Texas. *Wilson Bull.* **90**: 386–395.

Partridges, Grouse, Turkeys, and Old World Quail: Family Phasianidae

Badyaev, A., W. J. Etges, and T. E. Martin. 1996. Ecological and behavioral correlates of variation in seasonal home ranges of Wild Turkeys. *J. Wildl. Manage.* **60**: 154–164.

Buchholz, R. 1995. Female choice, parasite load, and male ornamentation in Wild Turkeys. *Anim. Behav.* **50**: 929–943.

———. 1996. Thermoregulatory role of the unfeathered head and neck in male Wild Turkeys. *Auk* **113**: 310–318.

Crawford, J. A., and E. G. Bolen. 1975. Spring lek activity of the Lesser Prairie Chicken in West Texas. *Auk* **92**: 808–810.

Dickson, J. G., ed. 1992. *The Wild Turkey: Biology and Management*. Harrisburg, Pa.: Stackpole.

Mateos, C., and J. Carranza. 1995. Female choice for morphological features of male Ring-necked Pheasants. *Anim. Behav.* **49**: 737–748.

———. 1997. Singles in intra-sexual competition between Ring-Necked Pheasant males. *Anim. Behav.* **53**: 471–485.

Peterson, M. J., and N. J. Silvy. 1996. Reproductive stages limiting productivity of the endangered Attwater's Prairie Chicken. *Conserv. Biol.* **10**: 1264–1276.

Phillips, J. B. 1990. Lek behaviour in birds: Do displaying males reduce nest predation? *Anim. Behav.* **39**: 555–565.

Thogmartin, W. E. 1999. Landscape attributes and nest-site selection in Wild Turkeys. *Auk* **116**: 912–923.

Whiteside, R. W., and F. S. Guthery. 1983. Ring-necked Pheasant movements, home ranges, and habitat use in West Texas. *J. Wildl. Manage.* **47**: 1097–1104.

New World Quail: Family Odontophoridae

Anderson, W. L. 1978. Vocalizations of Scaled Quail. *Condor* **80**: 49–63.

Cain, J. R., R. J. Lien, and S. L. Beasom. 1987. Phytoestrogen effects on reproductive performance of Scaled Quail. *J. Wildl. Manage.* **51**: 198–201.

Hagelin, J. C. 1998. Rare, local, little known, and declining North American breeders: A closer look, Montezuma Quail. *Birding* **30**: 406–414.

Hunt, J. L., and T. L. Best. 2001. Foods of Northern Bobwhites (*Colinus virginianus*) in southeastern New Mexico. *Southwest. Nat.* **46**: 239–245.

Kassinis, N. I., and F. Guthery. 1996. Flight behavior of Northern Bobwhites. *J. Wildl. Manage.* **60**: 581–585.

Stromberg, M. R. 1990. Habitat, movements, and roost characteristics of Montezuma Quail in southeastern Arizona. *Condor* **92**: 229–236.

Thomas, J. W. 1957. Anting performed by Scaled Quail. *Wilson Bull.* **69**: 280.

Williams, H. W., A. W. Stokes, and J. C. Wallen. 1968. The food call and display of the Bobwhite Quail (*Colinus virginianus*). *Auk* **85**: 464–476.

Wilson, M. H., and J. A. Crawford. 1987. Habitat selection by Texas Bobwhites and Chestnut-bellied Scaled Quail in South Texas. *J. Wildl. Manage.* **51**: 575–582.

Rails, Gallinules, and Coots: Family Rallidae
Cranes: Family Gruidae

Anderson, J. T., and A. M. Anderson. 1996. Unusual behavior of wintering Louisiana Clapper Rails. *Bull. Texas Ornithol. Soc.* **29**: 47–49.

Ballard, B. M., and J. E. Thompson. 2000. Winter diets of Sandhill Cranes from Central and Coastal Texas. *Wilson Bull.* **112**: 263–268.

Chavez-Ramirez, F., H. E. Hunt, R. D. Slack, and T. V. Stehn. 1996. Ecological correlates of Whooping Crane use of fire-treated upland habitats. *Conserv. Biol.* 10: 217–223.

Gullion, G. W. 1951. The frontal shield of the American Coot. *Wilson Bull.* 63: 157–166.

———. 1952. The displays and calls of the American Coot. *Wilson Bull.* 64: 83–95.

Meanly, B. 1957. Notes on the courtship behavior of the King Rail. *Auk* 74: 433–440.

Wauer, R. H. 1999. A Texas specialty: Whooping Crane. *Texas Birds* 1: 14–18.

Shorebirds, Gulls, Auks, and Allies: Order Charadriiformes

Davis, C. A., and L. Smith. 1998. Behavior of migrant shorebirds in playas of the Southern High Plains, Texas. *Condor* 100: 266–276.

———. 2001. Foraging strategies and niche dynamics of coexisting shorebirds at stopover sites in the Southern Great Plains. *Auk* 118: 484–495.

Lapwings and Plovers: Family Charadriidae

Drake, K. R., J. E. Thompson, and K. L. Drake. 2001. Movements, habitat use, and survival of nonbreeding Piping Plovers. *Condor* 103: 259–267.

Knopf, F. L., and B. J. Miller. 1994. *Charadrius montanus:* montane, grassland, or bare-ground plover? *Auk* 111: 504–506.

Paulson, D. R. 1990. Sandpiper-like feeding in Black-bellied Plovers. *Condor* 92: 245.

Staine, K. J., and J. Burger. 1994. Nocturnal foraging behavior of breeding Piping Plovers (*Charadrius melodus*) in New Jersey. *Auk* 111: 579–587.

Thibault, M., and R. McNeil. 1994. Day/night variation in habitat use by Wilson's Plovers in northeastern Venezuela. *Wilson Bull.* 106: 299–310.

Winton, B. R., D. M. Leslie Jr., and J. R. Rupert. 2000. Breeding ecology and management of Snowy Plovers in north-central Oklahoma. *J. Field Ornithol.* 71: 573–584.

Zharikov, Y., and E. Nol. 2000. Copulation behavior, mate guarding, and paternity in the Semipalmated Plover. *Condor* 102: 231–235.

Oystercatchers: Family Haematopodidae
Stilts and Avocets: Family Recurvirostridae
Jacanas: Family Jacanidae

Fleetwood, R. J. 1973. Jacana breeding in Brazoria County, Texas. *Auk* 90: 422–423.

Goss-Custard, J. D., J. T. Cayford, and S. E. G. Lea. 1998. The changing trade-off between food finding and food stealing in juvenile oystercatchers. *Anim. Behav.* 55: 745–760.

Winton, B. R., and D. M. Leslie Jr. 1997. Breeding ecology of American Avocets (*Recurvirostra americana*) in North-Central Oklahoma. *Bull. Oklahoma Ornithol. Soc.* 30: 25–32.

Sandpipers, Phalaropes, and Allies: Family Scolopacidae

Hamilton, W. J. Jr. 1959. Aggressive behavior in migrant Pectoral Sandpipers. *Condor* 61: 161–179.

Myers, J. P., P. G. Connors, and F. A. Pitelka. 1979. Territory size in wintering Sanderlings: The effects of prey abundance and intruder density. *Auk* 96: 551–561.

Roberts, G. 1997. How many birds does it take to put a flock to flight? *Anim. Behav.* 54: 1517–1522.

Rojas De Azuaje, L. M., S. Tai, and R. McNeil. 1993. Comparison of rod/cone ratio in three species of shorebirds having different nocturnal foraging strategies. *Auk* 110: 141–145.

Seyffert, K. D. 1985. A first nesting of the Wilson's Phalarope in Texas. *Bull. Texas Ornith. Soc.* 18: 27–29.

Williams, G. G. 1953. Wilson Phalaropes as commensals. *Condor* 55: 158.

Willis, E. O. 1994. Are *Actitis* sandpipers inverted flying fishes? *Auk* 111: 190–191.

Skuas, Gulls, Terns, and Skimmers: Family Laridae

Gonzalez-Solis, J., P. H. Becker, and H. Wendeln. 1999. Divorce and asynchronous arrival in Common Terns, *Sterna hirundo. Anim. Behav.* 58: 1123–1129.

Iacovides, S., and R. M. Evans. 1998. Begging as graded signals of need for food in young Ring-billed Gulls. *Anim. Behav.* 56: 79–85.

Palestis, B. G., and J. Burger. 1998. Evidence for social facilitation of preening in the Common Tern. *Anim. Behav.* 56: 1107–1111.

———. 1999. Individual sibling recognition in experimental broods of Common Tern chicks. *Anim. Behav.* 58: 375–381.

Schwagmeyer, P. L., D. W. Mock, T. C. Lamey, C. S. Lamey, and M. D. Beecher. 1991. Effects of sibling contact on hatch timing in an asynchronously hatching bird. *Anim. Behav.* 41: 887–894.

Smalley, A. E., G. B. Smalley, A. J. Mueller, and B. C. Thompson. 1991. Roof-nesting Gull-billed Terns in Louisiana and Texas. *J. Louisiana Ornithol.* 2: 18–20.

White, M. 1999. Texas review species: Black-headed Gull. *Texas Birds* 1: 29–33.

Pigeons and Doves: Family Columbidae

Blockstein, D. E. 1989. Crop milk and clutch size in Mourning Doves. *Wilson Bull.* 101: 11–25.

Bowman, R., and G. E. Woolfenden. 1997. Nesting chronology of the Common Ground-Dove in Florida and Texas. *J. Field Ornithol.* 68: 580–589.

Brush, T. 1998. Recent nesting and current status of Red-billed Pigeon along the lower Rio Grande in southern Texas. *Bull. Texas Ornith. Soc.* 31: 22–26.

Burley, N. 1980. Clutch overlap and clutch size: Alternative and complementary reproductive tactics. *Am. Nat.* 115: 223–246.

Cheng, M.-F. 1992. For whom does the female dove coo? A case for the role of vocal self-stimulation. *Anim. Behav.* **43**: 1035-1044.

Davies, S. J. 1970. Patterns of inheritance in the bowing display and associated behaviour of some hybrid *Streptopelia* doves. *Behaviour* **36**: 187-214.

Davis, J. M. 1975. Socially induced flight reactions in pigeons. *Anim. Behav.* **23**: 597-601.

Downing, R. L. 1959. Significance of ground nesting by Mourning Doves in northwestern Oklahoma. *J. Wildl. Manage.* **23**: 117-118.

Frankel, A. I., and T. S. Baskett. 1961. The effect of pairing on cooing of penned Mourning Doves. *J. Wildl. Manage.* **25**: 372-384.

Goforth, W. R., and T. S. Baskett. 1971. Social organization of penned Mourning Doves. *Auk* **88**: 528-542.

Hayslette, S. E., T. C. Tacha, and G. L. Waggerman. 2000. Factors affecting White-winged, White-tipped, and Mourning Dove reproduction in the Lower Rio Grande Valley. *J. Wildl. Manage.* **64**: 286-295.

Johnston, R. F. 1960. Behavior of the Inca Dove. *Condor* **62**: 7-24.

———. 1964. Remarks on the behavior of the Ground Dove. *Condor* **66**: 65-69.

———. 1992. Evolution in the Rock Dove: Skeletal morphology. *Auk* **109**: 530-542.

Johnston, R. F., and M. Janiga. 1995. *Feral Pigeons.* New York: Oxford University Press.

Kenward, R. E. 1978. Hawks and doves: Factors affecting success and selection in Goshawk attacks on wild pigeons. *J. Anim. Ecol.* **47**: 449-460.

Losito, M. P., and R. E. Mirarchi. 1991. Summertime habitat use and movements of hatching-year Mourning Doves in northern Alabama. *J. Wildl. Manage.* **55**: 137-146.

Lovell-Mansbridge, C., and T. R. Birkhead. 1998. Do female pigeons trade pair copulations for protection? *Anim. Behav.* **56**: 235-241.

MacMillan, R. E., and C. H. Trost. 1967. Nocturnal hypothermia in the Inca Dove, *Scardafella inca. Comp. Biochem. Physiol.* **23**: 243-253.

McClure, H. E. 1945. Reaction of the Mourning Dove to colored eggs. *Auk* **62**: 270-272.

Murton, R. K., N. J. Westwood, and R. J. P. Thearle. 1973. Polymorphism and the evolution of a continuous breeding season in the pigeon, *Columba livia. J. Reprod. and Fertility Suppl.* **19**: 563-577.

Passmore, M. F. 1981. Population biology of the Common Ground Dove and ecological relationships with Mourning and White-winged Doves in South Texas. Abstract in *Dissertation Abstracts International* **42**: 849B.

Peeters, H. J. 1962. Nuptial behavior of the Band-tailed Pigeon in the San Francisco Bay area. *Condor* **64**: 445-470.

Quay, W. B. 1982. Seasonal calling, foraging, and flocking of Inca Doves at Galveston, Texas. *Condor* **84**: 321-326.

Robertson, P. B., and A. F. Schnapf. 1987. Pyramiding behavior in the Inca Dove: Adaptive aspects of day-night differences. *Condor* 89: 185–187.

Sanders, T. A., and R. L. Jarvis. 2000. Do Band-tailed Pigeons seek a calcium supplement at mineral sites? *Condor* 102: 855–863.

Schroeder, M. A., and C. E. Braun. 1993. Movement and philopatry of Band-tailed Pigeons captured in Colorado. *J. Wildl. Manage.* 57: 103–112.

Small, M. F., R. A. Hilsenbeck, and J. F. Scudday. 1995. Utilization of undigested livestock feed grain by White-winged Doves in West Texas. *Texas J. Sci.* 47: 323–325.

Vega, J. H., and J. H. Rappole. 1994. Effects of scrub mechanical treatment on the nongame bird community in the Rio Grande Plain of Texas. *Wildl. Soc. Bull.* 22: 165–171.

Parakeets and Parrots: Family Psittacidae

Neck, R. W. 1986. Expansion of Red-crowned Parrot, *Amazona viridigenalis,* into southern Texas and changes in agricultural practices in northern Mexico. *Bull. Texas Ornith. Soc.* 19: 6–12.

Cuckoos, Roadrunners, and Anis: Family Cuculidae

Bowen, B. S., R. R. Koford, and S. L. Vehrencamp. 1991. Seasonal pattern of reverse mounting in the Groove-Billed Ani (*Crotophaga sulcirostris*). *Condor* 93: 159–163.

Cornett, J. W. 2001. *The Roadrunner.* Palm Springs, Calif.: Nature Trails Press.

Fischer, D. H. 1979. Black-billed Cuckoo (*Coccyzus erythropthalmus*) breeding in South Texas. *Bull. Texas Ornith. Soc.* 12: 25–27.

Holte, A. E., and M. A. Houck. 2000. Juvenile Greater Roadrunner (Cuculidae) killed by choking on a Texas Horned Lizard (Phrynosomatidae). *Southwest. Nat.* 45: 74–75.

Meinzer, W. 1993. *The Roadrunner.* Lubbock: Texas Tech University Press.

Rylander, M. K. 1972. Winter dormitory of the Roadrunner, *Geococcyx californicus,* in West Texas. *Auk* 89: 896.

Wilson, J. K. 2000. Additional observations on precopulatory behavior of Yellow-billed Cuckoos. *Southwest. Nat.* 45: 535–536.

Owls: Order Strigiformes

Gamel, C. M. 1997. Habitat selection, population density, and home range of the Elf Owl, *Microrathene whitneyi,* at Santa Ana National Wildlife Refuge, Texas. M.S. thesis. University of Texas–Pan American, Edinburg. [Cited in Brush, 1999.]

Gehlbach, F. R. 1995. *The Eastern Screech-Owl: Life History, Ecology, and Behavior in the Suburbs and Countryside.* College Station: Texas A&M Press.

Hayward, G. D. 1994. Review of technical knowledge: Boreal owls. USDA, Forest Serv. RM-GTR-253: 92–127.

Hogan, K. M., et al. 1996. Notes on the diet of Short-eared Owls (*Asio flammeus*) in Texas. *J. Raptor Res.* 30: 102–104.

Klatt, P. H., and G. Ritchison. 1993. The duetting behavior of Eastern Screech-Owls. *Wilson Bull.* 105: 483–489.

Martin, D. J. 1973. Selected aspects of Burrowing Owl ecology and behavior. *Condor* 75: 446–456.

McCallum, D. A. 1994. Review of technical knowledge: Flammulated owls. USDA, Forest Serv. RM-GTR-253: 14–46.

Otter, K. 1996. Individual variation in the advertising call of male Northern Saw-whet Owls. *J. Field Ornithol.* 67: 398–405.

Pavey, C. R., and A. K. Smyth. 1998. Effects of avian mobbing on roost use and diet of Powerful Owls, *Ninox strenua. Anim. Behav.* 55: 313–318.

Plumpton, D. L., and R. S. Lutz. 1994. Sexual size dimorphism, mate choice, and productivity of Burrowing Owls. *Auk* 111: 724–727.

Proudfoot, G. A., and S. L. Beasom. 1997. Food habits of nesting Ferruginous Pygmy-Owls in southern Texas. *Wilson Bull.* 109: 741–748.

Roberts, K. J., F. D. Yancey II, and C. Jones. 1997. Predation by Great-horned Owls on Brazilian Free-tailed Bats in North Texas. *Texas J. Sci.* 49: 215–218.

Shackelford, C. E., and F. C. Earley. 1996. Barred Owl nest in a natural hole in an earthen bank in eastern Texas. *J. Raptor Res.* 30: 41.

Goatsuckers: Family Caprimulgidae

Bayne, E. M., and R. M. Brigham. 1995. Prey selection and foraging constraints in Common Poorwills (*Phalaenoptilus nuttallii*: Aves: Caprimulgidae). *J. Zool.* 235: 1–8.

Brauner, J. 1953. Observations on the behavior of a captive Poor-will. *Condor* 55: 68–74.

Brigham, R. M., and R. M. R. Barclay. 1992. Lunar influence on foraging and nesting activity of Common Poorwills (*Phalaenoptilus nuttallii*). *Auk* 109: 315–320.

———. 1995. Prey detection by Common Nighthawks: Does vision impose a constraint? *Ecoscience* 2: 276–279.

Cowles, R. B., and W. R. Dawson. 1951. A cooling mechanism of the Texas Nighthawk. *Condor* 53: 19–22.

Firman, M. C., R. M. Brigham, and R. M. Barclay. 1993. Do free-ranging Common Nighthawks enter torpor? *Condor* 95: 157–162.

Freemyer, H. 1993. Call-counts of West Texas Common Poorwills. *Bull. Texas Ornith. Soc.* 26: 15–18.

———. 1996. Some observations on Common Poorwill foraging techniques. *Bull. Texas Ornith. Soc.* 29: 45–47.

Komar, O., and W. Rodriguez. 1997. Nesting of Lesser Nighthawks on beaches in El Salvador. *Wilson Bull.* 109: 167–168.

Mills, A. M. 1986. The influence of moonlight on the behavior of goatsuckers (Caprimulgidae). *Auk* 103: 370–378.

Todd, L. D., R. G. Poulin, and R. M. Brigham. 1998. Diet of Common Nighthawks (*Chordeiles minor:* Caprimulgidae) relative to prey abundance. *Am. Midl. Nat.* 139: 20–28.

Swifts: Family Apodidae

Davies, S. 2000. The North American Chimney Swift nest site research project. *Texas Birds* 2: 52–53.

Dell'omo, G., E. Alleva, and C. Carere. 1998. Parental recycling of nestling faeces in the Common Swift. *Anim. Behav.* 56: 631–637.

Hummingbirds: Family Trochilidae

Blem, C. R., L. B. Blem, and C. C. Cosgrove. 1997. Field studies of Rufous Hummingbird sucrose preference: Does source height affect test results? *J. Field Ornithol.* 68: 245–252.

Blem, C. R., L. B. Blem, J. Felix, and J. van Gelder. 2000. Rufous Hummingbird sucrose preference: Precision of selection varies with concentration. *Condor* 102: 235–238.

Ficken, M. S., K. M. Rusch, S. J. Taylor, and D. R. Powers. 2000. Blue-throated hummingbird song: A pinnacle of nonoscine vocalizations. *Auk* 117: 120–128.

Goldsmith, K. M., and T. H. Goldsmith. 1982. Sense of smell in the Black-chinned Hummingbird. *Condor* 84: 237–238.

Green, G. 1999. Noisy Buff-bellied Hummers. *Texas Birds* 1: 21–23.

Hurley, T. A., R. D. Scott, and S. D. Healy. 2001. The function of displays of male Rufous Hummingbirds. *Condor* 103: 647–651.

Kuban, J. F., and R. L. Neill. 1980. Feeding ecology of hummingbirds in the highlands of the Chisos Mountains, Texas. *Condor* 82: 180–185.

Phillips, J. N. 1998. A survey of wintering Rufous Hummingbirds in Texas. *Bull. Texas Ornith. Soc.* 31: 65–67.

Pytte, C., and M. S. Ficken. 1994. Aerial display sounds of the Black-chinned Hummingbird. *Condor* 96: 1088–1091.

Roberts, W. M. 1995. Hummingbird licking behavior and the energetics of nectar feeding. *Auk* 112: 456–463.

Rusch, K. M., C. L. Pytte, and M. S. Ficken. 1996. Organization of agonistic vocalizations in Black-chinned Hummingbirds. *Condor* 98: 557–566.

Slack, R. D., and K. A. Arnold. 1985. Nesting of the Magnificent Hummingbird in Jeff Davis County, Texas. *Bull. Texas Ornith. Soc.* 18: 27.

Stromberg, M. R., and P. B. Johnsen. 1990. Hummingbird sweetness preferences: Taste or viscosity? *Condor* 92: 606–612.

Sutherland, G. D., and C. L. Gass. 1995. Learning and remembering of spatial patterns by hummingbirds. *Anim. Behav.* 50: 1273–1286.

Kingfishers: Family Alcedinidae

Kelly, J. F. 1998. Behavior and energy budgets of Belted Kingfishers in winter. *J. Field Ornithol.* 69: 75–84.

Wauer, R. 2000. Green Kingfisher. *Texas Birds* 2: 11–12.

Woodpeckers and Allies: Family Picidae

Conner, R. N. 1982. Pileated Woodpecker feeds on horned passalus colony. *Bull. Texas Ornith. Soc.* 15: 15–16.

Conner, R. N., D. C. Rudolph, D. Saenz, and R. R. Schaefer. 1996. Red-cockaded Woodpecker nesting success, forest structure, and southern flying squirrels in Texas. *Wilson Bull.* 108: 697–711.

———. 1997. Species using Red-cockaded Woodpecker cavities in eastern Texas. *Bull. Texas Ornithol. Soc.* 30: 11–16.

Conner, R. N., D. C. Rudolph, R. R. Schaefer, and D. Saenz. 1997. Long-distance dispersal of Red-cockaded Woodpeckers in Texas. *Wilson Bull.* 109: 157–160.

Husak, M. S. 1995. Evidence of possible egg predation by Golden-fronted Woodpeckers. *Bull. Texas Ornithol. Soc.* 28: 55–56.

———. 1996. Breeding season displays of the Golden-fronted Woodpecker. *Southwest. Nat.* 41: 441–469.

———. 1997. Predation of Golden-fronted Woodpecker nestlings by a Texas rat snake. *Bull. Texas Ornithol. Soc.* 30: 17–18.

———. 2000. Seasonal variation in territorial behavior of Golden-fronted Woodpeckers in West-Central Texas. *Southwest. Nat.* 45: 30–38.

Jackson, J. A., B. J. Schardien, and R. Weeks. 1978. An evaluation of the status of some Red-cockaded Woodpecker colonies in East Texas. *Bull. Texas Ornith. Soc.* 11: 2–9.

Khan, M. Z., and J. R. Walters. 1997. Is helping a beneficial learning experience for Red-cockaded Woodpecker (*Picoides borealis*) helpers? *Behav. Ecol. Sociobiol.* 41: 69–73.

Kilham, L. 1958. Pair formation, mutual tapping, and nest hole selection of Red-bellied Woodpeckers. *Auk* 75: 318–329.

———. 1959. Behavior and methods of communication of pileated woodpeckers. *Condor* 61: 377–378.

———. 1960. Courtship and territorial behavior of Hairy Woodpeckers. *Auk* 77: 259–270.

Koenig, W. D., M. T. Stanback, and J. Haydock. 1999. Demographic consequences of incest avoidance in the cooperatively breeding Acorn Woodpecker. *Anim. Behav.* 57: 1287–1293.

Ligon, J. D. 1970. Behavior and breeding biology of the Red-cockaded Woodpecker. *Auk* 87: 255–278.

Lockwood, M. W., and C. E. Shackelford. 1998. The occurrence of Red-breasted

Sapsucker and suspected hybrids with Red-naped Sapsucker in Texas. *Bull. Tex. Ornith. Soc.* 31: 3–6.

Matthysen, E., T. C. Grubb Jr., and D. Cimprich. 1991. Social control of sex-specific foraging behaviour in Downy Woodpeckers, *Picoides pubescens. Anim. Behav.* 42: 515–517.

Neill, A. J., and R. G. Harper. 1990. Red-bellied Woodpecker predation on nestling House Wrens. *Condor* 92: 789.

Peters, W. D., and T. C. Grubb Jr. 1983. An experimental analysis of sex-specific foraging in the Downy Woodpecker, *Picoides pubescens. Ecology* 64: 1437–1443.

Reynolds, P. S., and S. L. Lima. 1994. Direct use of wings by foraging woodpeckers. *Wilson Bull.* 106: 408–411.

Rudolph, D. C., and R. N. Conner. 1994. Forest fragmentation and Red-cockaded Woodpecker population: An analysis at intermediate scale. *J. Field Ornithol.* 65: 365–375.

Saenz, E., R. N. Conner, C. E. Shackelford, and D. C. Rudolph. 1998. Pileated Woodpecker damage to Red-cockaded Woodpecker cavity trees in eastern Texas. *Wilson Bull.* 110: 362–367.

Shackelford, C. E., and R. N. Conner. 1997. Woodpecker abundance and habitat use in three forest types in eastern Texas. *Wilson Bull.* 109: 614–629.

Thomlinson, J. R. 1995. Landscape characteristics associated with active and abandoned Red-cockaded Woodpecker clusters in East Texas. *Wilson Bull.* 107: 603–614.

Volman, S. F., T. C. Grubb Jr., and K. C. Schuett. 1997. Relative hippocampal volume in relation to food-storing behavior in four species of woodpeckers. *Brain, Behav. Evol.* 49: 110–120.

Tyrant Flycatchers: Family Tyrannidae

Andrews, B. J., M. Sullivan, and J. D. Hoerath. 1996. Vermilion Flycatcher and Black Phoebe feeding on fish. *Wilson Bull.* 108: 377–378.

Archer, T. J. 1996. Observations on nesting and display flights of the Vermilion Flycatcher in western Texas. *Southwest. Nat.* 41: 443–444.

Baker, B. W. 1980. Commensal foraging of Scissor-tailed Flycatchers with Rio Grande Turkeys. *Wilson Bull.* 92: 248.

Beaver, D. L., and P. H. Baldwin, 1975. Ecological overlap and the problem of competition and sympatry in the Western and Hammond's Flycatchers. *Condor* 77: 1–13.

Blancher, P. J., and R. J. Robertson. 1984. Resource use by sympatric kingbirds. *Condor* 86: 305–313.

———. 1987. Effect of food supply on the breeding biology of Western Kingbirds. *Ecology* 68: 723–732.

Brush, T. 1999. Current status of Northern Beardless-tyrannulet and Tropical Parula in Bentsen–Rio Grande Valley State Park and Santa Ana National Wildlife Refuge, southern Texas. *Bull. Texas Ornith. Soc.* 32: 3–12.

Conrad, K. F., R. J. Robertson, and P. T. Boag. 1998. Frequency of extrapair young increases in second broods of Eastern Phoebes. *Auk* 115: 497–502.

Fitch, F. W. Jr. 1947. The roosting tree of the Scissor-tailed Flycatcher. *Auk* 64: 616.

Gorena, R. L. 1998. Notes on the feeding habits and prey of adult Great Kiskadees. *Bull. Texas Ornith. Soc.* 30: 18–19.

Husak, M. S., and J. F. Husak. 1997. Anting by a Scissor-tailed Flycatcher. *Southwest. Nat.* 42: 351–352.

Maigret, J. L., and M. T. Murphy. 1997. Costs and benefits of parental care in Eastern Kingbirds. *Behav. Ecol.* 8: 250–259.

Murphy, M. T., C. L. Cummings, and M. S. Palmer. 1997. Comparative analysis of habitat selection, nest site, and nest success by Cedar Waxwings (*Bombycilla cedrorum*) and Eastern Kingbirds (*Tyrannus tyrannus*). *Am. Midl. Nat.* 138: 344–356.

Nolte, K. R., and T. E. Fulbright. 1996. Nesting ecology of Scissor-tailed Flycatchers in South Texas. *Wilson Bull.* 108: 302–316.

Ohlendorf, H. M. 1976. Comparative breeding ecology of phoebes in Trans-Pecos Texas. *Wilson Bull.* 88: 255–271.

Palmer, P. C. 1986. Great Kiskadees observed feeding on reptiles. *Bull. Texas Ornith. Soc.* 19: 31–32.

Regosin, J. V., and S. Pruett-Jones. 1995. Aspects of breeding biology and social organization in the Scissor-tailed Flycatcher. *Condor* 97: 154–164.

———. 2001. Sexual selection and tail-length dimorphism in Scissor-tailed Flycatchers. *Auk:* 118: 167–175.

Russell, S. J., and L. D. Foreman. 1979. Use of artificial materials in nests of urban Scissor-tailed Flycatchers. *Bull. Texas Ornith. Soc.* 12: 22–25.

Sexton, C. W. 1999. The Vermilion Flycatcher in Texas. *Texas Birds* 1: 41–45.

Smith, S. M. 1977. Coral-snake pattern recognition and stimulus generalization by naive Great Kiskadees (Aves: Tyrannidae). *Nature* 265: 535–536.

Weeks, H. P. Jr. 1994. Pre-laying nest roosting in the Eastern Phoebe: An energy-conserving behavior? *J. Field Ornith.* 65: 52–57.

Woodard, J. D., and M. T. Murphy. 1999. Sex roles, parental experience, and reproductive success of Eastern Kingbirds, *Tyrannus tyrannus*. *Anim. Behav.* 57: 105–115.

Shrikes: Family Laniidae

Atkinson, E. C. 1997. Singing for your supper: Acoustical luring of avian prey by Northern Shrikes. *Condor* 99: 203–206.

Casto, S. D. 1992. Loggerhead Shrike kills Common Ground-Dove. *Bull. Texas Ornith. Soc.* 25: 27–28.

Chavez-Ramirez, F., D. E. Gawlik, F. G. Prieto, and R. D. Slack. 1994. Effects of habitat structure on patch use by Loggerhead Shrikes wintering in a natural grassland. *Condor* 96: 228–231.

Ingold, J. J., and D. A. Ingold. 1987. Loggerhead Shrike kills and transports a Northern Cardinal. *J. Field Ornithol.* 58: 66–68.

Smith, S. M. 1973. An aggressive display and related behavior in the Loggerhead Shrike. *Auk* 90: 287–298.

Tyler, J. D. 1992. Nesting ecology of the Loggerhead Shrike in southwestern Oklahoma. *Wilson Bull.* 104: 95–104.

Woods, C. P. 1995. Food delivery and food holding during copulation in the Loggerhead Shrike. *Wilson Bull.* 107: 762–764.

Yosef, R. 1993. Prey transport by Loggerhead Shrikes. *Condor* 95: 231–233.

Yosef, R., and T. C. Grubb Jr. 1994. Resource dependence and territory size in Loggerhead Shrikes (*Lanius ludovicianus*). *Auk* 111: 465–469.

Vireos: Family Vireonidae

Bates, J. M. 1992. Winter territorial behavior of Gray Vireos. *Wilson Bull.* 104: 425–433.

Bradley, R. A. 1980. Vocal and territorial behavior in the White-eyed Vireo. *Wilson Bull.* 92: 302–311.

Bryan, K. B., and M. W. Lockwood. 2000. Gray Vireo in Texas. *Texas Birds* 2: 19–24.

Grzybowski, J. A., D. J. Tazik, and G. D. Schnell. 1994. Regional analysis of Black-capped Vireo breeding habitats. *Condor* 96: 512–544.

Howes-Jones, D. 1985. Relationships among song activity, context, and social behavior in the Warbling Vireo. *Wilson Bull.* 97: 4–20.

———. 1985. The complex song of the Warbling Vireo. *Canad. J. Zool.* 63: 2756–2766.

McNair, D. B., and R. A. Forster. 1983. Heterospecific vocal mimicry by six oscines. *Canad. Field Nat.* 97: 321–322.

Crows and Jays: Family Corvidae

Balda, R. P., and G. C. Bateman. 1972. The breeding biology of the Pinon Jay. *Living Bird* 11: 5–42.

Bednekoff, P. A., and R. P. Balda. 1996. Observational spatial memory in Clark's Nutcrackers and Mexican Jays. *Anim. Behav.* 52: 833–839.

Blake, S. F. 1957. The function of the concealed throat-patch in the White-necked Raven. *Auk* 74: 95–96.

Brown, J. L., and E. R. Brown. 1998. Are inbred offspring less fit? Survival in a natural population of Mexican Jays. *Behav. Ecol.* 9: 60–63.

Brown, J. L., and S.-H. Li. 1995. Phylogeny of social behavior in *Aphelocoma* jays: A role for hybridization? *Auk* 112: 464–472.

Burt, D. B. 1996. Habitat-use patterns in cooperative and non-cooperative breeding birds: Testing predictions with Western Scrub Jays. *Wilson Bull.* 108: 712–727.

Cassamise, D. F., L. M. Reed, J. Romanowski, and P. C. Stouffer. 1997. Roosting behavior and group territoriality in American Crows. *Auk* **114**: 628–637.

Clayton, N. S., and A. Dickinson. 1999. Motivational control of caching behaviour in the Scrub Jay, *Aphelocoma coerulescens. Anim. Behav.* **57**: 435–444.

Clench, M. H. 1991. Another case of Blue Jay kleptoparasitism. *Florida Field Nat.* **19**: 109–110.

Conner, R. N., and J. H. Williamson. 1984. Food storing by American Crows. *Bull. Texas Ornith. Soc.* **17**: 13–14.

Freeman, B. 2000. Murder in the madrones [behavior of Mexican Jays]. *Texas Birds* **2**: 42–43.

Fritz, J., and K. Kotrschal. 1999. Social learning in Common Ravens, *Corvus corax. Anim. Behav.* **57**: 785–793.

Gayou, D. C. 1986. The social system of the Texas Green Jay. *Auk* **103**: 540–547.

Heinrich, B. 1995. An experimental investigation of insight in Common Ravens (*Corvus corax*). *Auk* **112**: 994–1003.

Heinrich, B., D. Kaye, T. Knight, and K. Schaumburg. 1994. Dispersal and association among Common Ravens. *Condor* **96**: 545–551.

Hendricks, P., and S. Schlang. 1998. Aerial attacks by common ravens, *Corvus corax,* on adult feral pigeons, *Columba livia. Canad. Field Nat.* **112**: 702–703.

Hickey, M. B., and M. C. Brittingham. 1991. Population dynamics of Blue Jays at a bird feeder. *Wilson Bull.* **103**: 401–414.

Marzluff, J. M., and B. Heinrich. 1991. Foraging by Common Ravens in the presence and absence of territory holders: An experimental analysis of social foraging. *Anim. Behav.* **42**: 755–770.

Marzluff, J. M., B. Heinrich, and C. S. Marzluff. 1996. Raven roosts are mobile information centres. *Anim. Behav.* **51**: 89–103.

McDonald, D. B., J. W. Fitzpatrick, and G. E. Woolfenden. 1996. Actuarial senescence and demographic heterogeneity in the Florida Scrub Jay. *Ecology* **77**: 2373–2381.

Pavey, C. R. 1998. Effects of avian mobbing on roost use and diet of Powerful Owls, *Ninox strenua. Anim. Behav.* **55**: 313–318.

Tarvin, K. A., and G. E. Woolfenden. 1997. Patterns of dominance and aggressive behavior in Blue Jays at a feeder. *Condor* **99**: 434–444.

Larks: Family Alaudidae

Beason, R. C. 1984. Altruism in the horned lark? *J. Field Ornith.* **55**: 489–490.

Beason, R. C., and E. C. Franks. 1974. Breeding behavior of the Horned Lark. *Auk* **91**: 65–74.

Swallows: Family Hirundinidae

Blem, C. R., and L. B. Blem. 1993. Do swallows sunbathe to control ectoparasites? An experimental test. *Condor* **95**: 728–730.

Brown, C. R. 1984. Vocalizations of the Purple Martin. *Condor* **86**: 433–442.

Brown, C. R., and M. B. Brown. 1991. Selection of high-quality host nests by parasitic Cliff Swallows. *Anim. Behav.* **41**: 457–465.

———. 1999. Nest spacing in relation to settlement time in colonial Cliff Swallows. *Anim. Behav.* **59**: 47–55.

———. 2001. Egg hatchability increases with colony size in Cliff Swallows. *J. Field Ornithol.* **72**: 113–123.

Brown, C. R., M. B. Brown, and M. L. Shaffer. 1991. Food-sharing signals among socially foraging Cliff Swallows. *Anim. Behav.* **42**: 551–564.

Casto, S. D. 1998. Winter mortality of Cave Swallows in South Texas. *Bull. Texas Ornithol. Soc.* **31**: 30–31.

De Lope, F., and A. P. Moeller. 1993. Female reproductive effort depends on the degree of ornamentation of their mates. *Evolution* **47**: 1152–1160.

Johnston, R. F., and J. W. Hardy. 1962. Behavior of the Purple Martin. *Wilson Bull.* **74**: 243–261.

Kopachena, J. G., A. J. Buckley, and G. A. Potts. 2000. Effects of the red imported fire ant (*Solenopsis invicta*) on reproductive success of Barn Swallows (*Hirundo rustica*) in northeast Texas. *Southwest. Nat.* **45**: 477–482.

Moeller, A. P. 1990. Male tail length and female mate choice in the monogamous swallow *Hirundo rustica*. *Anim. Behav.* **39**: 458–465.

Sanio, N., P. Galeotti, R. Sacchi, and A. P. Moeller. 1997. Song and immunological condition in male Barn Swallows (*Hirundo rustica*). *Behav. Ecol.* **8**: 364–371.

Smith, H. G., and R. Montgomerie. 1991. Sexual selection and the tail ornaments of North American Barn Swallows. *Behav. Ecol. Sociobiol.* **28**: 195–201.

Soler, J. J., J. J. Cuervo, A. P. Moeller, and F. De Lope. 1998. Nest building is a sexually selected behaviour in the Barn Swallow. *Anim. Behav.* **56**: 1435–1442.

Stutchbury, B. J. 1991. Floater behaviour and territory acquisition in male Purple Martins. *Anim. Behav.* **42**: 435–443.

Wagner, R. H., P. S. Davidar, M. D. Schug, and E. S. Morton. 1997. Do blood parasites affect paternity, provisioning, and mate-guarding in Purple Martins? *Condor* **99**: 520–523.

Chickadees and Titmice: Family Paridae

Brawn, J. D., and F. B. Samson. 1983. Winter behavior of Tufted Titmice. *Wilson Bull.* **95**: 222–232.

Clemmons, J. R., and M. M. Lambrechts. 1992. The waving display and other nest site anti-predator behavior of the Black-capped Chickadee. *Wilson Bull.* **104**: 749–756.

Cooper, S. J., and D. L. Swanson. 1994. Seasonal acclimatization of thermoregulation in the Black-capped Chickadee. *Condor* **96**: 638–646.

Healy, S. D., and J. R. Krebs. 1996. Food storing and the hippocampus in Paridae. *Brain, Behav. Evol.* **47**: 195–199.

Kleintjes, P. K., and D. L. Dahlsten. 1995. Within-season trends in the foraging behavior of the Mountain Chickadee. *Wilson Bull.* 107: 655–666.

Kroodsma, D. E., D. J. Albano, P. W. Houlihan, and J. A. Wells. 1995. Song development by Black-capped Chickadees (*Parus atricapillus*) and Carolina Chickadees (*P. carolinensis*). *Auk* 112: 29–43.

Lemon, R. W. 1968. Coordinated singing by Black-crested Titmice. *Canad. J. Zool.* 46: 1163–1167.

Popp, J. W., M. S. Ficken, and C. M. Weise. 1990. How are agonistic encounters among Black-capped Chickadees resolved? *Anim. Behav.* 39: 980–986.

Pravosudov, V. V., and T. C. Grubb Jr. 1998. Management of fat reserves in Tufted Titmice *Baelophus bicolor* in relation to risk of predation. *Anim. Behav.* 56: 49–54.

Sibley, C. G. 1955. Behavioral mimicry in the titmice (Paridae) and certain other birds. *Wilson Bull.* 67: 128–132.

Smulders, T. V., A. D. Sasson, and T. J. DeVoogd. 1995. Seasonal variation in hippocampal volume in a food-storing bird, the Black-capped Chickadee. *J. Neurobiol.* 27: 15–25.

Penduline Tits and Verdins: Family Remizidae
Long-tailed Tits and Bushtits: Family Aegithalidae

Austin, G. T. 1976. Behavioral adaptations of the Verdin to the desert. *Auk* 93: 245–262.

Carter, M. D. 1987. An incident of brood parasitism by the Verdin. *Wilson Bull.* 99: 136.

Chaplin, S. B. 1982. The energetic significance of huddling behavior in Common Bushtits (*Psaltriparus minimus*). *Auk* 99: 424–430.

Sloane, S. A. 1996. Incidence and origins of supernumeraries at Bushtit (*Psaltriparus minimus*) nests. *Auk* 113: 757–770.

Wolf, B. O., and G. E. Walsberg. 1996. Thermal effects of radiation and wind on a small bird and implications for microsite selection. *Ecology* 77: 2228–2236.

Wolf, B. O., K. M. Wooden, and G. E. Walsberg. 1996. The use of thermal refugia by two small desert birds. *Condor* 98: 424–428.

Nuthatches: Family Sittidae
Creepers: Family Certhiidae

Franzreb, K. E. 1985. Foraging ecology of Brown Creepers in a mixed-coniferous forest. *J. Field Ornithol.* 56: 9–16.

Hendricks, P. 1995. Ground-caching and covering of food by a Red-breasted Nuthatch. *J. Field Ornithol.* 66: 370–372.

Herb, A. Jr., and D. B. Burt. 2000. Influence of habitat use-patterns on cooperative breeding in the Brown-headed Nuthatch. *Bull. Texas Ornith. Soc.* 33: 25–36.

Kilham, L. 1972. Reproductive behavior of White-breasted Nuthatches, II: Court-ship. *Auk* **89**: 115–129.

———. 1973. Reproductive behavior of Red-breasted Nuthatches, I: Courtship. *Auk* **90**: 597–609.

Lima, S. L., and R. M. Lee III. 1993. Food caching and its possible origin in the Brown Creeper. *Condor* **95**: 483–484.

O'Halloran, K. A., and R. N. Conner. 1987. Habitat used by Brown-headed Nut-hatches. *Bull. Texas Ornith. Soc.* **20**: 7–13.

Ritchison, G. 1983. Vocalizations of the White-breasted Nuthatch. *Wilson Bull.* **95**: 440–451.

Wrens: Family Troglodytidae

Brenowitz, E. A., K. Lent, and D. E. Kroodsma. 1995. Brain space for learned song in birds develops independently of song learning. *J. Neurosci.* **15**: 6281–6286.

Brenowitz, E. A., B. Nalls, D. E. Kroodsma, and C. Horning. 1994. Female Marsh Wrens do not provide evidence of anatomical specializations of song nuclei for perception of male song. *J. Neurobiol.* **25**: 197–208.

Freeman, B. 1999. Of wasps and wrens [behavior of Carolina Wrens]. *Texas Birds* **1**: 33.

Gish, S. L., and E. S. Morton. 1981. Structural adaptations to local habitat acoustics in Carolina Wren songs. *Z. Tierpsychol.* **56**: 74–84.

Kroodsma, D. E., J. Sanchez, D. W. Stemple, E. Goodwin, M. L. Da Silvas, and J. M. E. Vielliard. 1999. Sedentary life style of Neotropical Sedge Wrens pro-motes song imitation. *Anim. Behav.* **57**: 855–863.

Merola, M. 1995. Observations on the nesting and breeding behavior of the Rock Wren. *Condor* **97**: 585–587.

Oppenheimer, S. D., and M. L. Morton. 2000. Nesting habits and incubation be-havior of the Rock Wren. *J. Field Ornithol.* **71**: 650–657.

Picman, J., S. Pribil, and A. K. Picman. 1996. The effect of intraspecific egg de-struction on the strength of Marsh Wren eggs. *Auk* **113**: 599–607.

Pogue, D. W., and G. D. Schnell. 1994. Habitat characterization of secondary cavity-nesting birds in Oklahoma. *Wilson Bull.* **106**: 203–226.

Simons, L. S., and L. H. Simons. 1990. Experimental studies of nest-destroying behavior by Cactus Wren. *Condor* **92**: 855–860.

Simpson, B. S. 1984. Tests of habituation to song repertoires by Carolina Wrens. *Auk* **101**: 244–254.

Kinglets: Family Regulidae
Old World Warblers and Gnatcatchers: Family Sylviidae

Keast, A., and S. Saunders. 1991. Ecomorphology of the North American Ruby-crowned (*Regulus calendula*) and Golden-crowned (*Regulus satrapa*) Kinglets. *Auk* **108**: 880–888.

Thrushes: Family Turdidae

Chavez-Ramirez, F., and R. D. Slack. 1996. Winter phenology and frugivory of American Robins and Cedar Waxwings on the Edwards Plateau of Central Texas. *Texas J. Sci.* **48**: 129–136.

Dickenson, J. L., W. D. Koenig, and F. A. Pitelka. 1996. Fitness consequences of helping behavior in the Western Bluebird. *Behav. Ecol.* **7**: 168–177.

McLean, I. G., J. N. M. Smith, and K. G. Stewart. 1986. Mobbing behaviour, nest exposure, and breeding success in the American Robin. *Behaviour* **96**: 171–186.

McRae, S. B., P. J. Weatherhead, and R. Montgomerie. 1993. American Robin nestlings compete by jockeying for position. *Behav. Ecol. Sociobiol.* **33**: 101–106.

Montgomerie, R., and P. J. Weatherhead. 1997. How robins find worms. *Anim. Behav.* **54**: 143–151.

Purcell, K. L., J. Verner, and L. W. Oring. 1997. A comparison of the breeding ecology of birds nesting in boxes and tree cavities. *Auk* **114**: 646–656.

Rivers, J. W., and D. E. Kroodsma. 2000. Singing behavior of the Hermit Thrush. *J. Field Ornithol.* **71**: 467–471.

Smith, H. G., and R. Montgomerie. 1991. Nestling American Robins compete with siblings by begging. *Behav. Ecol. Sociobiol.* **29**: 307–312.

Swihart, R. K., and S. G. Johnson. 1986. Foraging decisions of American Robins: Somatic and reproductive tradeoffs. *Behav. Ecol. Sociobiol.* **19**: 275–282.

Weatherhead, P. J., and S. B. McRae. 1990. Brood care in American Robins: Implications for mixed reproductive strategies by females. *Anim. Behav.* **39**: 1179–1188.

Whitney, C. L. 1985. Serial order in Wood Thrush song. *Anim. Behav.* **33**: 1250–1265.

Yong, W., and F. R. Moore. 1990. "Foot quivering" as a foraging maneuver among migrating *Catharus* thrushes. *Wilson Bull.* **102**: 542–545.

Mockingbirds and Thrashers: Family Mimidae

Boughey, M. J., and N. S. Thompson. 1981. Song variety in the Brown Thrasher (*Toxostoma rufum*). *Z. Tierpsychol.* **56**: 47–58.

Casto, S. D. 1999. Nest sites of Curve-billed Thrashers at a rural dwelling in southern Texas. *Bull. Texas. Ornith. Soc.* **32**: 44–46.

Fischer, D. H. 1978. Dew-bathing by Long-billed Thrashers (*Toxostoma longirostre*) and an Olive Sparrow (*Arremonops rufivirgatus*). *Bull. Texas Ornith. Soc.* **11**: 49–50.

———. 1980. Breeding biology of Curve-billed Thrashers and Long-billed Thrashers in southern Texas. *Condor* **82**: 392–397.

———. 1981. Factors affecting the reproductive success of the Northern Mockingbird in southern Texas. *Southwest. Nat.* **26**: 289–293.

———. 1981. Wintering ecology of thrashers in South Texas. *Condor* **83**: 340–346.

Hailman, J. P. 1963. The Mockingbird's "tail-up" display to mammals near the nest. *Wilson Bull.* **75**: 414–417.

Logan, C. A. 1985. Mockingbird use of chatbursts with neighbors versus strangers. *J. Field Ornithol.* **56**: 69–71.

———. 1991. Mate switching and mate choice in female Northern Mockingbirds: Facultative monogamy. *Wilson Bull.* **103**: 277–281.

Logan, C. A., L. E. Hyatt, and L. Gregorcyk. 1990. Song playback initiates nest building during clutch overlap in Mockingbirds, *Mimus polyglottos. Anim. Behav.* **39**: 943–953.

Rylander, M. K. 1981. Vocal mimicry in the Curve-billed Thrasher. *Bull. Texas Ornith. Soc.* **14**: 29–30.

Sutton, G. M. 1946. Wing-flashing in the Mockingbird. *Wilson Bull.* **58**: 206–209.

Starlings: Family Sturnidae

Clark, C. C., and L. Clark. 1990. "Anting" behavior by Common Grackles and European Starlings. *Wilson Bull.* **102**: 167–169.

Eens, M., and R. Pinxten. 1996. Female European Starlings increase their copulation solicitation rate when faced with the risk of polygyny. *Anim. Behav.* **51**: 1141–1147.

Ellis, C. R. Jr. 1966. Agonistic behavior in the male Starling. *Wilson Bull.* **78**: 208–223.

Gentner, T. Q., and S. H. Hulse. 1998. Perceptual mechanisms for individual vocal recognition in European Starlings, *Sturnus vulgaris. Anim. Behav.* **56**: 579–594.

Sandell, M. I., and M. Diemer. 1999. Intraspecific brood parasitism: A strategy for floating females in the European Starling. *Anim. Behav.* **57**: 197–202.

Wright, J., and I. Cuthill. 1990. Manipulation of sex differences in parental care: The effect of brood size. *Anim. Behav.* **40**: 462–471.

Wagtails and Pipits: Family Motacillidae

Freeman, B. 1999. Finding Sprague's Pipit in Texas. *Texas Birds* **1**: 50–51.

Robbins, M. B. 1998. Display behavior of male Sprague's Pipits. *Wilson Bull.* **110**: 435–438.

Waxwings: Family Bombycillidae
Silky Flycatchers: Family Ptilogonatidae

Avery, M. L., K. J. Goocher, and M. A. Cone. 1993. Handling efficiency and berry size preferences of Cedar Waxwings. *Wilson Bull.* **105**: 604–611.

Chavez-Ramirez, F., and R. D. Slack. 1996. Winter phenology and frugivory of American Robins and Cedar Waxwings on the Edwards Plateau of Central Texas. *Texas J. Sci.* **48**: 129.

King, D. I. 1996. Carnivory observed in the Cedar Waxwing. *Wilson Bull.* **108**: 381–382.

Mountjoy, D. J., and R. J. Robertson. 1988. Why are waxwings "waxy"? Delayed plumage maturation in the Cedar Waxwing. *Auk* 105: 61–69.

Mulvihill, R. S., K. C. Parkes, R. C. Leberman, and D. S. Wood. 1992. Evidence supporting a dietary basis for orange-tipped rectrices in the Cedar Waxwing. *J. Field Orinthol.* 63: 212–216.

Wood Warblers: Family Parulidae

Bay, M. D. 1999. Notes on the singing behavior and use of atypical songs in the Northern Parula Warbler, *Parula americana* (Aves: Emberizidae). *Texas J. Sci.* 51: 20–24.

Benson, R. H. 1995. The effect of roadway traffic noise on territory selection by Golden-cheeked Warblers. *Bull. Texas Ornith. Soc.* 28: 42–51.

Blem, C. R., and L. B. Blem. 1994. Composition and microclimate of Prothonotary Warbler nests. *Auk* 111: 197–200.

Bolsinger, J. S. 2000. Use of two song categories by Golden-cheeked Warblers. *Condor* 102: 539–552.

Brush, T. 1999. Current status of Northern Beardless-tyrannulet and Tropical Parula in Bentsen–Rio Grande Valley State Park and Santa Ana National Wildlife Refuge, Southern Texas. *Bull. Texas Ornith. Soc.* 32: 3–12.

Carrie, N. R. 1996. Swainson's Warblers nesting in early seral pine forests in East Texas. *Wilson Bull.* 108: 802–804.

Coldren, C. L. 1998. The effects of habitat fragmentation on the Golden-cheeked Warbler. Ph.D. diss. Texas A&M University, College Station.

Dudzinski, K. M., T. G. Frohoff, L. K. M. Shoda, and T. D. Sparks. 1991. The Pine Warbler song repertoire: A preliminary description and analysis. *Bull. Texas Ornith. Soc.* 24: 30.

Dunn, J. L., and K. L. Garrett. 1997. *A Field Guide to Warblers of North America.* New York: Houghton Mifflin.

Engels, T. M., and C. W. Sexton. 1994. Negative correlation of Blue Jays and Golden-cheeked Warblers near an urbanizing area. *Conservation Biol.* 8: 286–290.

Ficken, M. S. 1963. Courtship of the American Redstart. *Auk* 80: 307–317.

Gallucci, T. 1979. Successful breeding of Lucy's Warbler in Texas. *Bull. Texas Ornith. Soc.* 12: 37–41.

Gilbert, W. M. 1983. Flight song and song flight in the Orange-crowned Warbler. *Condor* 85: 113.

Gill, S. A., D. L. Neudorf, and S. G. Sealy. 1997. Host response to cowbirds near the nest: Cues for recognition. *Anim. Behav.* 53: 1287–1293.

Keast, A., L. Pearce, and S. Saunders. 1995. How convergent is the American Redstart (*Setophaga ruticilla*, Parulinae) with flycatchers (Tyrannidae) in morphology and feeding behavior? *Auk* 112: 310–325.

Kelly, J. P., and C. Wood. 1996. Diurnal, intraseasonal, and intersexual variation in foraging behavior of the Common Yellowthroat. *Condor* 98: 491–500.

King, D. I. 1996. A case of cooperative breeding in the Hooded Warbler. *Wilson Bull.* **108**: 382–384.

Lanning, D. V., J. T. Marshall, and J. T. Shiflett. 1990. Range and habitat of the Colima Warbler. *Wilson Bull.* **102**: 1–13.

Lockwood, M. W. 1996. Courtship behavior of Golden-cheeked Warblers. *Wilson Bull.* **108**: 591–592.

Luneau, G. 2001. An affection for the Swainson's Warbler. *Texas Birds* **3**: 4–7.

Morse, D. H. 1989. Song patterns of warblers at dawn and dusk. *Wilson Bull.* **101**: 26–35.

Morton, E. S., and K. Young. 1986. A previously undescribed method of song matching in a species with a single song "type," the Kentucky Warbler (*Oporornis formosus*). *Ethology* **73**: 334–342.

Morton, E. S., M. Van der Voort, and R. Greenburg. 1993. How a warbler chooses its habitat: Field support for laboratory experiments. *Anim. Behav.* **46**: 47–53.

Root, R. B. 1989. Dead-leaf-searching by the Orange-crowned Warbler in Louisiana in winter. *Wilson Bull.* **101**: 645–648.

Staicer, C. A. 1989. Characteristics, use, and significance of two singing behaviors in Grace's Warbler (*Dendroica graciae*). *Auk* **106**: 49–63.

Stutchbury, B. J. M. 1998. Extra-pair mating effort of male Hooded Warblers, *Wilsonia citrina*. *Anim. Behav.* **55**: 553–561.

Wauer, R. H. 2000. Colima Warbler: A Texas specialty. *Texas Birds* **2**: 45–47.

Weary, D. M., R. E. Lemon, and S. Perreault. 1994. Male Yellow Warblers vary use of song types depending on pairing status and distance from nest. *Auk* **111**: 727–729.

Tanagers: Family Thraupidae

Holmes, R. T. 1986. Foraging patterns of forest birds: Male-female differences. *Wilson Bull.* **98**: 196–213.

Leberg, P. L., T. J. Spengler, and W. C. Barrow Jr. 1996. Lipid and water depletion in migrating passerines following passage over the Gulf of Mexico. *Oecologia* **106**: 1–17.

Perez-Rivera, R. A. 1997. The importance of vertebrates in the diet of tanagers. *J. Field Ornithol.* **68**: 178–182.

Shy, E. 1985. Songs of Summer Tanagers (*Piranga rubra*): Structure and geographical variation. *Am. Midl. Nat.* **114**: 112–124.

Emberizids: Family Emberizidae

Adams, M. T., and K. B. Bryan. 1999. Botteri's Sparrow in Trans-Pecos Texas. *Texas Birds* **1**: 6–13.

Albrecht, D. J., and L. W. Oring. 1995. Song in Chipping Sparrows, *Spizella passerina:* Structure and function. *Anim. Behav.* **50**: 1233–1241.

Allaire, P. N., and C. D. Fisher. 1975. Feeding ecology of three resident sympatric sparrows in eastern Texas. *Auk* **92**: 260–269.

Anderson, J. T., and W. C. Conway. 2000. The flight song display of the Cassin's Sparrow (*Aimophila cassinii*): Form and possible function. *Bull. Texas Ornith. Soc.* **33**: 1–5.

Arnold, K. A., and N. C. Garza Jr. 1998. Populations and habitat requirements of breeding Henslow's Sparrows in Harris County, Texas. *Bull. Texas Ornith. Soc.* **31**: 42–49.

Baptista, L. F., P. W. Trail, B. B. Dewolfe, and M. L. Morton. 1993. Singing and its functions in female White-crowned Sparrows. *Anim. Behav.* **46**: 511–524.

Benkman, C. W., and H. R. Pulliam. 1988. The comparative feeding rates of North American sparrows and finches. *Ecology* **69**: 1195–1199.

Berthelsen, P. S., and L. M. Smith. 1995. Nongame bird nesting on CRP lands in the Texas Southern High Plains. *J. Soil Water Conserv.* **50**: 672–675.

Catchpole, C. K., and L. F. Baptista. 1988. A test of the competition hypothesis of vocal mimicry, using Song Sparrow imitation of White-crowned Sparrow song. *Behaviour* **106**: 119–128.

Cawthorn, J. M., D. L. Morris, E. D. Ketterson, and V. Nolan Jr. 1998. Influence of experimentally elevated testosterone on nest defence in Dark-eyed Juncos. *Anim. Behav.* **56**: 617–621.

Clement, P. 1993. *Finches and Sparrows: An Identification Guide.* Princeton, N.J.: Princeton University Press.

Cristol, D. A., V. Nolan Jr., and E. D. Ketterson. 1990. Effect of prior residence on dominance status of Dark-eyed Juncos, *Junco hyemalis. Anim. Behav.* **40**: 580–586.

Davis, J. 1957. Comparative foraging behavior of the Spotted and Brown Towhees. *Auk* **74**: 129–166.

DeVoogd, T. J., A. M. Houtman, and J. B. Falls. 1995. White-throated Sparrow morphs that differ in song production rate also differ in the anatomy of some song-related brain areas. *J. Neurobiol.* **28**: 202–213.

Eitniear, J. C., and T. Rueckle. 1995. Successful nesting of the White-collared Seed-eater in Zapata County, Texas. *Bull. Texas Ornith. Soc.* **28**: 20–22.

Elekonich, M. M. 1998. Song Sparrow males use female-typical vocalizations in fall. *Condor* **100**: 145–148.

Fischer, D. H. 1978. Dew-bathing by Long-billed Thrashers (*Toxostoma longirostre*) and an Olive Sparrow (*Arremonops rufivirgatus*). *Bull. Texas Ornith. Soc.* **11**: 49–50.

Greenberg, R. 1990. Feeding neophobia and ecological plasticity: A test of the hypothesis with captive sparrows. *Anim. Behav.* **39**: 375–379.

Greenlaw, J. S., C. E. Shackelford, and R. E. Brown. 1998. Call mimicry by Eastern Towhees and its significance in relation to auditory learning. *Wilson Bull.* **110**: 431–434.

Horning, C. L., M. D. Beecher, P. K. Stoddard, and S. E. Campbell. 1993. Song perception in the Song Sparrow: Importance of different parts of the song in song type classification. *Ethology* 94: 46–58.

Houtman, A. N., and J. B. Falls. 1994. Negative assortative mating in the White-throated Sparrow, *Zonotrichia albicollis:* The role of mate choice in intra-sexual competition. *Anim. Behav.* 48: 377–383.

Hurley, T. A., L. Ratcliffe, and R. Weisman. 1990. Relative pitch recognition in White-throated Sparrows, *Zonotrichia albicollis. Anim. Behav.* 40: 176–181.

Jarvis, E. D., H. Schwabl, S. Ribeiro, and C. V. Mello. 1997. Brain gene regulation by territorial singing behavior in freely ranging songbirds. *Neuroreport* 8: 2073–2077.

Joern, A. 1988. Foraging behavior and switching by the Grasshopper Sparrow *Ammodramus savannarum* searching for multiple prey in a heterogeneous environment. *Am. Midl. Nat.* 119: 225–234.

Jones, S. L., M. T. Green, and G. R. Geupel. 1997. Rare, local, little-known, and declining North American breeders: A closer look, Baird's Sparrow. *Birding* 29: 108–115.

Keys, G. C., and S. I. Rothstein. 1991. Benefits and costs of dominance and subordinance in White-crowned Sparrows and the paradox of status signaling. *Anim. Behav.* 42: 899–912.

Liu, W.-C., and D. E. Kroodsma. 1999. Song development by Chipping Sparrows and Field Sparrows. *Anim. Behav.* 57: 1275–1286.

Marler, P., and S. Peters. 1987. A sensitive period for song acquisition in the Song Sparrow, *Melospiza melodia:* A case of age-limited learning. *Ethology* 76: 89–100.

Maurer, B. A., E. A. Webb, and R. K. Bowers. 1989. Nest characteristics and nestling development of Cassin's and Botteri's Sparrows in southeastern Arizona. *Condor* 91: 736–738.

Nelson, B. S., and P. K. Stoddard. 1998. Accuracy of auditory distance and azimuth perception by a passerine bird in a natural habitat. *Anim. Behav.* 56: 467–477.

Nelson, D. A. 1988. Feature weighting in species song recognition by the Field Sparrow (*Spizella pusilla*). *Behaviour* 106: 158–182.

———. 1992. Song overproduction and selective attrition lead to song sharing in the Field Sparrow (*Spizella pusilla*). *Behav. Ecol. Sociobiol.* 30: 415–424.

Nelson, D. A., and L. J. Croner. 1991. Song categories and their functions in the Field Sparrow (*Spizella pusilla*). *Auk* 108: 42–52.

Nelson, D. A., C. Whaling, and P. Marler. 1996. The capacity for song memorization varies in populations of the same species. *Anim. Behav.* 52: 379–387.

O'Loghlen, A. L., and M. D. Beecher. 1997. Sexual preferences for mate song types in female Song Sparrows. *Anim. Behav.* 53: 835–841.

———. 1999. Mate, neighbour, and stranger songs: A female song sparrow perspective. *Anim. Behav.* 58: 13–20.

Plentovich, S., J. W. Tucker Jr., N. R. Holler, and G. E. Hill. 1998. Enhancing

Bachman's Sparrow habitat via management of Red-cockaded Woodpeckers. *J. Wildl. Manage.* **62**: 347–354.

Post, W. 1981. The influence of rice rats *Oryzomys palustris* on the habitat use of the Seaside Sparrow, *Ammospiza maritima*. *Behav. Ecol. Sociobiol.* **9**: 35–40.

Rising, J. D. 1996. Relationship between testis size and mating systems in American sparrows (Emberizinae). *Auk* **113**: 224–228.

Schnase, J. L., W. E. Grant, T. C. Maxwell, and J. J. Leggett. 1991. Time and energy budgets of Cassin's Sparrow (*Aimophila cassinii*) during the breeding season: Evaluation through modeling. *Ecol. Model.* **55**: 285–319.

Schnase, J. L., and T. C. Maxwell. 1989. Use of song patterns to identify individual male Cassin's Sparrows. *J. Field Ornithol.* **60**: 12–19.

Shackelford, C. E. 2000. Henslow's Sparrow in Texas. *Texas Birds* **2**: 39–44.

Slotow, R., J. Alcock, and S. I. Rothstein. 1993. Social status signaling in White-crowned Sparrows: An experimental test of the social control hypothesis. *Anim. Behav.* **46**: 977–989.

Titus, R. C. 1998. Short-range and long-range songs: Use of two acoustically distinct song classes by Dark-eyed Juncos. *Auk* **115**: 386–393.

White, M. 1999. Inland occurrences of Nelson's Sharp-tailed Sparrow. *Texas Birds* **1**: 34–39.

———. 2001. Texas review species: Snow Bunting. *Texas Birds* **3**: 18–21.

Wiley, R. H. 1991. Both high- and low-ranking White-throated Sparrows find novel locations of food. *Auk* **108**: 8–15.

Woodin, M. C., M. K. Skoruppa, G. W. Blacklock, and G. C. Hickman. 1999. Discovery of a second population of White-collared Seedeaters, *Sporophila torqueola* (Passeriformes: Emberizidae) along the Rio Grande of Texas. *Southwest. Nat.* **44**: 535–538.

Cardinals, Saltators, and Allies: Family Cardinalidae

Anderson, M. E., and R. N. Conner. 1985. Northern Cardinal song in three forest habitats in eastern Texas. *Wilson Bull.* **97**: 436–449.

Baird, J. 1964. Hostile displays of Rose-breasted Grosbeaks towards a Red Squirrel. *Wilson Bull.* **76**: 286–288.

Filliater, T. S., and R. Breitwisch. 1997. Nestling provisioning by the extremely dichromatic Northern Cardinal. *Wilson Bull.* **109**: 145–153.

Gould, P. J. 1961. Territorial relationships between Cardinals and Pyrrhuloxias. *Condor* **63**: 246–256.

Halkin, S. L. 1997. Nest-vicinity song exchanges may coordinate biparental care of Northern Cardinals. *Anim. Behav.* **54**: 189–198.

Hill, G. E. 1988. Age, plumage brightness, territory quality, and reproductive success in the Black-headed Grosbeak. *Condor* **90**: 379–388.

Kopachena, J. G., and C. J. Crist. 2000. Microhabitat features associated with the song perches of Painted and Indigo Buntings (Passeriformes: Cardinalidae) in Northeast Texas. *Texas J. Sci.* **52**: 133–144.

————. 2000. Macro-habitat features associated with Painted and Indigo Buntings in Northeast Texas. *Wilson Bull.* 112: 108–114.

Lanyon, S. M., and C. F. Thompson. 1984. Visual displays and their context in the Painted Bunting. *Wilson Bull.* 96: 396–407.

Lemon, R. E., and A. Herzog. 1969. The vocal behavior of Cardinals and Pyrrhuloxias in Texas. *Condor* 71: 1–15.

Linville, S. U., R. Breitwisch, and A. J. Schilling. 1998. Plumage brightness as an indicator of parental care in Northern Cardinals. *Anim. Behav.* 55: 119–127.

Lockwood, M. 1995. Rare, local, little-known, and declining North American breeders: A closer look, Varied Bunting. *Birding* 27: 110–113.

Nealen, P. M., and R. Breitwisch. 1997. Northern Cardinal sexes defend nests equally. *Wilson Bull.* 109: 269–278.

Ritchison, G. 1983. Possible "deceptive" use of song by female Black-headed Grosbeaks. *Condor* 85: 250–251.

Wolfenbarger, L. L. 1999. Is red coloration of male Northern Cardinals beneficial during the nonbreeding season? A test of status signaling. *Condor* 101: 655–663.

Blackbirds: Family Icteridae

Armstrong, T. A. 1995. Patterns of vocalization use by female Red-winged Blackbirds (Aves: Emberizidae, Icterinae): Variation and context. *Ethology* 100: 331–351.

Briskie, J. V., and S. G. Sealy. 1990. Evolution of short incubation periods in the parasitic cowbirds, *Molothrus* spp. *Auk* 107: 789–794.

Brush, T. 1998. Rare, local, little-known, and declining North American breeders: A closer look, Altamira Oriole. *Birding* 30: 46–53.

————. 2000. Audubon's Oriole. *Texas Birds* 2: 4–9.

Butcher, G. S. 1991. Mate choice in female Northern Orioles with a consideration of the role of the black male coloration in female choice. *Condor* 93: 82–88.

Carter, M. D. 1986. The parasitic behavior of the Bronzed Cowbird in South Texas. *Condor* 88: 11–25.

Clark, A. B., and W.-H. Lee. 1998. Red-winged Blackbird females fail to increase feeding in response to begging call playbacks. *Anim. Behav.* 56: 563–570.

Clotfelter, E. D. 1998. What cues do Brown-headed Cowbirds use to locate Red-winged Blackbird host nests? *Anim. Behav.* 55: 1181–1189.

Clotfelter, E. D., and T. Brush. 1995. Unusual parasitism by the Bronzed Cowbird. *Condor* 97: 814–815.

————. 1995. Courtship displaying and intrasexual competition in the Bronzed Cowbird. *Condor* 97: 816–818.

Davidson, A. H. 1994. Common Grackle predation on adult passerines. *Wilson Bull.* 106: 174–175.

Dearborn, D. C. 1996. Video documentation of a Brown-headed Cowbird nestling ejecting an Indigo Bunting nestling from the nest. *Condor* 98: 645–649.

Dixon, K. L. 1960. A courtship display of Scott's Oriole. *Auk* 77: 348–349.

Enstrom, D. A. 1993. Female choice for age-specific plumage in the Orchard Oriole: Implications for delayed plumage maturation. *Anim. Behav.* 45: 435–442.

Ficken, R. W. 1963. Courtship and agonistic behavior of the Common Grackle, *Quiscalus quiscula. Auk* 80: 52–72.

Flood, N. J. 1985. Incidences of polygyny and extrapair copulation in the Northern Oriole. *Auk* 102: 410–413.

———. 1990. Aspects of the breeding biology of Audubon's Oriole. *J. Field. Ornithol.* 61: 290–302.

Freeberg, T. M. 1998. The cultural transmission of courtship patterns in cowbirds, *Molothrus ater. Anim. Behav.* 56: 1063–1073.

Freeberg, T. M., S. D. Duncan, T. L. Kast, and D. A. Enstrom. 1999. Cultural influences on female mate choice: An experimental test in cowbirds, *Molothrus ater. Anim. Behav.* 57: 421–426.

Gill, S. A., D. L. Neudorf, and S. G. Sealy. 1997. Host responses to cowbirds near the nest: Cues for recognition. *Anim. Behav.* 53: 1287–1293.

Hahn, D. C., and R. C. Fleischer. 1995. DNA fingerprint similarity between female and juvenile Brown-headed Cowbirds trapped together. *Anim. Behav.* 49: 1577–1580.

Hamilton, K. S., A. P. King, D. R. Sengelaub, and M. J. West. 1998. Visual and song nuclei correlate with courtship skills in Brown-headed Cowbirds. *Anim. Behav.* 56: 973–982.

Horn, A., and J. B. Falls. 1988. Structure of Western Meadowlark (*Sturnella neglecta*) song repertoires. *Canad. J. Zool.* 66: 284–288.

Johnsen, T. S., J. D. Hengeveld, J. L. Blank, K. Yasukawa, and V. Nolan Jr. 1996. Epaulet brightness and condition in female Red-winged Blackbirds. *Auk* 113: 356–362.

Knapton, R. W. 1988. Nesting success is higher for polygynously mated females than for monogamously mated females in the Eastern Meadowlark. *Auk* 105: 325–329.

Mason, J. R., and R. F. Reidinger Jr. 1981. Effects of social facilitation and observational learning on feeding behavior of the Red-winged Blackbird (*Agelaius phoeniceus*). *Auk* 98: 778–784.

Nero, R. W. 1963. Comparative behavior of the Yellow-headed Blackbird, Red-winged Blackbird, and other icterids. *Wilson Bull.* 75: 376–411.

O'Loghlen, A. L., and S. I. Rothstein. 1993. An extreme example of delayed vocal development: Song learning in a population of wild Brown-headed Cowbirds. *Anim. Behav.* 46: 293–304.

Picman, J. 1997. Are cowbird eggs unusually strong from the inside? *Auk* 114: 66–73.

Post, W. 1992. Dominance and mating success in male Boat-tailed Grackles. *Anim. Behav.* 44: 917–929.

————. 1994. Redirected copulation by male Boat-tailed Grackles. *Wilson Bull.* 106: 770–771.

————. 1997. Mate choice and competition for mates in the Boat-tailed Grackle. *Anim. Behav.* 54: 525–534.

————. 1997. Dominance, access to colonies, and queues for mating opportunities by male Boat-tailed Grackles. *Behav. Ecol. Sociobiol.* 41: 89–98.

Pribil, S., and J. Picman. 1996. Polygyny in the Red-winged Blackbird: Do females prefer monogamy or polygamy? *Behav. Ecol. Sociobiol.* 38: 183–190.

Price, K. 1996. Begging as competition for food in Yellow-headed Blackbirds. *Auk* 113: 963–967.

Rothstein, S. I. 1977. The preening invitation or head-down display of parasitic cowbirds: I. Evidence for intraspecific occurrence. *Condor* 79: 13–23.

Searcy, W. A. 1990. Species recognition of song by female Red-winged Blackbirds. *Anim. Behav.* 40: 1119–1127.

Sherry, D. F., M. R. L. Forbes, M. Khurgel, and G. O. Ivy. 1993. Females have a larger hippocampus than males in the brood-parasitic Brown-headed Cowbird. *Proc. Natl. Acad. Sci. USA.* 90: 7839–7843.

Teather, K. L. 1993. Behavioral development of male and female Red-winged Blackbirds. *Wilson Bull.* 105: 159–166.

West, M. J., A. P. King, and T. M. Freeberg. 1998. Dual signaling during mating in Brown-headed Cowbirds (*Molothrus ater;* Family Emberizidae/Icterinae). *Ethology* 104: 250–267.

Fringilline and Cardueline Finches and Allies: Family Fringillidae

Belthoff, J. R., and S. A. Gauthreaux Jr. 1991. Aggression and dominance in House Finches. *Condor* 93: 1010–1013.

Bitterbaum, E., and L. F. Baptista. 1979. Geographical variation in songs of California House Finches (*Carpodacus mexicanus*). *Auk* 96: 462–474.

Clement, P. 1993. *Finches and Sparrows: An Identification Guide.* Princeton, N.J.: Princeton University Press.

Coutlee, E. L. 1967. Agonistic behavior in the American Goldfinch. *Wilson Bull.* 79: 89–107.

Hill, G. E. 1990. Female House Finches prefer colourful males: Sexual selection for a condition-dependent trait. *Anim. Behav.* 40: 563–572.

————. 1993. Male mate choice and the evolution of female plumage coloration in the House Finch. *Evolution* 47: 1515–1525.

Hooge, P. N. 1990. Maintenance of pair-bonds in the House Finch. *Condor* 92: 1066–1067.

McGraw, K. J., and G. E. Hill. 2000. Plumage brightness and breeding-season dominance in the House Finch: A negatively correlated handicap. *Condor* 102: 456–461.

Payne, P. N., L. L. Payne, and S. Whitesell. 1988. Interspecific learning and cultural transmission of song in House Finches. *Wilson Bull.* 100: 667–670.

Thompson, W. L. 1960. Agonistic behavior in the House Finch, Part I: Annual cycle and display patterns. *Condor* **62**: 245–271.

Tobias, M. C., and G. E. Hill. 1998. A test of sensory bias for long tails in the House Finch. *Anim. Behav.* **56**: 71–78.

Zahn, S. N., and S. J. Rothstein. 2001. House Finches are not just what they eat: A reply to Hill. *Auk* **118**: 260–266.

Old World Sparrows: Family Passeridae

Liker, A., and B. Zoltan. 2001. Male badge size predicts dominance against females in House Sparrows. *Condor* **103**: 151–157.

Veiga, J. P. 1990. Infanticide by male and female House Sparrows. *Anim. Behav.* **39**: 496–502.

———. 1993. Badge size, phenotypic quality, and reproductive success in the House Sparrow: A study of honest advertisement. *Evolution* **47**: 1161–1170.

———. 1995. Honest signaling and the survival cost of badges in the House Sparrow. *Evolution* **49**: 570–572.

Wade, V. E., and M. K. Rylander. 1982. Unrecorded foraging behaviour in the House Sparrow. *Bird Study* **29**: 166.

Whitekiller, R. R., D. F. Westneat, P. L. Schwagmeyer, and D. W. Mock. 2000. Badge size and extra-pair fertilizations in the House Sparrow. *Condor* **102**: 342–348.

INDEX

Printed in Australia
AUHW010140200120
322590AU00037B/143

9 780292 771208